Zinc in Soil-Plant Continuum

Sudhakar Srivastava • Harmanjit Kaur
Editors

Zinc in Soil-Plant Continuum

Springer

Editors
Sudhakar Srivastava
Plant Stress Biology Laboratory, Institute of Environment and Sustainable Development
Banaras Hindu University
Varanasi, Uttar Pradesh, India

Harmanjit Kaur
Department of Botany
University of Allahabad
Prayagraj, Uttar Pradesh, India

ISBN 978-981-96-4252-6 ISBN 978-981-96-4253-3 (eBook)
https://doi.org/10.1007/978-981-96-4253-3

This Springer imprint is published by the registered company Springer Nature Singapore Pte Ltd.
The registered company address is: 152 Beach Road, #21-01/04 Gateway East, Singapore 189721, Singapore

Preface

Zinc is an essential element required for the normal functioning of all the living beings on the earth owing to its involvement as a constituent of several enzymes related to various metabolic pathways. To get required trace quantities of zinc, plants have devised intricate system of transporters and other factors. Zinc homeostasis is of crucial importance to achieve good growth and yield in crop plants. However, plants depend on the bioavailable zinc present in the environment they live. To this end, it is important to note that several regions of earth have zinc-deficient levels, while others have an excess of zinc. Therefore, the crops growing in such conditions either fight with zinc deficiency or suffer from zinc toxicity. Since crops are a source of nutrition to humans, zinc deficiency is a prevailing problem crippling the health of millions of people on earth, especially those who are dependent on plant based diet. Contrastingly, crops can also be a provider of excessive zinc to humans, when they happen to grow in toxic zinc conditions. Through years of research, attempts for improving zinc nutrition of crop plants have been made and significant progress has been made.

This book aims to compile up-to-date knowledge related to zinc deficiency and excess responses in plants at one place. The book contains a total of 12 chapters divided into two sections. The first section of the book comprises five chapters that present an overview of zinc in soil and plant responses toward zinc deficiency and excess. Chapter 1 of the book discusses the factors regulating the bioavailability of zinc in soil. Chapter 2 reviews the significance of zinc for plants, animals, and humans and thoroughly presents various functions of zinc. Chapter 3 provides information related to uptake and transport of zinc in plants while Chap. 4 discusses about how plants respond to zinc deficiency and excess. Chapter 5 provides detailed information regarding molecular aspects of zinc homeostasis in plants both under zinc-deficient and excess conditions. The second section of the book delves around the strategies that are being targeted for managing the levels of zinc in plants. Chapter 6 gives an overall view of strategies that can be utilized to increase zinc uptake in plants so as to tackle zinc deficiency conditions. Chapters 7 and 8 discuss microbial and breeding related approaches to augment zinc nutrition of plants. Chapter 9 reviews various agronomic approaches targeted for zinc biofortification of crops. Chap. 10 provides understanding about how a few plants are able to hyper-accumulate zinc and how these plants can be a resource material to manage zinc in crop plants. Chapters 11 and 12 present views on microbe- and phytohormone-mediated approaches to tackle zinc toxicity in plants.

The book therefore comprises a unique combination of chapters on various aspects and will provide the reader a comprehensive view of zinc responses of plants and of ways to regulate zinc. This book would act as a guiding textbook for undergraduate and postgraduate students and as a mean to understand the latest research trends for doctoral students.

Varanasi, Uttar Pradesh, India Sudhakar Srivastava
Prayagraj, Uttar Pradesh, India Harmanjit Kaur

Contents

Editors and Contributors

About the Editor

Sudhakar Srivastava is an Assistant Professor in the Institute of Environment and Sustainable Development (IESD), Banaras Hindu University (BHU). He obtained Ph.D. in Botany from CSIR-National Botanical Research Institute (CSIR-NBRI) and University of Lucknow in 2008. He served as Scientific Officer in Bhabha Atomic Research Centre (BARC), Mumbai, from 2009 to 2014 before joining BHU. He has more than 19 years of research experience in the field of plant-metal interactions. He has published more than 130 research and review articles in high impact international and national journals. He has also published 25 book chapters and edited three books. He has supervised 7 Ph.D. and 3 M.Phil. students. Dr. Srivastava is NAAS Associate of National Academy of Agricultural Sciences. He has received Young Scientist Award of National Academy of Sciences, India (NASI), Allahabad, in 2011 and Young Scientist Award from Uttar Pradesh Council of Science & Technology (UPCST) in 2013–2014. Dr. Srivastava is also a Fellow of International Society of Environmental Botanists (ISEB), Lucknow.

Harmanjit Kaur currently works as an Assistant Professor at the Department of Botany, University of Allahabad, Prayagraj, Uttar Pradesh, India. Her research area revolves around Plant Stress Physiology, including influence of salinity and heavy metal stresses on legumes and their alleviation by using beneficial microbes and plant growth regulators. She topped in undergraduation as well as master's programs at college and university level, respectively. She was awarded Ph.D. degree in 2013 by Panjab University, Chandigarh, India. She received Young Scientist Award at National Conference on Biotechnology and Environment, Department of Biotechnology, Jamia Millia Islamia, New Delhi, India, in 2017. She is a life member of Indian Society for Plant Physiology (ISPP) since 2016. She has published around 20 research and review articles in various SCI and SCOPUS indexed international journals. She has published around 10 chapters in edited books. She has presented her research work in several national and international conferences. She has more than 9 years of teaching and research experience at both college and university level.

Contributors

M. Annapoorneshwari Nitte University Centre for Science Education and Research, Nitte (Deemed to be University), Mangaluru, Karnataka, India

Anjali Awasthi Department of Zoology, University of Rajasthan, Jaipur, Rajasthan, India

Garima Awasthi Department of Life Sciences, Vivekananda Global University, Jaipur, Rajasthan, India

Kumud Kant Awasthi Department of Life Sciences, Vivekananda Global University, Jaipur, Rajasthan, India

Maria Antonia Machado Barbosa Grupo de Estudos em Fitotecnia, Universidade do Estado de Minas Gerais, Ituiutaba, Minas Gerais, Brazil

Kuldeep Bauddh Institute of Environment and Sustainable Development, RGSC, Banaras Hindu University, Mirzapur, Uttar Pradesh, India

Erin L. Connolly Department of Plant Science, Pennsylvania State University, University Park, PA, USA

Viviane Kopp da Luz Universidade Federal de Pelotas, Fitotecnia, Pelotas, Brazil

Allan Klynger da Silva Lobato Núcleo de Pesquisa Vegetal Básica e Aplicada, Universidade Federal Rural da Amazônia, Paragominas, Pará, Brazil

Km Sandhya Devi Department of Botany, University of Allahabad, Prayagraj, Uttar Pradesh, India

Francesco Di Gioia Department of Plant Science, Pennsylvania State University, University Park, PA, USA

M. Divyashree Department of Environmental Health and Toxicology, Nitte University Centre for Science Education and Research (NUCSER), Paneer Campus, Nitte (Deemed to be University), Mangaluru, Karnataka, India

Komal Dubey Plant Stress Biology Laboratory, Institute of Environment and Sustainable Development, Banaras Hindu University, Varanasi, Uttar Pradesh, India

Smitha Hegde Nitte University Centre for Science Education and Research, Nitte (Deemed to be University), Mangaluru, Karnataka, India

Harmanjit Kaur Department of Botany, University of Allahabad, Prayagraj, Uttar Pradesh, India

Harpreet Kaur Department of Agriculture, Jain College of Pharmacy, Abohar, Punjab, India

P. M. Krupa Nitte University Centre for Science Education and Research, Nitte (Deemed to be University), Mangaluru, Karnataka, India

Camille Ferreira Maia Núcleo de Pesquisa Vegetal Básica e Aplicada, Universidade Federal Rural da Amazônia, Paragominas, Pará, Brazil

Rachel Carmelita Mathias Nitte University Centre for Science Education and Research, Nitte (Deemed to be University), Mangaluru, Karnataka, India

Vishnu Mishra National Institute of Plant Genome Research (NIPGR), New Delhi, India

Department of Plant and Soil Sciences, Delaware Biotechnology Institute (DBI), University of Delaware (UD), Newark, DE, USA

Neeksha Nitte University Center for Science Education and Research, Nitte (Deemed to be University), Derelakatte, Mangaluru, Karnataka, India

Camila Pegoraro Universidade Federal de Pelotas, Fitotecnia, Pelotas, Brazil

Ynglety Cascaes Pereira Núcleo de Pesquisa Vegetal Básica e Aplicada, Universidade Federal Rural da Amazônia, Paragominas, Pará, Brazil

Pradip Poudel Department of Plant Science, Pennsylvania State University, University Park, PA, USA

K. Shama Prakash K.S. Hegde Medical Academy, Nitte (Deemed to be University), Mangaluru, Karnataka, India

Dhanyashree Rai Nitte University Centre for Science Education and Research, Nitte (Deemed to be University), Derelakatte, Mangaluru, Karnataka, India

J. S. Rajatha Sriram Nitte University Centre for Science Education and Research, Nitte (Deemed to be University), Mangaluru, Karnataka, India

Swati Sachdev Department of Liberal Education, Era University, Lucknow, Uttar Pradesh, India

Anuj Sharma Department of Forensic Science, School of Sciences, ITM University, Gwalior, Madhya Pradesh, India

Ruhani Sharma Department of Zoology, Vivek P. G. College, Jaipur, Rajasthan, India

Department of Life Sciences, Vivekananda Global University, Jaipur, Rajasthan, India

Fizanaj Sheikh Institute of Environment and Sustainable Development, RGSC, Banaras Hindu University, Mirzapur, Uttar Pradesh, India

Pavana Shree Nitte University Centre for Science Education and Research, Nitte (Deemed to be University), Mangaluru, Karnataka, India

J. R. Shubha Department of Food Safety and Nutrition, Nitte University Centre for Science Education and Research (NUCSER), Nitte Deemed to be University, Mangaluru, Karnataka, India

Kavita Shukla Plant Stress Biology Laboratory, Institute of Environment and Sustainable Development, Banaras Hindu University, Varanasi, Uttar Pradesh, India

Pragya Shukla Molecular Biology and Biotechnology Division, CSIR-National Botanical Research Institute, Lucknow, Uttar Pradesh, India

Department of Botany, Institute of Science, Banaras Hindu University, Varanasi, Uttar Pradesh, India

Jawahar Singh University of Cambridge, Sainsbury Laboratory (SLCU), Cambridge, UK

Sudhakar Srivastava Plant Stress Biology Laboratory, Institute of Environment and Sustainable Development, Banaras Hindu University, Varanasi, Uttar Pradesh, India

Bhawna Sunkaria Yadavindra Department of Science, Punjabi University Guru Kashi Campus, Talwandi Sabo, Bathinda, Punjab, India

Vishal Varshney Department of Botany, Govt. Shaheed GendSingh College, Uttar Bastar Kanker, Chhattisgarh, India

Institut de Biologie Moléculaire des Plantes du CNRS, Strasbourg, France

Vívian Ebeling Viana Universidade Federal de Pelotas, Pelotas, Fitossanidade, Brazil

Ranjana Yadav Department of Botany, University of Allahabad, Prayagraj, Uttar Pradesh, India

Part I

An Overview of the Zinc in Soil and Plant Responses

Zinc in Soils and the Regulatory Roles of Soil Properties

Anuj Sharma, Ruhani Sharma, Anjali Awasthi, Kumud Kant Awasthi, and Garima Awasthi

Abstract

Zinc has been proven to be a good micronutrient for both plants and animals, and it has also been essential in several physiological and metabolic processes. In soils, zinc occurs in different forms, and its accessibility is greatly affected by soil pH, texture, organic matter content, and moisture. This chapter studies the properties of soils with a particular focus on zinc and how such properties affect the solubility, retention, and mobility of the zinc substance. Soil pH affects the bioavailability of zinc significantly, and it has been observed that this element is highly available in acidic soils in the alkaline soils. Water, together with moisture content and drainage systems, is responsible for the distribution of water in the porosity of the soils and therefore influences the leaching and accumulation of zinc in the soil profiles. This chapter further analyzes the issue of zinc uptake by the plants and points out the deficiency problem as an absolute necessity for conducting further research aimed at developing such a management approach allowing zinc to be used by crops in the most efficient way possible. Also, the

A. Sharma
Department of Forensic Science, School of Sciences, ITM University, Gwalior, Madhya Pradesh, India

R. Sharma
Department of Zoology, Vivek P. G. College, Jaipur, Rajasthan, India

Department of Life Sciences, Vivekananda Global University, Jaipur, Rajasthan, India

A. Awasthi
Department of Zoology, University of Rajasthan, Jaipur, Rajasthan, India

K. K. Awasthi · G. Awasthi (✉)
Department of Life Sciences, Vivekananda Global University, Jaipur, Rajasthan, India
e-mail: gariimaa21@gmail.com

S. Srivastava, H. Kaur (eds.), *Zinc in Soil-Plant Continuum*,
https://doi.org/10.1007/978-981-96-4253-3_1

topic touches contamination of zinc and its threat to the health of the population and environment and explains the need for regulation to decrease threats.

Keywords

Zinc contamination · Zinc in soil · Environmental health · Soil properties

1 Introduction

For plants and animals alike, zinc (Zn) is an essential element that is necessary for healthy growth, development, and metabolic processes (Chen et al. 2022). Despite its minimal need, zinc significantly influences physiological and cellular functions. Zinc is essential for several enzymatic and structural processes in plants, which directly affect their ability to grow, absorb nutrients, and withstand environmental stresses (Shukla et al. 2016). In a similar vein, zinc is necessary for many enzymatic processes that promote general health and survival in animals, as well as for immune system performance, reproduction, and DNA synthesis (Tso and Forde 2021). Researchers in the fields of agriculture, nutrition, and environmental sciences extensively study zinc because deficiencies in either plants or animals can have detrimental effects (Zargar et al. 2015).

Zinc has a number of important metabolic roles in plants, prominent among them being the cofactor of several enzymes, including RNA polymerase, carbonic anhydrase, and superoxide dismutase (Rudani et al. 2018). It promotes the production of hormones, proteins, and chlorophyll, which aid in the general growth of plants. Additionally, zinc is essential for maintaining the cell's structural integrity and membrane activities (Pellinen et al. 2022). The synthesis of auxin, a plant hormone necessary for cell elongation, tissue differentiation, and general growth, is one of zinc's most notable functions in plants. Sufficient zinc levels boost auxin production, which promotes healthy plant growth. However, a zinc shortage inhibits the production of auxin, resulting in slowed growth, smaller leaves, and shorter internodes (Prasad 2008). Zinc deficiency is particularly detrimental during critical plant development phases such as seed germination and blooming, when the requirement for micronutrients increases. Zinc, therefore, has a direct effect on crop quality and yield, affecting plant health as well as agricultural output (Nielsen 2012).

Zinc's involvement in disease resistance and stress tolerance in plants is another important function. Zinc, as a cofactor of antioxidant enzymes, is known to protect plants from oxidative damage (Khan et al. 2022). Superoxide dismutase is one of these enzymes that scavenge damaging reactive oxygen species (ROS) created under environmental stress conditions like high light intensity, harsh temperatures, or dryness (Prasad 2008). Zinc neutralizes ROS to prevent oxidative stress, which otherwise has the potential to harm DNA, proteins, and cell membranes. Zinc also aids in strengthening plant defenses against infections (Cakmak 2004a). Adequate zinc levels in plants frequently result in higher defenses against fungi and other diseases, enhancing overall resilience. Zinc is an essential component of sustainable agricultural techniques because it promotes both growth and stress tolerance (Cakmak 2004b). This is particularly true in areas where soils are inherently low in zinc or at risk of depleting as a result of intensive farming (Esposito et al. 2006; Agrawal et al. 2021; Sharma et al. 2022).

Animals also require zinc for a variety of biological processes that ensure healthy growth, development, and maintenance (Blasiak et al. 2020). More than 300 enzymes, including those involved in DNA synthesis, protein metabolism, and cell division, depend on zinc as a critical component (Lall and Kaushik 2021). Its role in DNA replication and repair is crucial for cell development and reproduction. Zinc, for example, is necessary for healthy immune function because it promotes the growth and activation of T-cells, a subset of white blood cells that are important for the immunological response (Skrajnowska and Bobrowska-Korczak 2019). Animals with zinc deficiencies may have weakened immune systems, which leaves them more vulnerable to illnesses and infections. This is especially important for cattle, since low zinc levels may result in higher disease incidence and decreased output, which can cause financial losses. Zinc is also essential for maintaining reproductive health (Wessels et al. 2021). Males require zinc for spermatogenesis and testosterone maintenance, while females require it for ovulation and fetal development (Garner et al. 2021; Vickram et al. 2021). Insufficient zinc levels during pregnancy may lead to difficulties in the form of stillbirths or congenital abnormalities in the baby (Uriu-Adams and Keen 2010).

Zinc is needed not only for immunological and reproductive functions but also for wound healing, skin integrity, and epithelial tissue structural integrity (Jeng and Chen 2022). Enzymes involved in collagen formation, which is required for tissue regeneration and repair, need zinc as a cofactor. Therefore, animals deficient in zinc may experience sluggish wound healing, skin lesions, and reduced development (Hagmeyer et al. 2015). Zinc also influences taste and smell, and zinc deficiency can lead to decreased appetite and anorexia in animals. This is especially problematic for cattle since low feed intake has a direct effect on growth rates and yields (Gupta et al. 2020; Awasthi et al. 2020; Agrawal et al. 2023).

Zinc deficiency, a common issue affecting both plants and animals, has severe consequences for animal health and agricultural output. Zinc shortage in plants is most common in calcareous, sandy, and heavily worn tropical soils, where elements like high pH or low organic matter content lower zinc bioavailability (Shukla et al. 2021). Plants that are zinc deficient frequently exhibit chlorosis or yellowing of the leaves; necrosis or areas of dead tissue; and poor root growth (Upadhayay et al. 2022). We often apply zinc fertilizers to soils, either through soil application or foliar sprays, to address zinc deficiency, restore zinc levels, and boost crop yields (Hamzah et al. 2022). In areas where low dietary zinc consumption is a result of zinc-deficient soils, biofortification—the process of breeding crops with increased zinc content—is also gaining traction as a long-term way to boost zinc nutrition in human diets (Das and Green 2016; Recena et al. 2021).

Animals suffering from zinc deficiency may exhibit stunted development, reduced appetite, skin lesions, an impaired immune system, and infertility (Wen et al. 2024). These health problems may affect livestock that are grown on pastures lacking in zinc or that are given diets low in zinc that are bioavailable (Grabrucker et al. 2016). To counteract zinc insufficiency, animal diets often include mineral supplements. However, it's crucial to properly manage zinc supplementation, as excessive zinc can be harmful and hinder the absorption of other vital minerals such as iron and copper (Awasthi et al. 2017).

2 Zinc in Soil

A survey of soils in England and Wales revealed that the average zinc (Zn) content was 97 mg/kg, with a median of 82.0 mg/kg. However, the results included polluted soils from metalliferous mining and heavy sewage sludge applications (McGrath and Loveland 1992). In France, Baize (1997) found that Zn contents varied with soil texture, with median values of 17.0 mg/kg for sandy soils, 63.5 mg/kg for loamy soils, and 132.0 mg/kg for heavy clay soils. Pendias (1992) found that Polish soils had mean Zn concentrations of 37 mg/kg for sandy soils, 60 mg/kg for loess soils, and 75 mg/kg for loamy soils.

In Sweden, Eriksson (2001) discovered that Zn contents in agricultural soils ranged from 6 to 152 mg/kg, with a mean of 65 mg/kg. Gorny et al. (2000) observed that median Zn values in German soils were 27.3 mg/kg in sandy soils, 59.2 mg/kg in silty soils, and 76.4 mg/kg in clay soils. These examples show a clear tendency of increased Zn content in soils with higher clay percentages.

2.1 Zinc Forms in Soil

2.1.1 Soluble or Dissolved Zinc

Zinc is more bioavailable to plants in its soluble form. In the soil solution, it may be found as free Zn3+ ions or as complex species with both organic and inorganic molecules (Silwamba et al. 2022). Plant roots may easily absorb this type of zinc via mass flow and diffusion mechanisms. Because zinc tends to precipitate or adsorb onto soil particles, the concentration of soluble zinc in the soil solution is usually low, commonly in the range of 1–100 micromolars (Gondek and Mierzwa-Hersztek 2023). The amount of soluble zinc that is available depends on many variables. One of the most important variables is soil pH; when soil pH rises, zinc availability falls (Alvarez et al. 1996). Zinc's solubility is decreased in alkaline soils because it tends to precipitate as zinc hydroxide ($Zn\,(OH)_2$) or zinc carbonate ($ZnCO_3$). On the other hand, zinc becomes more soluble and accessible to plants in acidic soils. Zinc, however, may become overly soluble and hazardous to plants in very acidic environments (Ortiz et al. 2023). Various zinc forms and their solubility are summarized in Table 1.

Another important factor influencing the availability of soluble zinc is organic matter. Zinc may combine with organic molecules to produce complexes, such as fulvic and humic acids. Depending on the nature of the complex, this can either increase or decrease zinc's mobility. Zinc may also combine with anions like sulphate (SO_4^{2-}), chloride (Cl^-), and carbonate (CO_3^{2-}) to produce soluble complexes, which may alter how zinc behaves in soil (Boguta and Sokołowska 2020).

Table 1 Research studies on zinc in soil and plants

S. no.	Zinc form	Solubility	Deficiency symptoms in plants	Reference
1.	Zinc sulfate ($ZnSO_4$)	Highly soluble in water	Chlorosis in younger leaves	Daccak et al. (2022)
2.	Zinc oxide (ZnO)	Poorly soluble in neutral to alkaline soils	Stunted growth	Singh et al. (2021)
3.	Zinc carbonate ($ZnCO_3$)	Low solubility in water	Reduced leaf size	Pimentel et al. (2023)
4.	Zinc chelate (Zn-EDTA)	Highly soluble and bioavailable	Interveinal chlorosis	Haroon et al. (2022)
5.	Zinc phosphate ($Zn_3(PO_4)_2$)	Poor solubility in acidic soils	Necrotic spots on leaves	Nath et al. (2024)
6.	Zinc chloride ($ZnCl_2$)	Highly soluble in water	Decreased seed yield	Hosamani et al. (2020)
7.	Zinc acetate ($Zn(C_2H_3O_2)_2$)	Moderate solubility in water	Leaf curling	Pal et al. (2022)
8.	Zinc nitrate ($Zn(NO_3)_2$)	Highly soluble in water	Delayed flowering	Farahzadi et al. (2020)
9.	Zinc silicate (Zn_2SiO_4)	Low solubility in water	Reduced root development	Luyckx et al. (2023)
10.	Zinc-humate complex	Moderately soluble	Lower chlorophyll content	Pirzada et al. (2024)
11.	Zinc sulfate monohydrate ($ZnSO_4{\cdot}H_2O$)	Highly soluble in water	Leaf yellowing	Suganya et al. (2020)
12.	Zinc oxide nanoparticles	Low solubility, improved with nanotechnology	Enhanced root architecture but chlorosis	Kashyap and Siddiqui (2022)
13.	Zinc ferrite ($ZnFe_2O_4$)	Insoluble in water, bioavailability enhanced in acidic conditions	Decreased fruit yield	Garg et al. (2024)
14.	Zinc-manganese oxide	Low solubility, co-applied with micronutrients	Necrosis in older leaves	Semenova et al. (2024)
15.	Zinc-amino acid complex	Highly soluble and bioavailable	Improved plant biomass with limited leaf discoloration	Saini et al. (2024)
16.	Zinc sulfate heptahydrate ($ZnSO_4{\cdot}7H_2O$)	Highly soluble	Reduced fruit size	Singh et al. (2023a, b)
17.	Zinc phosphate nanoparticles	Poor solubility, higher effectiveness in acidic soils	Stunted plant growth	Nath et al. (2024)
18.	Zinc malate ($Zn(C_4H_4O_5)$)	Moderately soluble	Leaf wilting	Montanha et al. (2020)

(continued)

Table 1 (continued)

S. no.	Zinc form	Solubility	Deficiency symptoms in plants	Reference
19.	Zinc citrate ($Zn_3(C_6H_5O_7)_2$)	Moderately soluble	Decreased chlorophyll production	Thiyagarajan (2021)
20.	Zinc-alginate complex	Low solubility, enhanced with organic complexing agents	Lowered seed germination	Selvapandian et al. (2024)
21.	Zinc borate ($Zn_3(BO_3)_2$)	Low solubility	Necrotic leaf margins	Petrová and Soudek (2022)
22.	Zinc sulfide (ZnS)	Insoluble in water, availability dependent on microbial activity	Chlorotic veins in leaves	Saini et al. (2021)
23.	Zinc-citrate complex	Moderate solubility	Poor root elongation	Thiyagarajan (2021)
24.	Zinc-iron complex	Low solubility	Reduced internode length	Cao et al. (2022a, b)
25.	Zinc-copper complex	Moderate solubility, co-applied with micronutrients	Leaf discoloration	Liščáková et al. (2022)
26.	Zinc-gluconate	Highly soluble in water	Delayed fruit ripening	Saini et al. (2024)
27.	Zinc-phytate complex	Poor solubility	Stunted plant growth	Shen et al. (2023)
28.	Zinc-nitrate nanoparticles	High solubility	Increased biomass but stunted growth without micronutrients	Palacio-Márquez et al. (2022)

2.1.2 Zinc Adsorbed

Zinc ions that are loosely attached to the surfaces of soil particles, especially clay minerals, organic matter, and metal oxides like aluminum (Al) and iron (Fe) oxides, are referred to as adsorbed zinc. Although plant roots may absorb dissolved zinc and reduce its concentration in the soil solution, this form of zinc is not immediately accessible to plants. Instead, it can become available by desorption into the soil solution (Schlattl et al. 2021). Several variables, including soil pH, cation exchange capacity (CEC), and the availability of competing cations such as calcium (Ca^{2+}) and magnesium (Mg^{2+}), affect the adsorption of zinc into particles (Cao et al. 2022a, b). More zinc may be adsorbed onto soil particles in soils with high CEC, which would decrease the amount of zinc available. More zinc is absorbed as soil pH rises, which further lowers zinc's bioavailability to plants (Szerement et al. 2022).

2.1.3 Zinc in Mineral Forms

Additionally, zinc may be found in soil as a component of primary and secondary minerals. Because they are comparatively insoluble, primary zinc minerals like sphalerite (ZnS) and smithsonite ($ZnCO_3$) are usually of geologic origin and have less effect on the immediate availability of zinc for plant absorption (Szerement et al. 2022). These minerals may be gradually broken down by weathering processes over lengthy periods, releasing zinc into the soil solution (Singh et al. 2024).

Under some circumstances, secondary minerals, such as hydroxides and oxides of zinc, may occur in soils, especially those with high pH levels. Even while these minerals are more soluble than primary minerals, they nonetheless make up a small portion of the zinc that is immediately accessible (Adamczyk-Szabela et al. 2015).

2.2 Zinc Deficiency in Soil

Zn deficiency in agricultural soils is regarded as the most geographically widespread micronutrient deficiency constraint limiting crop production (yield losses can exceed 40%), whereas its excess in soil can be geological or anthropogenic in origin (Alloway 2008; Singh et al. 2005; Sadeghzadeh 2013). Zn deficiency is more common in calcareous (high pH) soils that have been levelled by machines to provide consistent irrigation coverage. This is because field levelling, particularly in calcareous soils, destroys topsoil and micronutrient-rich organic debris (Alloway 2008). Furthermore, deficits arise in soils with low total Zn concentration (e.g., sandy soils) or soils with relatively high accessible phosphorus content (Singh et al. 2005).

However, overall Zn concentration is not a reliable indicator of the soil's ability to deliver Zn for plant uptake. The soil solution contains a modest amount of Zn ($<$ 1 mg/kg), which crops can absorb (Kabata-Pendias 2000). The main causes of Zn deficiency in crops are mainly soil-related: low Zn availability (high pH, calcareous, and sodic soils); low total soil Zn content (especially in sandy, sodic, and calcareous soils); loss of soil organic matter; presence of nitrogen, sodium, calcium, magnesium, and phosphates; restricted root exploration due to soil compaction or high water table; and climate factors (Alloway 2008; Cakmak 2008; Alloway 2009).

As a result, Zn deficiency in crops can develop in a variety of soil types, including highly weathered soils, calcareous high pH soils (semiarid locations), sandy soils, and acid-leached soils with low total Zn concentration (Takkar and Walker 1993). Deficiencies can also be attributed to soil parent material inheritance, as demonstrated in Northern Greece, where Zn deficiency was detected in 92.8% of the acidic soils investigated (Noulas et al. 2018). Plant accessible Zn in Crete, Greece ($n = 80$), varied between 0.2 and 12.3 mg/kg due to variances in soil parent materials and farming practices (Vavoulidou et al. 2009).

3 Soil Properties Influencing Zinc Availability

3.1 Soil pH

Soil pH is a major factor in determining the mobility, chemical forms, and solubility of zinc (Zn) in soils. Zinc is a very important element for plants in micronutrient values, and its availability is dependent on the soil pH which affects the forms in which zinc is found in soils. Obviously, depending on the pH of the soil, zinc could be found in exchangeable form, organically complexed, or tightly bound to iron and manganese oxide minerals as mentioned in Table 2. These forms are responsible for whether zinc is in a form that is easy for plant uptake or whether it gets fixed by the soil (Martınez and Motto 2000). Zinc also tends to be in an exchangeable form which is within easy reach of the plants in acidic soils (low pH). Research depicts that in soils with PH less than 5.5, zinc has been observed to be present dominated

Table 2 Role of soil properties influencing zinc uptake in plants

S. no.	Soil property	Effect on zinc uptake	Plant type	Study findings	References
1.	Soil pH	High pH reduces zinc availability, leading to deficiency	Wheat (*Triticum aestivum*)	Alkaline soils (>7.5 pH) restrict zinc solubility, reducing uptake	Elhaj Baddar and Unrine (2021)
2.	Organic matter	High organic matter enhances zinc complexation and mobility	Maize (*Zea mays*)	Organic amendments increase zinc uptake due to higher microbial activity and chelation	Abedi et al. (2022)
3.	Soil texture	Clay soils tend to fix zinc, reducing its availability	Rice (*Oryza sativa*)	Clay-rich soils show lower zinc uptake due to zinc adsorption on clay particles	Sipos et al. (2021)
4.	Soil moisture	Excess moisture can lead to zinc leaching or reduced uptake	Soybean (*Glycine max*)	Waterlogged soils reduce zinc uptake due to lower root aeration	Costa and Coutinho (2022)
5.	Cation exchange capacity (CEC)	High CEC soils can hold more zinc but may also bind it tightly, reducing availability	*Zea mays*	High CEC leads to better retention of zinc, but excessive retention decreases uptake	Suganya et al. (2020)
6.	Soil salinity	High salinity reduces zinc uptake due to ionic competition	Barley (*Hordeum vulgare*)	Saline soils reduce zinc absorption by roots due to competition with sodium and chloride ions	Noreen et al. (2021)
7.	Soil compaction	Compacted soils reduce root growth, limiting zinc uptake	Corn (*Zea mays*)	Compacted soils reduce root access to zinc, especially in the subsoil layers	Costa and Coutinho (2022)

(continued)

Table 2 (continued)

S. no.	Soil property	Effect on zinc uptake	Plant type	Study findings	References
8.	Phosphorus levels	High phosphorus levels in soil antagonize zinc uptake	Spinach (*Spinacia oleracea*)	Excessive phosphorus reduces zinc availability due to formation of zinc-phosphate complexes	Singh et al. (2023a, b)
9.	Soil aeration	Poorly aerated soils reduce zinc bioavailability	Wheat (*Triticum aestivum*)	Lack of oxygen in compacted or waterlogged soils reduces zinc solubility	Steffens et al. (2005)
10.	Carbonates	Calcareous soils bind zinc, making it less available for plant uptake	Alfalfa (*Medicago sativa*)	Calcareous soils (>15% $CaCO_3$) limit zinc mobility, leading to plant deficiency	Grewal and Williams (1999)
11.	Soil microbial activity	Microbial activity enhances zinc solubility and availability	Peas (*Pisum sativum*)	High microbial activity increases zinc availability via mineralization of organic matter	Sharma et al. (2023)
12.	Soil temperature	Higher soil temperatures increase zinc availability	Cotton (*Gossypium hirsutum*)	Warm soil temperatures increase microbial activity, enhancing zinc uptake	Ramazanoglu et al. (2024)
13.	Soil redox potential	Low redox potential in waterlogged soils reduces zinc solubility	Rice (*Oryza sativa*)	Anaerobic conditions reduce zinc uptake in paddy soils due to precipitation of zinc as insoluble forms	Gao et al. (2012)
14.	Soil structure	Well-structured soils promote root growth and zinc uptake	Sorghum (*Sorghum bicolor*)	Soils with good structure (aggregates) enhance root access to available zinc	Dimkpa et al. (2019)
15.	Fertilizer application	Zinc fertilizers improve uptake in deficient soils	Wheat (*Triticum aestivum*)	Zinc application (e.g., zinc sulfate) enhances uptake, especially in deficient soils	Awasthi et al. (2017)

(continued)

Table 2 (continued)

S. no.	Soil property	Effect on zinc uptake	Plant type	Study findings	References
16.	Sulfur content	Adequate sulfur levels increase zinc uptake by improving root metabolism	Brassica (*Brassica oleracea*)	Co-application of sulfur improves zinc uptake and overall plant health	Shah et al. (2022)
17.	Soil acidity	Acidic soils improve zinc availability but may cause toxicity at very low pH	Beans (*Phaseolus vulgaris*)	Zinc availability increases with low pH (<5.5), but excessive acidity may lead to toxicity	Magomere et al. (2024)
18.	Soil aluminum levels	High aluminum levels reduce zinc uptake due to antagonism	Wheat (*Triticum aestivum*)	Aluminum toxicity in acidic soils reduces zinc uptake by altering root structure	Abedi and Mojiri (2020)
19.	Soil iron content	High iron levels compete with zinc for plant uptake	Rice (*Oryza sativa*)	Excessive iron in soils can block zinc uptake by roots	Shrestha et al. (2020)
20.	Soil potassium levels	High potassium levels reduce zinc absorption	Tomato (*Solanum lycopersicum*)	Excessive potassium application reduces zinc uptake by competing at the root surface	Ali et al. (2022)
21.	Soil magnesium content	High magnesium levels reduce zinc uptake due to ionic interactions	Peas (*Pisum sativum*)	Magnesium excess reduces zinc absorption due to competition at root uptake sites	Liščáková et al. (2022)
22.	Soil organic carbon	Organic carbon enhances zinc solubility by forming chelates	Wheat (*Triticum aestivum*)	Organic carbon enhances zinc mobility and availability in the soil, promoting better uptake	Abedi and Mojiri (2020)
23.	Soil depth	Zinc availability decreases with increasing soil depth	Maize (*Zea mays*)	Zinc is more available in topsoil than subsoil, requiring deeper rooting for higher uptake	Bossolani (2022)

(continued)

Table 2 (continued)

S. no.	Soil property	Effect on zinc uptake	Plant type	Study findings	References
24.	Soil sodium adsorption ratio (SAR)	High SAR levels reduce zinc uptake due to salt stress	Barley (*Hordeum vulgare*)	High sodium adsorption reduces zinc uptake by altering root physiology	Khan et al. (2024)
25.	Soil heavy metals	Heavy metals such as cadmium reduce zinc uptake due to competition	Lettuce (*Lactuca sativa*)	Heavy metal contamination (e.g., cd) reduces zinc uptake due to ionic competition	Li et al. (2020)

in its exchangeable fraction. This improves the utilization of zinc by plants. On the other hand, soil pH has to be avoided at low levels so that excess solubility of zinc does not occur leading to adverse effects on some crops. However, in the case of alkaline soils (higher pH), zinc reduces solubility; hence it gets precipitated in organic matters or oxides and is not easily absorbed by plants (McBride and Blasiak 1979). As pH goes up above 6.0, zinc moves from the exchangeable soluble form to the bound form with iron oxides or chelation with organic matter, making it less accessible to crops. In soils whose pH levels are above 7.0, implementable zinc opportunities lie extremely low, therefore causing deficiency diseases regardless of adequate total zinc levels in the soil (Sims 1986). Over-dominance of exchangeable form makes zinc that is significantly available in low pH soils. It has been shown that in surroundings with a pH of less than 5.2, exchangeable zinc dominates and thereby encourages bean uptake. For example, in a study from Delaware on poorly buffered coastal plain soils, maximum solubility of zinc was noted in the acid range of the soils, and crops responded positively to micronutrient contents (Xian and In Shokohifard 1989). Unfortunately, too much soil acidity leads to the toxication of zinc. The application of poultry manure (PM) to low pH soil has been noted to promote retention of considerable proportions of zinc that are not exchangeable, making them less available in the short term but possibly preventing plants that are sensitive to high levels from being exposed to high levels of zinc. Reasons behind the limitations in the uptake of zinc from Carats for plants, especially at elevated soil pH's, stems from the processes involving the fixation of zinc by organic matter and iron oxide. As explained in the same Delaware study, coastal plain soils are prone to overliming, which refers to adding excessive lime to increase the soil pH which in turn causes the zinc to immobilize (McBride et al. 1997). Past 7.0, lime soil helps farmers in NH, but the solubility of zinc is frequently reduced, and the compounds leached off and/or permanently secured in iron oxides. This severely results in problems for Delaware among other areas whereby over-application of lime and/or manure increases the soil's relative pH to levels where zinc deficiency occurs even when the soil contains enough zinc (Xian and In Shokohifard 1989, Römkens and Salomons 1998). Soil pH is a key factor in determining zinc availability to plants. Zinc is more soluble and readily available in acidic soils (low pH)

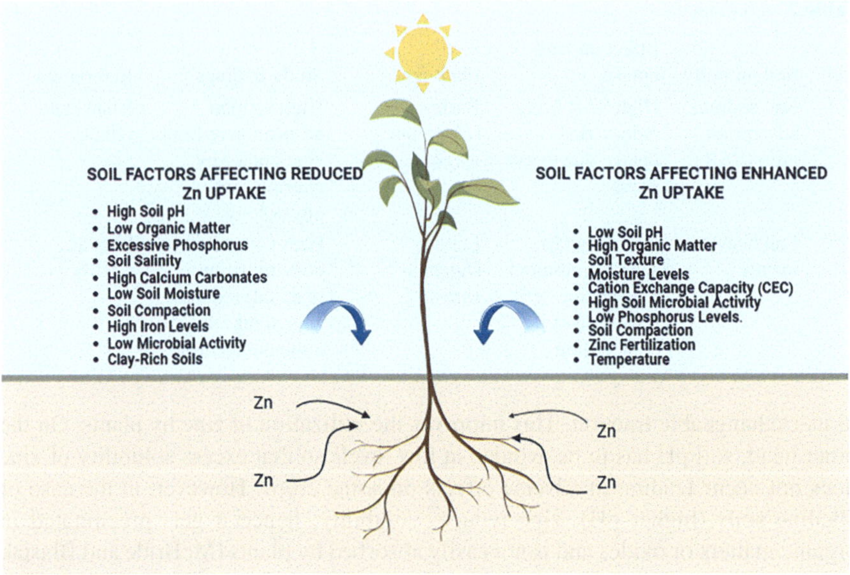

Fig. 1 Soil factors that affect zinc uptake in plants

compared to alkaline soils (high pH) (Hosamani et al. 2020). In alkaline soils, zinc can form insoluble compounds, such as zinc hydroxides or zinc carbonates, reducing its bioavailability to plants. Conversely, in acidic soils, zinc remains in its soluble form, allowing for greater uptake by plant roots (Rengel 2015). Figure 1 is showing the factors affecting Zn uptake in plants.

3.2 Soil Texture

Soil texture characterized by the constituents of sand, silt, and clay is an important factor for the retention and movement of zinc in soil. The physical properties such as the size and surface area of soil particles affect the behavior of zinc when dispersed in the soil environment. Sandy soils which are coarse in texture have large particles but with less surface area and cation exchange capacity (CEC) (Azouzi et al. 2015). Therefore, the amount of zinc retained will be very little and hence available in the soil for displacement. However, in terms of the retention of zinc, this will not be the case with soils that have finer textures such as clay. This is because soils that have clay in composition are dominated by smaller particles and in this case have bigger CEC; hence absorption of zinc increases, making its retention swift (Angelaki et al. 2022). In these soils, zinc when present can be retained either in solution intercalated in the clay lattices or adsorbed on organic matter which reduces leachability but in turn may decrease the immediate supply to plants. This is consistent where it has been shown that there is zinc retention in small particles as clay while little zinc retention happens in large particles such as sand (Duplay et al. 2014). According to a study of long-term soil management, it was noted that, for particular retention of metals such as zinc, the smaller particle size fractions (clay > silt > fine sand > coarse sand) were much more effective than the bigger ones. This is because the smaller particles being more attractive in sorption are usually rich in surface area and iron and manganese oxides (Lair et al. 2007). While confirming the dynamic relationship between soil texture and zinc mobility, one study compared the extent of zinc derived from various sources (Zn-amino acids and Zn-DTPA-HEDTA-EDTA) in weakly acidic and neutral soils and zinc distribution to different soil layers. There was low zinc mobility for sandy soils of poor permeability because the coarse particles and the little surface area were unable to hold the metal (Alvarez 2007). Highly complex organic molecules mainly restrained the mobility of zinc exogenously administrated to high clay soils. However, as more basic zinc fertilizer as synthetic chelates was subsequently applied, it leached out less in these soils, emphasizing the role of texture in controlling the zinc nutrition status of crops when grown. Also, organic matter content and variation in particle size are important factors affecting zinc sorption. In a prolonged field study, the capacity of different soils treated with farmyard manure or mineral fertilizer to retain metals was found to be quite different (Najafi et al. 2012). Clays and clayey soils with greater amounts of organic matter retained zinc, copper, and cadmium, more especially in their clay fractions, due to their high surface area and carbon content. However, in sandy soils with lower organic contents, zinc resided in more leachable fractions posing a risk of loss through leaching. Therefore, the influence of soil texture on the retention and mobility of zinc is dependent on the size of the soil particles and their arrangement. Soils that are rich in clay and organic matter can hold more zinc, while sandy soils have more ease with the mobility of the metals. Such interactions are important in the vicinity of zinc fertilization management approaches and the soil over a long period as they enhance the availability of zinc for plant absorption as well as the potential loss of zinc through leaching (Brazauskiene et al. 2008; Santos et al. 2019).

3.3 Organic Matter Content

The binding and mobility of zinc in soil particularly those of high organic matter content are highly influenced by zinc and its interaction with organic matter. Organic matter, be it particulate organic matter or dissolved organic matter, plays an essential role in binding Zn and other heavy metals, thereby affecting the bioavailability of plants (Pérez-Novo et al. 2008). Organic matter-rich soils normally do complexation with metals effectively, and only a fraction becomes bulk phase, and most is not obtainable; thus the metals are non-mobile. Organic matter contains dedicated proportions of certain elements such as humic substances and protein-like components which affect the binding of zinc to the organic matter (Jiang et al. 2015). POM has been suggested to act as a dryer for metals in the soil particularly paddy soils. The porous structure of the POM, as well as the functional groups such as hydroxyl and carboxyl, makes these organic polymers hold metals such as zinc (Latrille et al.

2003). Studies have shown that POM with its high absorption ability for metals is even more improved in zinc absorption by bacteria present at the root region which roughens and increases the surfaces of the cells. It is in such an environment where zinc is mainly in bound form with the organic substance of Zn-humic complexes which constitutes the major contribution of POM and zinc interaction. However, in some cases, the stability of these chelates may alter leading to a relative increase in the fraction of zinc otherwise entrenched in softer or more mobile types such as zinc sulfide. In cases accompanied by POM, dissolved organic matter (DOM) is also instrumental when it comes to the binding of zinc, especially for water body sediments (Shi et al. 2018). Researches conducted at Lake Dongping in China demonstrate how different DOM components react with zinc and what substances are responsible for DOM zinc binding. Humic-like and protein-like components present in DOM are bound to zinc but with different fates (Lü et al. 2020). For instance, the binding of zinc to protein-like fractions which the tyrosine and tryptophan are in is more forthright than to humic-like fractions. This points out that the amount and type of organic matter in the soils or even sediments could affect how zinc is held within the medium and how far it can move. It has been reported that zinc does promote the humification of DOM which in turn leads to a higher retention of zinc later on making it less available to plants in the short period. The effect of organic matter on zinc availability is evident in the observation whereby the protein-like components were suppressed while enhancing human-like components when zinc was added (Cheng and Allen 2006). This shift toward humic-like complexes would mean that soils with a higher organic matter content would be more effective in holding zinc since humic substances are associated with strong metal binding. Organic matter retention of zinc can be beneficial for plants growing in zinc-deficient areas as it holds relatively more zinc. However, this is double corn because organic matter improves zinc complexation, so availability for plants to absorb is less as the metal is mostly in stable complexes (Pinto et al. 2004). In those soil management practices over a long period, it has been observed that organic matter addition such as that of farm yard manure enhances the retention of metals like zinc in soil. Mineral soils supplemented with organic matter, especially farm yard manure or compost, also have a higher zinc sorption ability even in small particle size fractions that are believed to constrain organic matter. This enhanced retention, however, may lead to decreased mobility of zinc, thereby preventing its leaching to groundwater but, in the process, reducing its potential availability for uptake by crops (Jouvin et al. 2009).

3.4 Soil Moisture and Drainage

Soil wetness and drainage have an important impact on the bioavailability and mobility of zinc (Zn) in the soil. Once the degree of soil moisture increases, zinc usually becomes more available since it dissolves easily in water. Nevertheless, excessive moisture contents in soils, especially poorly drained ones, often lead to the leaching of zinc out and not in. For South Florida, high water holding sandy soils may easily leach out zinc, due to flooding results in its movement beyond the rooting depth/zone, therefore making it unavailable to the plants (Min et al. 2013; Angelaki et al. 2022). Boom and Chiorinis in the Indian River area of Florida show how long-standing nut-saturated conditions, such as after rainstorms, help load zinc into water even more readily (He et al. 2006). Zinc leaching is mainly from the carbonate-bound portion which is high in solubility and mobility when the moisture level is high (Bauer and Lindsay 1965). The effect of drainage conditions on zinc mobility is also of great importance, especially in poorly drained soils or flood-prone areas. Where there is inappropriate drainage in the soil, there is less oxygen present. This type of condition can change the forms of zinc in the soil, making it more leachable (Udovic and Lestan 2009). Using column leaching experiments performed in South Florida, it was determined that zinc leaches easily with decreasing pH and the leachability decreases as the pH level increases from 3.0 to 9.0. Effective drainage control will assist in that it will reduce the leaching of zinc compounds, especially in soils that are acidic and which contain more ion-exchangeable zinc (Udovic and Lestan 2009). When adequate drainage facilities are provided, zinc remains largely immobilized within soils, thereby limiting loss from such soils to water bodies. Water and air location also control the zinc's solubility by altering the concentration of these ions in the calcium-saturated soil. Soil characteristics are further connected to cations in the film and soil solution as cation exchange creates a cation-anion interface (Beesley et al. 2010). The study revealed that upon adding cations (calcium, Ca; potassium, K; sodium, Na; ammonium, NH4+; etc.), cation-exchangeable zinc release was enhanced. This is especially relevant in lowland regions and within soils that are frequently waterlogged. Sodium in particular disrupts the soil's colloids, thereby enhancing the release of zinc from the soil, while calcium enhances the structure of organic matter minimizing the mobility of zinc (Gonzalez et al. 2008).

4 Zinc Uptake by Plants

Zn is absorbed by the roots of the plant, which take it from the soil and distribute it to various parts of the plant for growth. Zinc is taken up by plants from the soil in the form of Zn^{2+} and moved to the shoot through the xylem (Clemens 2001; Hart et al. 1998). Zinc is moved from the soil into the root cell by the plasma membrane (Krishna et al. 2017). Zinc can move from root epidermal cells to cortical cells and then to the xylem through the symplast pathway, before being pumped into the stellar apoplast (Caldelas and Weiss 2017). Zinc is taken up by the plant from the soil via transporters bound to the membrane (Hacisalihoglu and Kochian 2003). These transporters are involved in the absorption of Zn from the soil, transport within the plant, xylem loading and unloading, vacuolar sequestration, and remobilization from the vacuole (Krishna et al. 2017). The difference is caused by variations in the levels of metal transporter proteins (MTPs, ZIPs, VITs) and internal binding sites in a specific organ. Various factors like the type of plant, its growth stage, amount of biomass, and soil zinc concentration, among others, also play a role in affecting zinc accumulation levels (Gupta et al. 2016; Rudani et al. 2018).

4.1 Factors Influencing Plant Uptake of Zinc

Zn accumulation is affected by several variables, including species and biotic and abiotic environmental influences. The morphological characteristics of plant organs greatly influence the plant's ability to absorb and accumulate zinc. Root biomass, age, root cover in the soil, root length, and lateral root density all impact how much zinc is absorbed by roots. The amount of zinc in leaves, stems, flowers, and fruits is regulated by the biomass of the shoot, the quantity and area of leaves, and the distribution of zinc among metal sinks. Zn accumulation also differs throughout plant development stages (Gupta et al. 2016). A recent study on *Solanum nigrum* found that the post-flowering stage of the roots' cortical parenchyma had greater zinc buildup (Samardjieva et al. 2015). Reduced Zn translocation to aerial parts and strong constitutive production of organic acids in root tissue contributed to the post-flowering stage's additional Zn tolerance. The amount of zinc in leaves varies depending on where they are on the stem. Zn accumulation in older leaves is lower than in younger leaves. Zn levels change from the top to the bottom of the stem of the plant (Vymazal and Březinová 2015; Gupta et al. 2016).

4.2 Role of Soil Properties in Influencing Plant Zinc Uptake

The pH of the soil affects how well Zn fertilizer is absorbed by the soil. Zinc is more readily available in soil with an acidic pH. With every pH unit increase, soil solubility drops 100 times (Lindsay 1991). In addition to pH, complexation with organic matter, blockage in minerals, diffusion into micropores and interparticle space, solid-phase diffusion, and co-precipitation with other metals are further variables that affect zinc fixation (Prasad et al. 2014; Sparks 1998; Tye et al. 2003). Since Zn primarily enters plant roots by diffusion, the process is limited by low soil moisture and organic matter, which lowers plant Zn availability (Alloway 2009). Restrictive moisture conditions and poor organic matter are frequently seen in agricultural areas; as a result, it is anticipated that these circumstances will limit plant roots' absorption of zinc (Graham et al. 1992). In soils with limited zinc availability, soil moisture is crucial. In the diffusion process, soil moisture serves as a channel for the movement of zinc from the soil to the root (Cakmak 2008). Therefore, the detrimental effects of a zinc shortage are more lethal in rainfed environments than in irrigated ones. The ideal zinc content for fertile, healthy soil is between 1 and 200 mg/kg. It would be wise to examine the food that is produced in your soil if the concentration of nickel in it is more than 7400 mg/kg. You should also consult an expert to see whether any particular remedial action is necessary if the levels are high (Rengel 2002). Low plant-available zinc soils are prevalent in tropical and temperate climates, although they are most prevalent in areas with Mediterranean temperatures (Sillanpää and Vlek 1985; Rudani et al. 2018).

The following factors may impact zinc levels and availability in soil:

- pH: A higher pH reduces zinc availability by lowering zinc levels.
- Phosphorus: Zinc availability is reduced by high phosphorus levels.
- Organic matter: Organic matter can enhance zinc availability and provide zinc.
- Nitrogen: Low amounts of nitrogen might hinder a plant's capacity to absorb zinc.
- Moisture: Plant absorption of zinc is impeded by very damp soils.
- Copper: Plants absorb copper and zinc through the same process, so when one is abundant, plants don't absorb enough of the other.
- Magnesium: Magnesium may help in zinc absorption.
- Arsenic: Elevated arsenic levels can prevent the absorption of zinc (Rudani et al. 2018).

4.3 Strategies for Managing Zinc Availability for Optimal Plant Nutrition

Numerous global research initiatives related to zinc biofortification have already been launched to address nutritional deficiencies. Since cereal grains are the primary staple food in poor nations, they are the obvious option for these biofortification initiatives. Both the agronomic and genetic biofortification strategies for treating zinc deficiency are covered by these biofortification challenge programs (Fageria and Press 2009; White and Broadley 2009; Zou et al. 2012). The efficiency of plants' absorption of zinc is a determining factor in agronomic biofortification, which refers to the use of agronomic techniques such as fertilization, soil alkalinity remediation, crop rotation, and introduction of beneficial microorganisms to increase Zn phytoavailability in soil (Phattarakul et al. 2012).

The amount of zinc that plants may use is increased by applying fertilizers to the soil or by spraying exogenous zinc on leaves. However, because nutrients are lost during root absorption, foliar treatment is frequently thought to be preferable to soil application (Gupta et al. 2016).

Apart from agronomic fortification, genetic modification offers a broad opportunity to increase Zn content in crops like rice, wheat, corn, chickpeas, brassica, etc. The wild germplasm of these crops has significant genetic variability for several characteristics, such as zinc uptake through the roots and zinc accumulation in edible tissues (leaves, grains, fruits, etc.) (Graham et al. 1999; Bänziger and Long 2000; Cakmak 2004b; Gupta et al. 2016).

A plant's capacity to absorb mineral nutrients from the soil is influenced by the growth of its roots. Therefore, when there is a specific zinc deficit, the root architecture needs to be modified to acquire zinc. Additionally, a plant's capacity to absorb mineral nutrients from the soil is influenced by the development of its roots (Verma et al. 2021). For example, this can lead to an increase in the length and density of lateral root hairs, an increase in the area of the active root surface, and the ability to explore a larger volume of soil, which may improve the efficiency of zinc absorption (Hamzah et al. 2022). To support the sustainability and resilience of agriculture, plant biostimulators are a crucial component of modern integrated crop management systems (Hamid et al. 2021). This may be accomplished in various ways, including by providing biostimulants or applying a variety of plant nutrition techniques, such as adding trace components to meals that are generally absent from a regular human diet (Yakhin et al. 2017).

One kind of microbial biostimulant employed in agricultural biofortification is arbuscular mycorrhizal fungus (Verma et al. 2021). By incorporating mycorrhizal uptake by roots, these beneficial soil fungi establish mutualistic relationships with most land plant species and generate additional mycelia networks, which help expand the area of soil that has been studied and facilitate the transport of both macro- and micronutrients to plant tissues (Hamid et al. 2021). It is believed that maintaining soil microbial diversity is essential to ensuring that agricultural systems can sustainably produce food. However, there is still much to learn about the connections between microbial diversity and environmental processes (Kumar et al. 2017). The place where roots and soil interact and where the roots alter the soil is referred to as the "rhizosphere." The microbial activity in the soil surrounding a plant's rhizome is influenced by its roots. Bacteria that colonize plant roots and positively impact plant development through direct or indirect means are known as plant growth-promoting bacteria (Fahsi et al. 2021; Hamzah et al. 2022).

5 Environmental and Health Implications

5.1 Impact of Soil Zinc Levels on Plant Health and Crop Yields

Zinc is an essential element for plant growth and development. Depending on its form and concentration, zinc (Zn) can have both positive and negative impacts on a plant's health and growth. It is important to comprehend these impacts to optimize farming techniques.

5.1.1 Essentiality of Zinc and Its Deficiency Effects on Plants

Plant development and reproduction require zinc for proper, healthy growth. This element is necessary in trace levels to maintain the structural and functional integrity of membranes and to facilitate the regular operation of various essential physiological pathways in plants (Hafeez et al. 2013). Zinc plays critical roles in the regulation of growth, activation of enzymes, regulation of gene expression, activity of phytohormones, protein synthesis, photosynthesis, carbohydrate metabolism, fertility, seed development, and disease resistance (Marschner 1995). Inadequate zinc levels can hinder these physiological processes, jeopardizing plant well-being and efficiency, resulting in significant declines in growth, reduced crop yields, and substandard agricultural output (Sadeghzadeh 2013). Zinc is the only metal required in all six enzyme classes (oxidoreductases, transferases, hydrolases, lyases, isomerases, and ligases) (Singh et al. 2005).

Zinc is crucial for protein metabolism. Its role in stabilizing and enabling the functionality of genetic information is the primary function of zinc in protein synthesis (Singh et al. 2005). Zinc is essential for gene expression, DNA/RNA metabolism, and chromatin structure. Protein synthesis is inhibited by zinc deficiency due to RNA degradation (Cakmak et al. 1989), RNA polymerase inactivity, ribosomal deformation, and a reduction in ribosome count (Sadeghzadeh 2013; Suganya et al. 2020).

5.1.2 Impact of Excess Zinc on Physiological Functions of Plants

Elevated zinc concentrations in soils are phytotoxic, leading to various anatomical and functional defects that impair crop growth. These reactions differ depending on the plant species and its growth stage (Khudsar et al. 2004; Vaillant et al. 2005). The most apparent physiological responses to zinc toxicity include reductions in germination vigor and biomass, which ultimately lead to lower yields and product quality (Garg and Kaur 2013; Garg and Singh 2018). Excess zinc inhibits amylase activity and the transport of sugars to the embryo axis, reducing germination percentages (Sharma et al. 2009).

Zinc toxicity reduces the relative water content (RWC) in plant tissues or alters water relations (Garg and Singh 2018). This may result from decreased water transporters and ration rates, as well as severe dehydration caused by zinc accumulation in aerial tissues (Ghnaya et al. 2010). Reduced RWC due to excessive zinc exposure might be due to xylem blockage, disrupting the sap ascent process and decreasing the absorbing surface of roots or vascular systems (Koleva 2010). Zinc toxicity also decreases photosynthesis and respiration, creates redox imbalances, and weakens the defense system (Kaur and Garg 2021).

5.2 Potential Environmental and Human Health Concerns Related to Zinc Contamination

Zinc can occur in one of five forms: water-soluble, complexed, chelated, adsorbed, or exchangeable. These forms differ in terms of uptake, dissemination, and strength, which are regulated by factors such as pH and the concentration of other ions like Fe and manganese (Deb 1992). Zinc persists in the environment due to anthropogenic sources, such as smelting mines, chemical industries, and household wastes, and natural sources (Hanahan and Weinberg 2000). These sources can lead to serious contamination of water bodies, affecting marine life and microbes. Furthermore, wastewater discharged into irrigated fields and soils can result in zinc seeping into soils, impacting pH and microbial diversity (Gautam et al. 2016).

Excessive concentrations of zinc have a phytotoxic effect, which can negatively impact crop quality and yield. Zinc's bioavailability as ionic zinc and as a component of organo-metallic complexes (Pueyo et al. 2004; Anju and Banerjee 2011; Fedotov et al. 2012; Agnieszka et al. 2014; Kim et al. 2015; Baran et al. 2018) also contributes to the inhibition and alteration of soil microorganisms (Olaniran et al. 2013; Hussain et al. 2022).

Zinc can accumulate in the human body through inhalation, skin contact, or ingestion (Roney 2005). Inhalation of zinc-containing smoke can cause metal fume fever (MFF), an acute syndrome mostly caused by inhalation of zinc oxide. This occupational disease occurs during zinc smelting or welding, with symptoms like fever, muscle soreness, nausea, fatigue, chest pain, cough, and dyspnea (Vogelmeier et al. 1987). Small amounts of zinc are essential for survival, with the recommended dietary allowance (RDA) being 11 mg/day for men and 8 mg/day for women (Trumbo et al. 2001). Symptoms of this reversible syndrome begin generally a few hours after acute exposure and include fever, muscle soreness, nausea, fatigue, and respiratory effects like chest pain, cough, and dyspnea (Rohrs 1957). The respiratory symptoms are accompanied by an increase in bronchiolar leukocytes (Vogelmeier et al. 1987).

Lower zinc intake is recommended for infants (2–3 mg/day) and children (5–9 mg/day) because of their lower average body weights (Trumbo et al. 2001). This is significantly below the LD50 value, which has been estimated to be 27 g zinc/day in humans based on comparison with equivalent studies in rats and mice (Roney 2005). In general, uptake of such an amount is unlikely, because approximately 225–400 mg of zinc has been determined to be an emetic dose (Brown et al. 1964). Immediate symptoms after uptake of toxic amounts of zinc include abdominal pain, nausea, and vomiting. Additional effects include lethargy, anemia, and dizziness (Porea et al. 2000). In the case of dermal exposure, Zn can only act as an irritant; it can be concluded that dermal exposure does not produce a significant toxicological health risk.

Other responses may include focal neuronal deficits, respiratory disorders, gastrointestinal tract irritation, and elevated risk of prostate cancer in case of excess zinc consumption and neurosensory disorders, mental lethargy, decreased nerve conduction, thymus atrophy, skin lesions, acrodermatitis, infertility, hypogonadism, etc. when a deficiency of zinc occurs in the body (Plum et al. 2010).

5.3 Regulatory Standards and Guidelines for Zinc in Soils

In acidic soils, a DTPA-extractable Zn content above 10 mg/kg is considered potentially harmful. As discussed earlier "total Zn" contents in soils usually fall in the range 10–300 mg/kg, with contents above 150 mg/kg regarded as high (Landon 2014) and likely to result in reduced plant growth. Soil quality guidelines for national/regional jurisdictions vary in their consideration of bioavailability and application to different land uses.

Zn is also found in drinking water in the form of salts or organic complexes. Normal levels of Zn in surface water and groundwater do not exceed 0.01 and 0.05 mg/L, respectively. Joint FAO/WHO committee on food additives proposed a daily dietary requirement of Zn of 0.3 mg/kg of body weight and a provisional maximum tolerable daily intake of 1.0 mg/kg of body weight (Joint FAO/WHO Expert Committee on Food Additives 1988). Daily human adult requirement is 15–22 mg Zn/day; however; values of 5–22 mg Zn/day have been reported in studies on the average daily intake of Zn in different areas. The recommended maximum content of Zn in irrigation water is 2.0 mg/L, since above this limit Zn is toxic to many plant species. Its toxicity can be reduced at pH > 6.0 and in fine textured or organic soils.

Zn allows critical physiological pathways to function normally, and these pathways have important roles in photosynthesis and sugar formation, protein synthesis, fertility and seed production, growth regulation, and immunity to disease. Most plants contain between 30 and 100 mg Zn/kg dry matter, with contents above 300 mg Zn/kg to be considered generally as toxic (Marschner 1995; Noulas et al. 2018).

6 Conclusion

Zinc is an important and, indeed, an indispensable micronutrient, for both plants and animals, and the content of zinc in soils depends on several soil characteristics. What's more important for agricultural and environmental management is the relationship between zinc and factors like soil pH, texture, organic matter content, and moisture. Acid soils are very good at dissolving and making zinc available, and alkaline soils allow to precipitation of zinc making plants uptake less. Different soils have varying abilities to hold and or move zinc; clay soils, for instance, retain zinc more than sandy soils because of the small particle size of the clay. Organic matter has been shown to have positive and negative effects on zinc, for example, it could hold zinc, rendering it unavailable for plant uptake, but it could also leach out without any losses. Availability and drainage of soil water also have a bearing on zinc's movement in the soil, since too much water can cause leaching, especially in soils that do not drain well.

Being able to regulate zinc availability to plants to improve overall plant nutrition is critical in managing yield losses due to zinc deficiency and the risks of zinc deficiency. It does bear noting, however, that excessive wet and calcic soil zinc can also be toxic to the ecosystem and the population which necessitates compliance with given thresholds. Unlike most other metals, advances in soil science have made it possible to step jump the barriers to understanding and reacting to zinc to soil management in a more reasonable way while strengthening plant process and stability without compromising soil health.

References

Abedi T, Mojiri A (2020) Cadmium uptake by wheat (Triticum aestivum L.): an overview. Plan Theory 9(4):500

Abedi T, Gavanji S, Mojiri A (2022) Lead and zinc uptake and toxicity in maize and their management. Plan Theory 11(15):1922

Adamczyk-Szabela D, Markiewicz J, Wolf WM (2015) Heavy metal uptake by herbs. IV. Influence of soil pH on the content of heavy metals in Valeriana officinalis L. Water Air Soil Pollut 226(4). https://doi.org/10.1007/s11270-015-2360-3

Agnieszka B, Tomasz C, Jerzy W (2014) Chemical properties and toxicity of soils contaminated by mining activity. Ecotoxicology 23:1234–1244. https://doi.org/10.1007/s10646-014-1266-y

Agrawal A, Sharma A, Awasthi G, Awasthi A, Awasthi KK (2021) Toxicity assessment and antibacterial activity of ZnO nanoparticles. In: Nanostructured zinc oxide. Elsevier, pp 511–552

Agrawal A, Sharma R, Sharma A, Awasthi KK, Awasthi K, Awasthi A (2023) ZnO nanostructures for latent fingerprints. In: Friction ridge analysis: applications of nanoparticles for latent fingerprint development. Singapore, Springer, pp 113–127

Ali M, Parveen A, Malik Z, Kamran M, Saleem MH, Abbasi GH et al (2022) Zn alleviated salt toxicity in Solanum lycopersicum L. seedlings by reducing Na+ transfer, improving gas exchange, defense system and Zn contents. Plant Physiol Biochem 186:52–63

Alloway BJ (2008) Zinc in soils and crop nutrition. International Zinc Association and International Fertilizer Association, p 16

Alloway BJ (2009) Soil factors associated with zinc deficiency in crops and humans. Environ Geochem Health 31(5):537–548

Alvarez JM (2007) Influence of soil type on the mobility and bioavailability of chelated zinc. J Agric Food Chem 55(9):3568–3576

Alvarez JM, Obrador A, Rico MI (1996) Effects of chelated zinc, soluble and coated fertilizers, on soil zinc status and zinc nutrition of maize. Commun Soil Sci Plant Anal 27(1–2)

Angelaki A, Dionysidis A, Sihag P, Golia EE (2022) Assessment of contamination management caused by copper and zinc cations leaching and their impact on the hydraulic properties of a sandy and a loamy clay soil. Land 11(2):290

Anju M, Banerjee DK (2011) Associations of cadmium, zinc, and lead in soils from lead and zinc mining area as studied by single and sequential extractions. Environ Monit Assess 176:67–85. https://doi.org/10.1007/s10661-010-1567-4

Awasthi A, Bansal S, Jangir LK, Awasthi G, Awasthi KK, Awasthi K (2017) Effect of ZnO nanoparticles on germination of Triticum aestivum seeds. Macromol Symp 376(1):1700043

Awasthi A, Sharma P, Jangir L, Awasthi G, Awasthi KK, Awasthi K (2020) Dose dependent enhanced antibacterial effects and reduced biofilm activity against Bacillus subtilis in presence of ZnO nanoparticles. Mater Sci Eng C 113:111021

Azouzi R, Charef A, Hamzaoui AH (2015) Assessment of the effect of pH, temperature and organic matter on zinc mobility in a hydromorphic soil. Environ Earth Sci 74:2967–2980

Baize D (1997) Teneurs totales en éléments traces métalliques dans les sols (France): Références et stratégies d'interprétation. Programme ASPITET

Bänziger M, Long J (2000) The potential for increasing the iron and zinc density of maize through plant-breeding. Food Nutr Bull 21(4):397–400

Baran A, Wieczorek J, Mazurek R, Urbański K, Klimkowicz-Pawlas A (2018) Potential ecological risk assessment and predicting zinc accumulation in soils. Environ Geochem Health 40:435–450. https://doi.org/10.1007/s10653-017-9924-7

Bauer A, Lindsay WL (1965) The effect of soil temperature on the availability of indigenous soil zinc. Soil Sci Soc Am J 29(4):413–416

Beesley L, Moreno-Jimenez E, Clemente R, Lepp N, Dickinson N (2010) Mobility of arsenic, cadmium and zinc in a multi-element contaminated soil profile assessed by in-situ soil pore water sampling, column leaching and sequential extraction. Environ Pollut 158(1):155–160

Blasiak J, Pawlowska E, Chojnacki J, Szczepanska J, Chojnacki C, Kaarniranta K (2020) Zinc and autophagy in age-related macular degeneration. Int J Mol Sci 21(14). https://doi.org/10.3390/ijms21144994

Boguta P, Sokołowska Z (2020) Zinc binding to fulvic acids: assessing the impact of pH, metal concentrations and chemical properties of fulvic acids on the mechanism and stability of formed soluble complexes. Molecules 25(6). https://doi.org/10.3390/molecules25061297

Bossolani JW, Crusciol CAC, Momesso L, Portugal JR, Moretti LG, Garcia A et al (2022) Surface liming triggers improvements in subsoil fertility and root distribution to boost maize crop physiology, yield and revenue. Plant Soil 477(1):319–341

Brazauskiene D-M, Paulauskas V, Sabiene N (2008) Speciation of Zn, Cu, and Pb in the soil depending on soil texture and fertilization with sewage sludge compost. J Soils Sediments 8:184–192
Brown MA, Thom JV, Orth GL, Cova P, Juarez J (1964) Food poisoning involving zinc contamination. Archives of Environmental Health: J Int 8(5):657–660
Cakmak I (2004a) Zinc deficiency as a practical problem in plant and human nutrition in Turkey: a NATO science series. Springer, Dordrecht, pp 13–19
Cakmak I (2004b) Identification and correction of widespread zinc deficiency in Turkey-a success story (a NATO-Science for Stability project) (No. 552), pp 1–26
Cakmak I (2008) Zinc deficiency in wheat in Turkey. In: Micronutrient deficiencies in global crop production. Springer, Dordrecht, pp 181–200
Cakmak I, Marschner H, Bangerth F (1989) Effect of zinc nutritional status on growth, protein metabolism and levels of indole-3-acetic acid and other phytohormones in bean (Phaseolus vulgaris L.). J Exp Bot 40(3):405–412
Caldelas C, Weiss DJ (2017) Zinc homeostasis and isotopic fractionation in plants: a review. Plant Soil 411:17–46
Cao HW, Li C, Zhang BQ, Rono JK, Yang ZM (2022a) A metallochaperone HIPP33 is required for rice zinc and iron homeostasis and productivity. Agronomy 12(2):488
Cao J, Zhang D, Yue Y, Chanajaree R, Wang S, Han J, Zhang X, Qin J, Huang Y (2022b) Regulating solvation structure to stabilize zinc anode by fastening the free water molecules with an inorganic colloidal electrolyte. Nano Energy 93. https://doi.org/10.1016/j.nanoen.2021.106839
Chen F, Muhammad FG, Khan ZI, Ahmad K, Malik IS, Ashfaq A, Naeem M, Nadeem M, Ma J, Awan MUF, Mahpara S, Mehmood S (2022) Bioaccumulation and transfer of zinc in soil plant and animal system: a health risk assessment for the grazing animals. Environ Sci Pollut Res 29(2). https://doi.org/10.1007/s11356-021-15808-z
Cheng T, Allen HE (2006) Comparison of zinc complexation properties of dissolved natural organic matter from different surface waters. J Environ Manag 80(3):222–229
Clemens S (2001) Molecular mechanisms of plant metal tolerance and homeostasis. Planta 212:475–486
Costa MCG, Coutinho ÍAC (2022) Root systems of agricultural crops and their response to physical and chemical subsoil constraints. In: Subsoil constraints for crop production. Springer, Cham, pp 225–261
Daccak D, Lidon FC, Pessoa CC, Luís IC, Coelho ARF, Marques AC et al (2022) Enrichment of grapes with zinc-efficiency of foliar fertilization with ZnSO4 and ZnO and implications on winemaking. Plan Theory 11(11):1399
Das S, Green A (2016) Zinc in crops and human health. In: Biofortification of food crops. https://doi.org/10.1007/978-81-322-2716-8_3
Deb DL (1992) Development of soil and plant analytical methods for micronutrients and sulphur Sri Lanka
Dimkpa CO, Singh U, Bindraban PS, Elmer WH, Gardea-Torresdey JL, White JC (2019) Zinc oxide nanoparticles alleviate drought-induced alterations in sorghum performance, nutrient acquisition, and grain fortification. Sci Total Environ 688:926–934
Duplay J, Semhi K, Errais E, Imfeld G, Babcsanyi I, Perrone T (2014) Copper, zinc, lead and cadmium bioavailability and retention in vineyard soils (Rouffach, France): the impact of cultural practices. Geoderma 230:318–328
Elhaj Baddar Z, Unrine JM (2021) Effects of soil pH and coatings on the efficacy of polymer coated ZnO Nanoparticulate fertilizers in wheat (Triticum aestivum). Environ Sci Technol 55(20):13532–13540
Eriksson J (2001) Concentrations of 61 trace elements in sewage sludge, farmyard manure, mineral fertiliser, precipitation and in oil and crops, vol 5159. Swedish Environmental Protection Agency, Stockholm
Esposito S, Baglivo I, Malgieri G, Russo L, Zaccaro L, D'Andrea LD, Mammucari M, Di Blasio B, Isernia C, Fattorusso R, Pedone PV (2006) A novel type of zinc finger DNA binding domain

in the agrobacterium tumefaciens transcriptional regulator Ros. Biochemistry 45(34). https://doi.org/10.1021/bi060697m
Fageria NK, Press C (2009) The use of nutrients in crop plants. Cereal Res Commun 37:149–150
Fahsi N, Mahdi I, Mesfioui A, Biskri L, Allaoui A (2021) Plant growth-promoting Rhizobacteria isolated from the jujube (Ziziphus lotus) plant enhance wheat growth, Zn uptake, and heavy metal tolerance. Agriculture 11(4):316
Farahzadi HN, Arbabian S, Majd A, Tajadod G (2020) Long-term effect different concentrations of Zn (NO3) 2 on the development of male and female gametophytes of Capsicum annuum L. var California wonder. Caryologia 73(1)
Fedotov PS, Koerdel W, Miro M, Peijnenburg WJGM, Wennrich R, Huang PM (2012) Extraction and fraction methods for exposure assessment of trace metals, metalloids and hazardous organic compounds in terrestrial environments. Crit Rev Environ Sci Technol 42:1117–1171. https://doi.org/10.1080/10643389.2011.556544
Gao X, Hoffland E, Stomph T, Grant CA, Zou C, Zhang F (2012) Improving zinc bioavailability in transition from flooded to aerobic rice. A review. Agron Sustain Dev 32:465–478
Garg N, Kaur H (2013) Response of antioxidant enzymes, phytochelatins and glutathione production towards Cd and Zn stresses in Cajanus cajan (L.) Millsp. Genotypes colonized by Arbuscular Mycorrhizal fungi. J Agron Crop Sci 199(2):118–133
Garg N, Singh S (2018) Mycorrhizal inoculations and silicon fortifications improve rhizobial symbiosis, antioxidant defense, trehalose turnover in pigeon pea genotypes under cadmium and zinc stress. Plant Growth Regul 86(1):105–119
Garg J, Chiu MN, Krishnan S, Kumar R, Rifah M, Ahlawat P et al (2024) Emerging trends in zinc ferrite nanoparticles for biomedical and environmental applications. Appl Biochem Biotechnol 196(2):1008–1043
Garner TB, Hester JM, Carothers A, Diaz FJ (2021) Role of zinc in female reproduction. Biol Reprod 104(5). https://doi.org/10.1093/biolre/ioab023
Gautam PK, Gautam RK, Banerjee S, Chattopadhyaya MC, Pandey JD (2016) Heavy metals in the environment: fate, transport, toxicity and remediation technologies. Nova Science 60:101–130
Ghnaya AB, Hourmant A, Cerantola S, Kervarec N, Cabon JY, Branchard M, Charles G (2010) Influence of zinc on soluble carbohydrate and free amino acid levels in rapeseed plants regenerated in vitro in the presence of zinc. Plant Cell Tissue Organ Cult 102:191–197
Gondek K, Mierzwa-Hersztek M (2023) Effect of soil supplementation with mineral-organic mixtures on the amount of maize biomass and the mobility of trace elements in soil. Soil Tillage Res 226
Gonzalez D, Novillo J, Rico M, Alvarez J (2008) Leaching and efficiency of six organic zinc fertilizers applied to navy bean crop grown in a weakly acidic soil of Spain. J Agric Food Chem 56(9):3214–3221
Gorny A, Uterman J, Eckelmann W (2000) Germany: in heavy metal (trace element) and organic matter contents of European Soils. European Commission. CEN Soil Team, p 30
Grabrucker S, Boeckers TM, Grabrucker AM (2016) Gender dependent evaluation of autism like behavior in mice exposed to prenatal zinc deficiency. Front Behav Neurosci 10(Mar)
Graham RD, Ascher JS, Hynes SC (1992) Selecting zinc-efficient cereal genotypes for soils of low zinc status. Plant Soil 146:241–250
Graham R, Senadhira D, Beebe S, Iglesias C, Monasterio I (1999) Breeding for micronutrient density in edible portions of staple food crops: conventional approaches. Field Crop Res 60(1–2):57–80
Grewal HS, Williams R (1999) Alfalfa genotypes differ in their ability to tolerate zinc deficiency. Plant Soil 214:39–48
Gupta N, Ram H, Kumar B (2016) Mechanism of Zinc absorption in plants: uptake, transport, translocation and accumulation. Rev Environ Sci Biotechnol 15:89–109
Gupta S, Brazier AKM, Lowe NM (2020) Zinc deficiency in low- and middle-income countries: prevalence and approaches for mitigation. J Hum Nutr Diet 33(5). https://doi.org/10.1111/jhn.12791

Hacisalihoglu G, Kochian LV (2003) How do some plants tolerate low levels of soil zinc? Mechanisms of zinc efficiency in crop plants. New Phytol 159(2):341–350

Hafeez BMKY, Khanif YM, Saleem M (2013) Role of zinc in plant nutrition-a review. Am J Exp Agric 3(2):374–391

Hagmeyer S, Haderspeck JC, Grabrucker AM (2015) Behavioral impairments in animal models for zinc deficiency. Front Behav Neurosci

Hamid B, Zaman M, Farooq S, Fatima S, Sayyed RZ, Baba ZA et al (2021) Bacterial plant biostimulants: a sustainable way towards improving growth, productivity, and health of crops. Sustain For 13(2856):1–24. https://doi.org/10.3390/su13052856

Hamzah Saleem M, Usman K, Rizwan M, Al Jabri H, Alsafran M (2022) Functions and strategies for enhancing zinc availability in plants for sustainable agriculture. Front Plant Sci 13:1033092

Hanahan D, Weinberg RA (2000) The hallmarks of cancer. Cell 100(1):57–70

Haroon M, Khan ST, Malik A (2022) Zinc-solubilizing bacteria: an option to increase zinc uptake by plants. In: Microbial biofertilizers and micronutrient availability: the role of zinc in agriculture and human health, pp 207–238

Hart JJ, Norvell WA, Welch RM, Sullivan LA, Kochian LV (1998) Characterization of zinc uptake, binding, and translocation in intact seedlings of bread and durum wheat cultivars. Plant Physiol 118(1):219–226

He Z, Zhang M, Yang X, Stoffella P (2006) Release behavior of copper and zinc from sandy soils. Soil Sci Soc Am J 70(5):1699–1707

Hosamani V, Yalagi M, Sasvihalli P, Hosamani V, Nair KS, Harlapur VK et al (2020) Importance of micronutrients (Zinc) in crop production: a review. Int J Chem Stud 8(1):1060–1064

Hussain S, Khan M, Sheikh TMM, Mumtaz MZ, Chohan TA, Shamim S, Liu Y (2022) Zinc essentiality, toxicity, and its bacterial bioremediation: a comprehensive insight. Front Microbiol 13:900740

Jeng SS, Chen YH (2022) Association of zinc with anemia. Nutrients 14(22). https://doi.org/10.3390/nu14224918

Jiang C, Aiken GR, Hsu-Kim H (2015) Effects of natural organic matter properties on the dissolution kinetics of zinc oxide nanoparticles. Environ Sci Technol 49(19):11476–11484

Joint FAO/WHO Expert Committee on Food Additives. Meeting, & World Health Organization (1988) Toxicological evaluation of certain food additives, vol 31. Cambridge University Press, Cambridge

Jouvin D, Louvat P, Juillot F, Maréchal CN, Benedetti MF (2009) Zinc isotopic fractionation: why organic matters. Environ Sci Technol 43(15):5747–5754

Kabata-Pendias A (2000) Trace elements in soils and plants. CRC Press

Kashyap D, Siddiqui ZA (2022) Effect of zinc oxide nanoparticles and rhizobium leguminosarum on growth, photosynthetic pigments and blight disease complex of pea. Gesunde Pflanzen 74(1):29–40

Kaur H, Garg N (2021) Zinc toxicity in plants: a review. Planta 253(6):129

Khan ST, Malik A, Alwarthan A, Shaik MR (2022) The enormity of the zinc deficiency problem and available solutions; an overview. Arab J Chem 15(3). https://doi.org/10.1016/j.arabjc.2021.103668

Khan WA, Penrose B, Yun P, Zhou M, Shabala S (2024) Exogenous zinc application mitigates negative effects of salinity on barley (Hordeum vulgare) growth by improving root ionic homeostasis. Funct Plant Biol 51(6)

Khudsar T, Mahmooduzzafar, Iqbal M, Sairam RK (2004) Zinc-induced changes in morpho-physiological and biochemical parameters in *Artemisia annua*. Biol Plant 48:255–260

Kim RY, Yoon JK, Kim TS, Yang JE, Owens G, Kim KR (2015) Bioavailability of heavy metals in soils: definitions and practical implementation: a critical review. Environ Geochem Health 37:1041–1061. https://doi.org/10.1007/s10653-015-9695-y

Koleva L (2010) Mineral nutrients content in zinc and cadmium treated durum wheat plants with similar growth inhibition. Gene Appl Plant Physiol 36:60–63

Krishna TA, Ceasar SA, Maharajan T, Ramakrishnan M, Duraipandiyan V, Al-Dhabi NA, Ignacimuthu S (2017) Improving the zinc-use efficiency in plants: a review

Kumar V, Yadav AN, Verma P, Sangwan P, Saxena A, Kumar K, Singh B (2017) β-Propeller phytases: diversity, catalytic attributes, current developments and potential biotechnological applications. Int J Biol Macromol 98:595–609

Lair G, Gerzabek M, Haberhauer G (2007) Retention of copper, cadmium and zinc in soil and its textural fractions influenced by long-term field management. Eur J Soil Sci 58(5):1145–1154

Lall SP, Kaushik SJ (2021) Nutrition and metabolism of minerals in fish. Animals 11(9). https://doi.org/10.3390/ani11092711

Landon JR (2014) Booker tropical soil manual: a handbook for soil survey and agricultural land evaluation in the tropics and subtropics. Routledge

Latrille C, Denaix L, Lamy I (2003) Interaction of copper and zinc with allophane and organic matter in the B horizon of an Andosol. Eur J Soil Sci 54(2):357–364

Li J, Qiu Y, Zhao Q, Chen D, Wu Z, Peng AA et al (2020) Lead and copper-induced hormetic effect and toxicity mechanisms in lettuce (Lactuca sativa L.) grown in a contaminated soil. Sci Total Environ 741:140440

Lindsay WL (1991) Inorganic equilibria affecting micronutrients in soils. Micronutrients Agric 4:89–112

Liščáková P, Nawaz A, Molnárová M (2022) Reciprocal effects of copper and zinc in plants. Int J Environ Sci Technol 19(9):9297–9312

Lü W, Yao X, Ren H, Deng H, Yao M, Zhang B (2020) Characterizing the interactions between sediment dissolved organic matter and zinc using multispectroscopic techniques. Environ Pollut 261:113644

Luyckx M, Hausman JF, Guerriero G, Lutts S (2023) Silicon reduces zinc absorption and triggers oxidative tolerance processes without impacting growth in young plants of hemp (Cannabis sativa L.). Environ Sci Pollut Res 30(1):943–955

Magomere KM, Nchimbi-Msolla S, Tryphone GM (2024). Effects of soil pH levels on iron and zinc concentrations of common bean (Phaseolus vulgaris L.) genotypes

Marschner H (1995) Mineral nutrition of higher plants, 2nd edn. Institute of Plant Nutrition University of Hohenheim: Germany

Martınez C, Motto H (2000) Solubility of lead, zinc and copper added to mineral soils. Environ Pollut 107(1):153–158

McBride M, Blasiak J (1979) Zinc and copper solubility as a function of pH in an acid soil. Soil Sci Soc Am J 43(5):866–870

McBride M, Sauve S, Hendershot W (1997) Solubility control of Cu, Zn, Cd and Pb in contaminated soils. Eur J Soil Sci 48(2):337–346

McGrath SP, Loveland PJ (1992) Soil geochemical atlas of England and Wales

Min XB, Xie XD, Chai LY, Liang YJ, Mi L, Yong K (2013) Environmental availability and ecological risk assessment of heavy metals in zinc leaching residue. Trans Nonferrous Metals Soc China 23(1):208–218

Montanha GS, Rodrigues ES, Romeu SL, de Almeida E, Reis AR, Lavres J Jr, de Carvalho HWP (2020) Zinc uptake from ZnSO4 (aq) and Zn-EDTA (aq) and its root-to-shoot transport in soybean plants (Glycine max) probed by time-resolved in vivo X-ray spectroscopy. Plant Sci 292:110370

Najafi-Ghiri M, Rezaei M, Sameni A (2012) Zinc sorption–desorption by sand, silt and clay fractions in calcareous soils of Iran. Arch Agron Soil Sci 58(9):945–957

Nath S, Dey S, Kundu R, Paul S (2024) Phosphate and zinc interaction in soil and plants: a reciprocal cross-talk. Plant Growth Regul:1–25

Nielsen FH (2012) History of zinc in agriculture. Adv Nutr 3(6). https://doi.org/10.3945/an.112.002881

Noreen S, Sultan M, Akhter MS, Shah KH, Ummara U, Manzoor H et al (2021) Foliar fertigation of ascorbic acid and zinc improves growth, antioxidant enzyme activity and harvest index in barley (Hordeum vulgare L.) grown under salt stress. Plant Physiol Biochem 158:244–254

Noulas C, Tziouvalekas M, Karyotis T (2018) Zinc in soils, water and food crops. J Trace Elem Med Biol 49:252–260

Olaniran AO, Balgobind A, Pillay B (2013) Bioavailability of heavy metals in soil: impact on microbial biodegradation of organic compounds and possible improvement strategies. Int J Mol Sci 14:10197–10228. https://doi.org/10.3390/ijms140510197

Ortiz R, Gascó G, Méndez A, Sanchez-Martín L, Obrador A, Almendros P (2023) Comparative study of traditional and environmentally friendly zinc sources applied in alkaline fluvisol soil: lettuce biofortification and soil zinc status. Agronomy 13(12). https://doi.org/10.3390/agronomy13123014

Pal K, Khurana SP, Debnath N, Das S (2022) Micronutrient nanoparticles: synthesis, properties and application in agriculture. In: Synthesis and applications of nanoparticles. Singapore, Springer, pp 337–356

Palacio-Márquez A, Ramírez-Estrada CA, Sanchez E, Ojeda-Barrios DL, Chavez-Mendoza C, Sida-Arreola JP (2022) Biofortification with nanoparticles and zinc nitrate plus chitosan in green beans: effects on yield and mineral content. Notulae Botanicae Horti Agrobotanici Cluj-Napoca 50(2):12672–12672

Pellinen T, Päivärinta E, Isotalo J, Lehtovirta M, Itkonen ST, Korkalo L, Erkkola M, Pajari AM (2022) Replacing dietary animal-source proteins with plant-source proteins changes dietary intake and status of vitamins and minerals in healthy adults: a 12-week randomized controlled trial. Eur J Nutr 61(3). https://doi.org/10.1007/s00394-021-02729-3

Pendias, H. (1992). Trace elements in soils and plants

Pérez-Novo C, Pateiro-Moure M, Osorio F, Nóvoa-Muñoz J, López-Periago E, Arias-Estévez M (2008) Influence of organic matter removal on competitive and noncompetitive adsorption of copper and zinc in acid soils. J Colloid Interface Sci 322(1):33–40

Petrová Š, Soudek P (2022) Ecotoxicity study of additives composed of zinc and boron. Toxics 10(12):795

Phattarakul N, Rerkasem B, Li LJ, Wu LH, Zou CQ, Ram H et al (2012) Biofortification of rice grain with zinc through zinc fertilization in different countries. Plant Soil 361:131–141

Pimentel Guerra C, Pina Martínez CM, Müller N, Lara LA, Melo Rodriguez G, Orlando F, ..., Fernández V (2023) Mineral particles in foliar fertilizer formulations can improve the rate of foliar uptake

Pinto A, Mota AD, De Varennes A, Pinto F (2004) Influence of organic matter on the uptake of cadmium, zinc, copper and iron by sorghum plants. Sci Total Environ 326(1–3):239–247

Pirzada T, Solangi AG, Talpur MMA, Chandio W, Memon K (2024) Metal binding studies of soil extracted humic acid with K (I), Zn (II) and Fe (III) ions. J Chem Soc Pak 46(2)

Plum LM, Rink L, Haase H (2010) The essential toxin: impact of zinc on human health. Int J Environ Res Public Health 7(4):1342–1365

Porea TJ, Belmont JW, Mahoney Jr DH (2000) Zinc-induced anemia and neutropenia in an adolescent. J pediatrics 136(5):688–690

Prasad AS (2008) Zinc in human health: effect of zinc on immune cells. Mol Med 14:353–357

Prasad R, Shivay YS, Kumar D (2014) Agronomic biofortification of cereal grains with iron and zinc. Adv Agron 125:55–91

Pueyo M, Lopez-Sanchez JF, Rauret G (2004) Assessment of CaCl2, NaNO3, and NH4NO3 extraction procedures for the study of Cd, Cu, Pb and Zn extractability in contaminated soils. Anal Chim Acta 504:217–226. https://doi.org/10.1016/j.aca.2003.10.047

Ramazanoglu E, Cun S, Cevheri CI, Beyyavas V, Sakin E (2024) Cottonseed meal soil application, nutrient uptake, biochemical attributes of cotton (Gossypium hirsutum L.) and activities of soil enzymes in saline soil. J Elem 29(3)

Recena R, García-López AM, Delgado A (2021) Zinc uptake by plants as affected by fertilization with Zn sulfate, phosphorus availability, and soil properties. Agronomy 11(2). https://doi.org/10.3390/agronomy11020390

Rengel Z (2002) Role of pH in availability of ions in soil. In: Handbook of plant growth pH as the master variable. CRC Press, pp 317–342

Rengel Z (2015) Availability of Mn, Zn and Fe in the rhizosphere. J Soil Sci Plant Nutr 15(2):397–409

Rohrs LC (1957) Metal-fume fever from inhaling zinc oxide. AMA Arch Intern Med 100(1):44–49

Römkens PF, Salomons W (1998) Cd, cu and Zn solubility in arable and forest soils: consequences of land use changes for metal mobility and risk assessment. Soil Sci 163(11):859–871
Roney N (2005) Toxicological profile for zinc. US Department of Health and Human Services, Public Health Service, Agency for Toxic Substances and Disease Registry
Rudani K, Vishal P, Kalavati P (2018) The importance of zinc in plant growth-a review. Int Res J Nat Appl Sci 5(2):38–48
Sadeghzadeh B (2013) A review of zinc nutrition and plant breeding. J Soil Sci Plant Nutr 13(4):905–927
Saini P, Nagpal S, Saini P, Kumar A, Gani M (2021) Microbial mediated zinc solubilization in legumes for sustainable agriculture. In: Phytomicrobiome interactions and sustainable agriculture, pp 254–276
Saini S, Kumar P, Sharma DP, Sharma NC, Chauhan A, Shandil D (2024) Organic Zn and nano-Zn amino acids chelates modulate quality growth attributes and antioxidants activity for biofortified apple (Malus x domestica Borkh.) production. Sci Hortic 337:113594
Samardjieva KA, Gonçalves RF, Valentão P, Andrade PB, Pissarra J, Pereira S, Tavares F (2015) Zinc accumulation and tolerance in Solanum nigrum are plant growth dependent. Int J Phytoremediation 17(3):272–279
Santos WO, de Aquino PM, Cuesta RLC, Saldanha IMH, Mattiello EM, Vergütz L (2019) Zinc diffusion and availability affected by different sources in soils of contrasting textures. J Agric Sci 11:259–270
Schlattl M, Buffler M, Windisch W (2021) Clay minerals affect the solubility of Zn and other bivalent cations in the digestive tract of ruminants in vitro. Animals 11(3). https://doi.org/10.3390/ani11030877
Selvapandian S, Muthusamy N, Kathirvelan C, Kumaresan G, Vasanthakumar P, Senthilkumar RP (2024) Effects of dietary zinc alginate complex supplementation on growth performance, nutrient utilisation, Tibia bone characteristics, ileum morphology, carcass characteristics, Meat quality and footpad health in broiler chicken
Semenova NA, Burmistrov DE, Shumeyko SA, Gudkov SV (2024) Fertilizers based on nanoparticles as sources of macro-and microelements for plant crop growth: a review. Agronomy 14(8):1646
Shah SH, Islam S, Mohammad F (2022) Sulphur as a dynamic mineral element for plants: a review. J Soil Sci Plant Nutr 22(2):2118–2143
Sharma S, Verma TS, Chander G (2009) Effect of long term lantana addition on soil zinc fractions and crop yield in rice-wheat cropping. Trop Agric 83(1):5–10
Sharma R, Agrawal A, Sharma A, Kumar S, Sharma PK, Awasthi KK et al (2022) Effect of biosynthesized ZnO nanoparticles on wheat seedlings. Mater Today Proc 69:1587–1595
Sharma M, Shilpa, Kaur M, Sharma AK, Sharma P (2023) Influence of different organic manures, biofertilizers and inorganic nutrients on performance of pea (Pisum sativum L.) in North Western Himalayas. J Plant Nutr 46(4):600–617
Shen Y, Wiita E, Nghiem AA, Liu J, Haque E, Austin RN et al (2023) Zinc localization and speciation in rice grain under variable soil zinc deficiency. Plant Soil 491(1):605–626
Shi J, Wu Q, Zheng C, Yang J (2018) The interaction between particulate organic matter and copper, zinc in paddy soil. Environ Pollut 243:1394–1402
Shrestha J, Kandel M, Subedi S, Shah KK (2020) Role of nutrients in rice (Oryza sativa L.): a review. Agri 9(1):53–62
Shukla AK, Tiwari PK, Abhijit P, Chandra P (2016) Zinc and iron in soil, plant, animal and human health. Indian J Fertil 12(11)
Shukla AK, Behera SK, Prakash C, Tripathi A, Patra AK, Dwivedi BS, Trivedi V, Rao CS, Chaudhari SK, Das S, Singh AK (2021) Deficiency of phyto-available sulphur, zinc, boron, iron, copper and manganese in soils of India. Sci Rep 11(1). https://doi.org/10.1038/s41598-021-99040-2
Sillanpää M, Vlek PLG (1985) Micronutrients and the agroecology of tropical and Mediterranean regions. In: Micronutrients in tropical food crop production. Springer, Dordrecht, pp 151–167

Silwamba M, Ito M, Hiroyoshi N, Tabelin CB, Hashizume R, Fukushima T, Park I, Jeon S, Igarashi T, Sato T, Nyambe I, Nakata H, Nakayama S, Ishizuka M (2022) Alkaline leaching and concurrent cementation of dissolved Pb and Zn from zinc plant leach residues. Fortschr Mineral 12(4). https://doi.org/10.3390/min12040393

Sims JT (1986) Soil pH effects on the distribution and plant availability of manganese, copper, and zinc. Soil Sci Soc Am J 50(2):367–373

Singh B, Natesan SKA, Singh BK, Usha K (2005) Improving zinc efficiency of cereals under zinc deficiency. Curr Sci:36–44

Singh K, Madhusudanan M, Verma AK, Kumar C, Ramawat N (2021) Engineered zinc oxide nanoparticles: an alternative to conventional zinc sulphate in neutral and alkaline soils for sustainable wheat production. 3Biotech 11(7):322

Singh SK, Barman M, Sil A, Prasad JP, Kundu S, Bahuguna RN (2023a) Nutrient biofortification in wheat: opportunities and challenges. Cereal Res Commun 51(1):15–28

Singh S, Singh J, Dhaliwal SS (2023b) Optimizing the Zn, Mn, and Fe mineral dose as tank mix foliar application for improvement of fruit yield, quality, and uptake of nutrients in the Kinnow mandarin. Trends Hortic 6(2):3527

Singh K, Gupta S, Singh AP (2024) Review: nutrient-nutrient interactions governing underground plant adaptation strategies in a heterogeneous environment. Plant Sci 342

Sipos P, Kis VK, Balázs R, Tóth A, Németh T (2021) Effect of pedogenic iron-oxyhydroxide removal on the metal sorption by soil clay minerals. J Soils Sediments 21:1785–1799

Skrajnowska D, Bobrowska-Korczak B (2019) Role of zinc in immune system and anti-cancer. Nutrients 11(10)

Sparks DL (1998) Kinetics of soil chemical phenomena: future directions. In: Future prospects for soil chemistry, vol 55, pp 81–101

Steffens D, Hutsch BW, Eschholz T, Losak T, Schubert S (2005) Water logging may inhibit plant growth primarily by nutrient deficiency rather than nutrient toxicity. Plant Soil Environ 51(12):545

Suganya A, Saravanan A, Manivannan N (2020) Role of zinc nutrition for increasing zinc availability, uptake, yield, and quality of maize (Zea mays L.) grains: an overview. Commun Soil Sci Plant Anal 51(15):2001–2021

Szerement J, Szatanik-Kloc A, Mokrzycki J, Mierzwa-Hersztek M (2022) Agronomic biofortification with Se, Zn, and Fe: an effective strategy to enhance crop nutritional quality and stress defense – a review. J Soil Sci Plant Nutri 22(1). https://doi.org/10.1007/s42729-021-00719-2

Takkar PN, Walker CD (1993) The distribution and correction of zinc deficiency. In: Zinc in soils and plants: proceedings of the international symposium on 'zinc in soils and plants' held at the University of Western Australia, 27–28 September, 1993. Springer, Dordrecht, pp 151–165

Thiyagarajan C (2021) Organo Zinc chelates for improving the yield and Zinc nutrition of hybrid tomato on calcareous soil under drip fertigation. J Soil Sci Plant Nutr:1–10

Trumbo P, Yates AA, Schlicker S, Poos M (2001) Dietary reference intakes. J Am Diet Assoc 101(3):294–294

Tso R, Forde CG (2021) Unintended consequences: nutritional impact and potential pitfalls of switching from animal- to plant-based foods. Nutrients 13(8). https://doi.org/10.3390/nu13082527

Tye AM, Young SD, Crout NMJ, Zhang H, Preston S, Barbosa-Jefferson VL et al (2003) Predicting the activity of Cd2+ and Zn2+ in soil pore water from the radio-labile metal fraction. Geochim Cosmochim Acta 67(3):375–385

Udovic M, Lestan D (2009) Pb, Zn and Cd mobility, availability and fractionation in aged soil remediated by EDTA leaching. Chemosphere 74(10):1367–1373

Upadhayay VK, Singh AV, Khan A, Sharma A (2022) Contemplating the role of zinc-solubilizing bacteria in crop biofortification: an approach for sustainable bioeconomy. Front Agron 4. https://doi.org/10.3389/fagro.2022.903321

Uriu-Adams JY, Keen CL (2010) Zinc and reproduction: effects of zinc deficiency on prenatal and early postnatal development. Birth Defects Res B Dev Reprod Toxicol 89(4). https://doi.org/10.1002/bdrb.20264

Vaillant N, Monnet F, Hitmi A, Sallanon H, Coudret A (2005) Comparative study of responses in four Datura species to a zinc stress. Chemosphere 59(7):1005–1013
Vavoulidou E, Charoulis A, Soulis K, Karyotis T, Kavvadias V (2009) Soil survey for improvement of farming practices in Malia municipality, Greece. Communications in soil science and plant analysis, 40(1–6):1020–1033
Verma PK, Verma S, Chakrabarty D, Pandey N (2021) Biotechnological approaches to enhance zinc uptake and utilization efficiency in cereal crops. J Soil Sci Plant Nutr 21(3):2412–2424
Vickram S, Rohini K, Srinivasan S, Veenakumari DN, Archana K, Anbarasu K, Jeyanthi P, Thanigaivel S, Gulothungan G, Rajendiran N, Srikumar PS (2021) Role of zinc (Zn) in human reproduction: a journey from initial spermatogenesis to childbirth. Int J Mol Sci 22(4). https://doi.org/10.3390/ijms22042188
Vogelmeier C, König G, Bencze K, Fruhmann G (1987) Pulmonary involvement in zinc fume fever. Chest 92(5):946–948
Vymazal J, Březinová T (2015) Heavy metals in plants in constructed and natural wetlands: concentration, accumulation and seasonality. Water Sci Technol 71(2):268–276
Wen C, Wang J, Sun Z, Zhong R, Li M, Shen X, Ye Q, Qin K, Peng X (2024) Dietary zinc ameliorates TNBS-induced colitis in mice associated with regulation of Th1/Th2/Th17 balance and NF-κB/NLRP3 signaling pathway. Biol Trace Elem Res 202(2). https://doi.org/10.1007/s12011-023-03715-y
Wessels I, Fischer HJ, Rink L (2021) Dietary and physiological effects of zinc on the immune system. Annu Rev Nutr 41. https://doi.org/10.1146/annurev-nutr-122019-120635
White PJ, Broadley MR (2009) Biofortification of crops with seven mineral elements often lacking in human diets–iron, zinc, copper, calcium, magnesium, selenium and iodine. New Phytol 182(1):49–84
Xian X, In Shokohifard G (1989) Effect of pH on chemical forms and plant availability of cadmium, zinc, and lead in polluted soils. Water Air Soil Pollut 45:265–273
Yakhin OI, Lubyanov AA, Yakhin IA, Brown PH (2017) Biostimulants in plant science: a global perspective. Front Plant Sci 7:2049
Zargar SM, Mahajan R, Farhat S, Nazir M, Mir RA, Nazir M, Salgotra RK, Mallick SA (2015) Understanding the role of iron and zinc in animals and crop plants from genomics perspective. Curr Trends Biotechnol Pharm 9(2)
Zou CQ, Zhang YQ, Rashid A, Ram H, Savasli E, Arisoy RZ et al (2012) Biofortification of wheat with zinc through zinc fertilization in seven countries. Plant Soil 361:119–130

Physiological Significance of Zinc for Plants, Animals, and Humans

Harmanjit Kaur, Km Sandhya Devi, and Ranjana Yadav

Abstract

Zinc (Zn) is an essential nutrient for plants as well as animals. It is required for diverse critical plant physiological and metabolic processes, for instance, photosynthesis, chlorophyll synthesis, carbohydrate metabolism, and starch and sucrose synthesis. Zinc plays a key role in gene expression and regulation, protein synthesis, nitrogen fixation, and plant defense against numerous biotic and abiotic stressors. Even though the food availability has increased, its nutritional characteristics have decreased. In this context, the significance of micronutrients, including Zn, in dietary food has become a subject of universal interest having health-related issues. There are several important animal and human physiological processes involving Zn. Zn deficiency leads to loss of several millions of dollars every year in the form of additional medical expenses and claims lives of millions of people, mostly affecting kids. It is believed that around 17% of the human population on earth are facing the threat of Zn deficiency. Zinc deficit is linked to difficulty in pregnancy, under-developed growth, early deaths, immune system impairment, and several neurobehavioral ailments in human beings. The present chapter highlights the significance of Zn in several metabolic processes of plants and animals (including humans) in order to understand the problem of Zn paucity with respect to its influence on crop growing and human well-being, taking into account the available literature.

Keywords

Carbohydrate metabolism · Cardiovascular system · Nitrogen fixation · Phytohormones · Protein synthesis · Trichome development

H. Kaur (✉) · K. S. Devi · R. Yadav
Department of Botany, University of Allahabad, Prayagraj, Uttar Pradesh, India
e-mail: harmanjitkaur@allduniv.ac.in

S. Srivastava, H. Kaur (eds.), *Zinc in Soil-Plant Continuum*,
https://doi.org/10.1007/978-981-96-4253-3_2

1 Introduction

The availability of nutrients is the foremost feature involved in the optimum growth and well-being of plants. Plants are inherently in need of several nutrients, broadly categorized into macronutrients such as sulfur (S), calcium (Ca), phosphorus (P), nitrogen (N), and potassium (K), along with micronutrients such as zinc (Zn), iron (Fe), boron (B), and manganese (Mn) based on the amount of specific nutrients they require (White and Brown 2010). All these nutrients impact diverse physiological and biochemical processes taking place inside the plants. Further, they help against several diseases that negatively influence plant growth (Toor et al. 2021; Gondal et al. 2021). Out of these nutrients, Zn is the most crucial metal for the living world, being the 23rd most widespread element present in earth's crust, with atomic number of 30 and atomic weight of 65.37 (Khan et al. 2022). Natural Zn is a lustrous, bluish-white colored metal having amphoteric nature (Escobedo Monge et al. 2019). Some researchers have argued that Zn is the most imperative trace element subsequent to macronutrients N, P, and K. Raulin discovered the significance of Zn and that it is necessary for *Aspergillus niger* growth in 1869 (Raulin 1869). Similarly, necessity of Zn as a trace element for rats was reported in 1934 (Todd et al. 1933). Furthermore, the role of Zn in humans was first reported in 1961 (Prasad et al. 1961). Zinc ranks second in importance among essential metals for fish, only surpassed by Fe in terms of amount (Watanabe et al. 1997). Zinc requirement for the diet of fish ranges between 230 and 460 mol kg^{-1} dry mass of diet (Gatlin and Wilson 1983).

Living beings need Zn as a micronutrient at minimum levels for normal metabolic functions (Sturikova et al. 2018). In fact about 6% of proteins in the proteome of prokaryotes and roughly 9% of proteins in the proteome of eukaryotes are estimated to bind with Zn (Andreini et al. 2006). In plants, Zn is required for the metabolism of carbohydrates since it activates majority of the enzymes involved (Taheri et al. 2011). Additionally, Zn is an integral part of many enzymes where it functions as a structural, regulatory, and physical co-factor for several enzymes (Grotz and Guerinot 2002; Akay 2011). Additional roles of Zn include DNA transcription, regulation of chlorophyll, auxin synthesis, growth-modulatory substances, and assisting in the catalization of oxidation process in plants (Singh et al. 2012). Because Zn is necessary for the production of both proteins and starch, low Zn levels in plant cells cause build-up of amino acids along with a decline in sugar levels (Taheri et al. 2011). Besides plants, Zn is also crucial for human and animal health. In human beings, its involvement in immunity, neuro-transmission, and good brain functioning is becoming gradually more evident (Frederickson et al. 2005). Besides this, it supports the defense system against oxidative stress as well as production, storage, and release of insulin. Zinc binds to several proteins, important enzymes, and transcription factors, all of which are believed to be dependent on Zn for their respective roles (Jamsheer and Kumar 2021). Zinc is a key component of numerous metabolic reactions that sustain life, including cell oxygen consumption, expression of DNA and RNA, cell membrane integrity preservation, free oxygen radical sequestration, as well as defense against lipid peroxidation (Skrajnowska and Bobrowska-Korczak

2019; Costa et al. 2023). Zinc is necessary for metalloenzymes, carboxypeptidase, lactate dehydrogenase, and DNA and RNA polymerases in animals, including humans (Cousins 1998; Brown et al. 2001; Erdman et al. 2012). Therefore, Zn is rightly referred to as a "metal of life" (Khan et al. 2022). In light of this, the current chapter aims to summarize diverse physiological roles of Zn in flora and fauna (including humans).

2 Zinc in Plants

Zinc (Zn) is an important nutrient in plant system, playing a vital role in numerous life and physico-chemical processes (Noman et al. 2019; Zaheer et al. 2022). In majority of crops, the normal foliar Zn concentration necessitated for optimal growth ranges from 15 to 20 mg kg^{-1} of dry weight (Broadley et al. 2007). Zinc affects the activity, structure, and proper folding of many proteins, serving as a key component or catalyst in various plant enzymes (Castillo-González et al. 2018; Zaheer et al. 2020). It is considered that interplay between phospholipids, sulfhydryl groups, and Zn ions in cell membrane proteins improves membrane integrity and stability (Bhatla et al. 2018; Kareem et al. 2022). Plants contain several zinc (Zn)-dependent proteins, including enzymes, for example, Zn-finger DNA-binding proteins, carbonic anhydrases, Zn/Cu superoxide dismutases (SOD), and aldehyde dehydrogenases (Sofo et al. 2018; Al Jabri et al. 2022). Moreover, Zn can deactivate certain proteins via attaching to their functional groups or dislocating other cations from their sites, thereby alleviating the negative impacts of several substances on plants (Natasha et al. 2022). Zinc also participates in regulating the synthesis of tryptophan, which is a primary precursor of auxins in plants. Additionally, it is involved in signal transduction via mitogen-activated protein kinases (Bhantana et al. 2021; Kaur and Garg 2021; Ahmad et al. 2022). Diverse functions of Zn in planta have been considered and discussed in detail in subsequent sub-sections:

2.1 Carbohydrate Metabolism

Carbonic anhydrase (CA), a Zn requiring metalloenzyme, participates in numerous processes in plants including regulation of pH, ion exchange, transport of carbon dioxide (CO_2), CO_2 fixation in photosynthesis, respiration, and closing of stomata. It helps in the quick conversion of CO_2 and water (H_2O) to bicarbonate ions (HCO_3^-) and a proton:

$$CO_2 + \mathrm{H_2O} ------> HCO_3^- + \mathrm{H}^+$$

The catalysis performed by CA relies on Zn which acts as a cofactor through its binding to H_2O, thereby activating the enzyme catalytic site (Fig. 1). Numerous pieces of evidence indicate that Zn-deficit condition decreases the amount of CA in certain plants (Escudero-Almanza et al. 2012). In dicotyledonous plants, CA

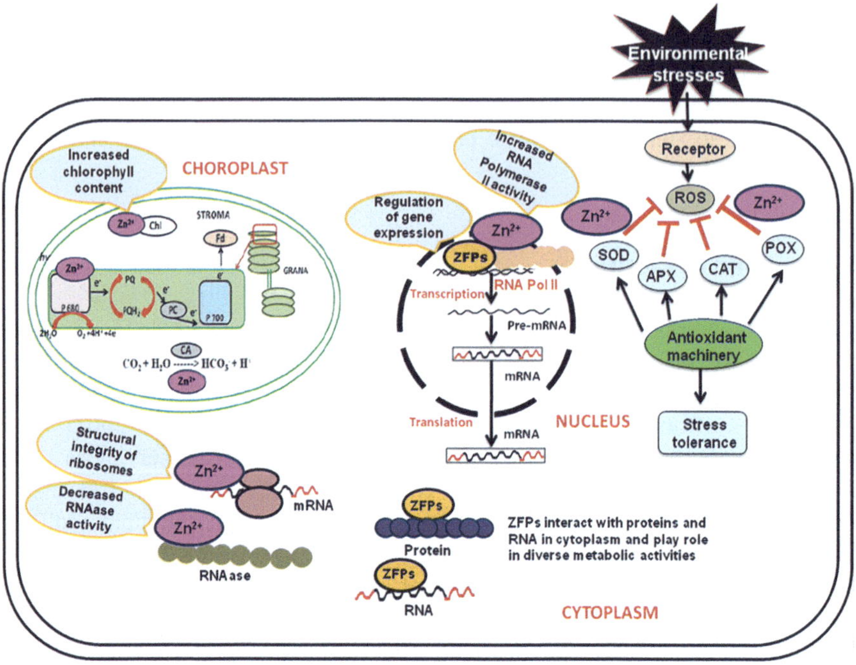

Fig. 1 Diverse roles of zinc (Zn) in plant cell. *APX*, ascorbate peroxidase, *CA* carbonic anhydrase, *CAT* catalase, *Chl* chlorophyll, *Fd* ferredoxin, *PC* plastocyanin, *POX* peroxidase, *ROS* reactive oxygen species, *SOD* superoxide dismutase, *ZFP* zinc finger proteins, Zn^{2+} zinc

comprises of six subunits, six Zn atoms per molecule, and molecular mass of 180 kDa. It is found in both chloroplasts and cytoplasm. Any alteration in its activity directly impacts the CO_2 fixation during photosynthesis under CO_2 deficiency. Among the enzymes involved in photosynthetic carbon metabolism across several plants, CA is the sole one whose functionality fluctuates with change in atmospheric CO_2 levels. Carbonic anhydrase plays a pivotal role in photosynthetic CO_2 fixation. For instance, in C3 plants, β-CA increases the CO_2 levels in chloroplast, which acts as a substrate for RuBisCO. This enhances carboxylation, enabling the incorporation of C2 into carbohydrates in photosynthetic process, wherein only CO_2 is utilized as a source of carbon. In CAM and C4, the hydration of CO_2 to HCO_3^- provides a substrate for the carboxylase phosphoenolpyruvate (PEPC). Carbonic anhydrase facilitates the distribution of CO_2 in plasma membrane as well as chloroplast. It is also involved in the transport of CO_2 across the membrane via converting CO_2 to HCO_3^- (Tiwari 2005). In light-limiting conditions, alternative sources of energy are utilized to fix CO_2, whereas respiration keeps on continuing, causing a rise in CO_2 concentration. This increase in CO_2 provides more substrate for CA, resulting in the production of more HCO_3^- and H^+. The increased production of H^+ lowers the pH, which inactivates amylase, thereby averting glucose and starch hydrolysis.

Consequently, the osmotic gradient is reduced, leading to water loss in guard cell and causing stomatal closure (Casson and Gray 2008).

It has been documented that external supplementation of Zn fertilizer improves the photochemical reactions and increases quantum efficiency of electron transport in PSII (Eisvand et al. 2018), thereby indicating probable involvement of Zn in light reactions of photosynthesis. Moreover, the foliar application of Zn has been shown to elevate the levels of photosynthetic pigments, viz., chlorophyll a, chlorophyll b, total chlorophyll, along with carotenoids in foliage of plants (Zhao et al. 2024), suggesting its crucial role in photosynthesis. In plants, Zn is involved in carbohydrate metabolism wherein it acts as a cofactor for enzymes, most important being fructose 1, 6-bisphosphate (commonly known as aldolase). Aldolase is among the six enzymes of Calvin cycle which lacks post-translational regulation. Additionally, the mathematical model (Zhu et al. 2007) indicated that non-regulatory enzymes of Calvin cycle, like aldolase, influence the flow of carbon through photosynthesis. Furthermore, aldolase activity is believed to speed up the regeneration of RuBP, thereby enhancing photosynthetic potential, growth, and biomass production (Uematsu et al. 2012). Aldolase facilitates the cleavage of fructose 1,6-bisphosphate into two triose sugars, glyceraldehyde 3-phosphate and dihydroxyacetone phosphate. Additionally, aldolase catalyzes reversible cleavage of fructose 1,6-phosphate into dihydroxyacetone and glyceraldehyde phosphate. There are two isozymes of aldolase known in higher plants: Class I-aldolase A and Class II-aldolase B. These isozymes are encoded by diverse genes which express different forms during developmental course. Structural differences between the two isozymes lead to varying activities on the substrates, namely, fructose 1-phosphate and fructose 1, 6-bisphosphate. Aldolase B shows no preference for either substrate and catalyzes both the reactions, whereas aldolase A specifically operates on fructose 1, 6-bisphosphate. Aldolase A belonging to Class I possesses a tetrahedral protein configuration of 160 kDa and is found within the plastids. Its catalytic approach involves generation of a Schiff base intermediate linking substrate with lysine residue within enzyme's active site. In contrast, Class II aldolase has a dimeric protein structure of 80 kDa and is located in the cytoplasm. In contrast to class I, class II needs divalent cations, for example, Zn^{2+}, Fe^{2+}, or Ca^{2+} as cofactors for stabilizing the carbon ion intermediate produced in the reaction process (Uematsu et al. 2012). It has been reported that under low aldolase, starches and sugars tend to build up in Zn-deficit plants. However, when Zn is restored after 24 h, activity of Hill reaction and sugar levels return to normal (Kirkby and Römheld 2004; Barker and Pilbeam 2015).

2.2 Protein Synthesis and Genetic Material

In plants with Zn deficiency, synthesis of proteins along with protein levels significantly declines, leading to an accumulation of amino acids. However, Zn supplementation to deficient plants results in resumption of protein synthesis in a rapid manner (Patel et al. 2007). The lesser protein content coupled with elevated amino acid levels in plants experiencing Zn shortage is attributed not just to decreased transcription and translation but to increased RNA degradation as well. Zinc deficiency typically leads to heightened activity of ribonuclease (RNase). There exists a clear inverse relationship between Zn availability and RNase activity (Fig. 1) as well as between protein content and RNase activity (Elsby and Roberts 2008). Elevated activity of RNase often precedes visible signs of Zn shortage, for instance, sluggish growth as well as alterations in leaf structure. Zinc serves as a structural constituent of ribosomes and is necessary for maintaining their configurational integrity. In unicellular algae like *Euglena,* ribosomal RNA contains Zn in amounts ranging between 650 and 1280 $\mu g\ g^{-1}$ RNA. Nevertheless, in higher plants experiencing Zn deficiency, this content drops to a range of 300–380 $\mu g\ g^{-1}$ RNA. In Zn-limiting conditions, ribosomes disorganize; however, their structure can be restored when Zn supply is replenished (Mendoza-Cózatl et al. 2006). Additionally, there is a significant demand for Zn in the centers where synthesis of proteins occurs inside pollen tubes (Elsby and Roberts 2008). An excess or deficiency of Zn can lead to its pro-oxidant behavior (Kinraide et al. 2011) As a result, plants tightly regulate the cellular concentration and compartmentalization of mobile Zn using various proteins that sense, transport, buffer, store, and release it.

2.3 Trichome Development

It has been reported that ZFP5 (zinc finger protein 5) regulates initiation of trichome via signalling pathway involving gibberellins (GA) (Zhou et al. 2011). ZFP5 is crucial for regulating origination as well as extension of root hairs. It also mediates the response of root hair development to signals from cytokinin and ethylene (An et al. 2012). ZFP5 features a characteristic C2H2 domain, common in the TFIIIA transcription factor family, as well as a conserved motif QALGGH which is plant-specific (Gan et al. 2006, 2007). C2H2-type Zn finger proteins are essential for vital functions of a plant cell, such as regulating transcription, development, defending against pathogens, and responding to stress (Laity et al. 2001; Englbrecht et al. 2004; Ciftci-Yilmaz and Mittler 2008).

2.4 Regulation of Nitrogen Fixation

Nitrogen homeostasis in legumes is achieved by maintaining a balance between the nitrogen obtained from soil and the nitrogen which is fixed by symbiotic bacteria in nodules present on roots (Okamoto et al. 2009: Lin et al. 2018). Zinc serves as an intracellular secondary messenger, connecting ecological changes with regulation of metabolic activities in nodules via transcription factors. FUN (fixation under nitrate) gene encodes a protein belonging to TGA family of transcription factors, most similar to transcription factor of *Arabidopsis*, i.e., PERIANTHIA (PAN) (Running and Meyerowitz 1996; Maier et al. 2011). The TGA family, classified as group D bZIP (basic leucine zipper) transcription factors (Dröge-Laser et al. 2018), is distinguished by a bZIP DNA-binding domain located at N terminal end and a

DOG1 domain having anonymous role at C terminal end (Tomaž et al. 2022). The FUN sensor domain shows a remote similarity to metal-binding proteins (Trepreau et al. 2011). The transcriptional regulator of FUN functions as a sensor, with Zn modulating the shift between an inert filament-like structure and a functional transcriptional regulator. When Zn levels decrease in the nodule, typically in response to increased soil nitrate levels, the filament dissociates, activating FUN. Once active, FUN quickly directs many pathways, leading to nodule disintegration. This Zn-reliant filamentation process provides a concentration-dependent signal that helps the plant in fine-tuning of nodule function in line with the nitrogen availability in the environment (Lin et al. 2024).

2.5 Plant Reproduction

It has been demonstrated that pollen tubes have a large demand for Zn in areas where synthesis of proteins occurs (Elsby and Roberts 2008). Applying Zn is highly beneficial for crop reproductive yield, as it enhances male and female gametogenesis, leading to an increased number of flowers per plant. Furthermore, application of Zn promotes the production of sporogenous tissue, which increases the number of pollen grains in anthers. Additionally, Zn enhances pollen-stigma interactions by improving receptiveness of stigma and its function and pollen survivance. These effects together add to the proper germination of pollen grains, normal development, and improvements in yield parameters such as number, volume, and mass of seeds and pods (Pandey and Gupta 2012: Borah et al. 2021).

2.6 Environmental Stresses

The role of nutrients like copper (Cu), manganese (Mn), zinc (Zn), and iron (Fe) in plant defense is well-documented (Dordas 2008; Fones and Preston 2013). Several reports have indicated vital role of Zn in aiding the plants to cope with diverse abiotic stressors such as salinity (Al-Zahrani et al. 2021; Hussein and Abu-Baker 2018), drought (Hassan et al. 2020), metal toxicity (ul Hassan et al. 2017), and various biotic stresses. The mechanisms by which plants use Zn to defend against pests and pathogens are diverse, and the results of these interactions can vary. The success of Zn-related defenses in repelling invaders, combined with the ability of pests or pathogens to overcome these defenses, as well as other environmental factors, all influences the outcome. Numerous studies indicate that Zn fertilization typically reduces the adversity of biotic stress symptoms in planta (Li et al. 2016; Machado et al. 2018). Nevertheless, while a certain level of Zn might protect a plant against specific pathogens, it can also make the plant more susceptible to other pathogens (Helfenstein et al. 2015). In this context, Zn proteins have a twofold function in the defense system of plant, as they can both support the plant and assist its invaders simultaneously.

Zinc nutrition aids in defending plants against viral attack by enhancing their growth, development, vigor, and robustness. In this context, external application of $ZnSO_4$ considerably decreased the severeness of cotton leaf curl virus (CLCuV) and made the conditions hard for CLCuV vector, i.e., whitefly (Kalsoom et al. 2019). Similarly, leaf supplementation of $ZnSO_4$ diminished the warning signs and amount of potato viruses (PLRV and PVY) by enhancing tolerance to viral diseases. This increased resistance was due to the elevated levels of phenolic compounds, which assist in developing the plant's defense machinery (Ibrahim et al. 2016). Since Zn plays a vital role in protein and starch synthesis, hence Zn-deficit plants tend to build up reducing sugars and amino acids in their cells (Dutta et al. 2017). Because of amino acid accumulation, plant becomes more attractive for providing nourishment and for breeding of sucking insects. From a vector perspective, elevation in feeding potency and reproduction can enhance transmission of virus and development of disease (Dutta et al. 2017). Recent advancements have introduced several Zn-based nanoparticles (Zn-NPs) for plant nutrition, which also exhibit less environmental toxicity. ZnO-NPs serve as an environment-friendly and cost-effective antiviral agent against TMV, thereby reducing oxidative injury by lowering the levels of MDA (malondialdehyde), O_2, and H_2O_2 (Sofy et al. 2021). Cai et al. (2019) showed that foliar spray of ZnO-NPs directly targeted TMV and boosted the plant's defense system. They observed that applying ZnO-NPs induced the production of ROS, which activated the antioxidant defense system and upregulated resistance-associated genes, PR1 and PR2, through salicylic acid signalling.

Zinc nutrition is also effective in managing plant bacterial diseases, such as ginseng bacterial leaf spot triggered by *Pseudomonas cichorii* (Dutta et al. 2017). The study conducted by Quaglia et al. (2021) reported shielding influence of $ZnSO_4$ against *Pseudomonas syringae pv.* tomato, the bacterium responsible for bacterial speck disease in tomatoes. Further, in vitro assays revealed that elevated Zn concentrations had direct antimicrobial activity. Additionally, the presence of $ZnSO_4$ triggered defense mechanisms, including callose deposition and stimulation of SA-dependent pathogenesis-linked protein PR1b1 gene (Quaglia et al. 2021). Callose, a β-(1,3)-D-glucan polysaccharide, is accumulated between cell membrane and cell wall at contagion sites. This deposition helps to impede the attack as well as proliferation of pathogens (Wang et al. 2021). Lentils infected with *Xanthomonas axonopodis pv. phaseoli* and *P. syringae pv. syringae*, which are responsible for bacterial leaf spots and bacterial blight, were reported to benefit from application of ZnO-NPs in terms of mitigation of these diseases (Siddiqui et al. 2018).

Zinc plays a pivotal task in alleviating the impact of plant pathogenic fungal infections as well. Adequate Zn levels can strengthen cellular integrity and enhance cell membrane permeability, serving as a defense mechanism against fungal invasions (Machado et al. 2018). Foliar application of Zn effectively diminished the acuteness of tan spot disease caused by *Pyrenophora tritici-repentis* in durum wheat (Simoglou and Dordas 2006). Similarly, Zn nutrition plays a key role in mitigating *Fusarium* root rot disease in wheat (Khoshgoftarmanesh et al. 2010). Further, Zn enhances plant resistance to infection through stabilization of root cell membranes, which decreases permeability of membranes, ultimately promoting root growth and overall plant vitality (Khoshgoftarmanesh et al. 2010). In a pot experiment, the Zn application to soil in combination with NPK fertilizers effectively controlled *Alternaria solani*-induced early blight infection in tomatoes (Awan et al. 2019). The control of early blight was linked to the enhanced antioxidant enzymatic activities such as superoxide dismutase and defense-related proteins. Likewise in another experiment, Zn compounds were effective in decreasing the symptoms of *Helminthosporium* leaf spot disease in sesame, under in vitro as well as field conditions (El-Fawy and El-Said 2018). The anti-fungal and anti-mycotoxin effects of two different compounds of Zn were tested against virulent strains of *Penicillium citrinum*, *Fusarium graminearum*, and *Aspergillus flavus* by Savi et al. (2013). The authors observed that $Zn(ClO_4)_2$ and $ZnSO_4$ fully hampered fungal growth as well as mycotoxin production by disrupting fungal cellular metabolism, affecting conidia production and altering hyphal morphology. In *Arabidopsis thaliana*, the application of Zn at high but non-phytotoxic concentrations enhanced the plant's defense response against the fungal pathogen *Alternaria brassicicola* through the activation of JA/ET-reliant defense signaling pathway (Martos et al. 2016). $ZnSO_4$ inhibited the in vitro development of *Alternaria alternata* in a concentration-dependent fashion (Shoaib et al. 2021). In addition, application of $ZnSO_4$ to leaves in vivo decreased *Alternaria* leaf spot infection in plants of chilli pepper. This effect was achieved by enhancing resistance through the activation of antioxidative enzymes (CAT and POX) as well as defense-related compounds [phenylalanine ammonia lyase (PAL) and polyphenol oxidase (PPO)] (Shoaib et al. 2021). Generally, in response to fungal infection, enzyme activities, for example, tyrosine ammonia lyase (TAL), POX, PPO, and PAL, are elevated as a part of plant's defense mechanism which further gets boosted by supplementation of Zn (Wadhwa et al. 2014). Peroxidase facilitates creation of an obstruction at infection site via lignification, whereas TAL and PAL are crucial for phenylpropanoid metabolism and defensive responses. The severe symptoms of coffee rust disease caused by *Hemileia vastatrix* decreased with increasing Zn concentrations varying from 0.25 to 2 mg L^{-1}, although the reduction was less pronounced at levels outside this range (Wadhwa et al. 2014). *Arabidopsis thaliana* Col-0 (wild type) exhibited increased susceptibility to *Botrytis cinerea* at both deficient and excessive Zn concentrations, likely due to inadequate activation of PDF 1.2, a gene associated with pathogen tolerance (Quaglia et al. 2022). Additionally, susceptibility to *Golovinomyces cichoracearum*, which is a biotrophic fungus, rose with higher Zn doses because the elevated Zn levels impaired the reduction of pathogen development and hindered the oxidative burst. This was because of the induction of antioxidant defense machinery at these high Zn levels, which led to increased ROS scavenging (Quaglia et al. 2022). An oxidative burst, characterized by a swift accumulation of reactive oxygen species (ROS) at the infection region, is a key defense response employed by plants against diseases (Zhang et al. 2020). Reduced vulnerability to *G. cichoracearum* in response to Zn deficiency could be attributed to increased oxidative burst, which creates an inhospitable environment for biotrophic fungi. According to Qualiga et al. (2022), oxidative stress in Zn-deficit plants may explicate the lesser vulnerability to biotrophic pathogens coupled with greater susceptibility to necrotrophic ones. This is for the reason that oxidative stress leads to cellular damage that benefits necrotrophic fungi that thrive on dead tissues (Quaglia et al. 2022). Additionally, ZnO-NPs application has been reported to play a vital role against *Fusarium proliferatum* (Pan et al. 2023).

Managing nematode infestations with Zn nutrition may be linked to improved cellular integrity. Zinc boosts the activity of SOD, which helps detoxify excessive ROS. Without this detoxification, high ROS levels could lead to lipid peroxidation, decreased membrane integrity, and increased root permeability, making the roots more susceptible to nematode infection (Couto et al. 2016). Zinc deficiency can inhibit protein synthesis, leading to an accumulation of free amino acids. These amino acids may leak into root exudates, which attract nematodes and consequently increase the extent of nematode infestation (Couto et al. 2016). When Zn was used in conjunction with various strains of *Rhizobium* (which produce potent nematicidal compounds) against *Meloidogyne incognita* in tomatoes, the mortality of the nematodes increased in vitro with higher Zn concentrations (Shaukat and Siddiqui 2003). Zinc does not directly impact the embryonic growth or young hatching of *Meloidogyne incognita*. Nevertheless, it does reduce the formation of eggs and galls. Zinc considerably decreases the number of nematodes gm^{-1} and, to some extent, the overall population of nematodes per root system. The decrease was notably more effective when Zn was combined with mannan-oligosaccharide obtained from cell walls of *Saccharomyces cerevisiae* (Couto et al. 2016). Application of Zn to soil, either singly or combined with *Pseudomonas aeruginosa* IE-65+, significantly decreased the invasion of *Meloidogyne javanica* nematodes.

Previous studies on plants have indicated that non-noxious levels of Zn could help alleviate diverse abiotic stresses. Role of Zn in responding to and mitigating salinity stress may depend on its ability to enhance uptake and transport of water in plants, thus lessening the harmful impacts of interim salinity and heat stress (Tavallali et al. 2010; Peck and McDonald 2010). Additionally, Zn induces ascorbate peroxidase (APX), catalase (CAT), peroxidase (POD), and polyphenol oxidase (PPO) activities and increases proline levels, which impacts the configurational integrity as well as permeability of membranes under salt stress (Weisany et al. 2012; Hussein and Alva 2014; Tolay 2021). A study on *Rosmarinus officinalis* (*L.*) raised under NaCl stress in modified Hoagland's nutrient solution showed that application of nano-Zn sulfate to leaves enhanced salinity tolerance and increased the plant biomass accumulation and physiological attributes (Hassanpouraghdam et al. 2020). Furthermore, several reports have revealed that plants mitigate drought stress through Zn-mediated mechanisms, including increased synthesis of hormones (auxins, melatonin, and gibberellins), antioxidant defense system, improved photosynthetic potential, cell membrane integrity, plant water relations, regulation of stomata, water uptake, and accretion of osmolytes (Hassan et al. 2020). In another experiment, Zn was shown to counteract the harmful effects of copper-generated oxidative burst by inducing antioxidant enzymes in rice seedlings raised in Hoagland solution (Thounaojam et al. 2014).

2.7 Zn Proteins or Enzymes Involved in Plant Defense

2.7.1 Zinc Finger Proteins

Zinc participates in gene expression regulation and stimulation or suppression of DNA elements through metalloproteins known as zinc finger proteins (ZFPs) (Sinclair and Kramer 2012) (Fig. 1). These motifs consist of a polypeptide chain typically comprising 11–13 amino acid residues, which interact specifically with DNA sequence (Kirkby and Römheld 2004). These proteins, characterized by a wide range of structures and functions, possess a minute, openly coiled functional domain which depends on Zn ions for structural stability (Laity et al. 2001). Zinc finger binding domains are found in tolerance proteins, for example, nucleotide binding site-leucine rich repeat proteins (NBS-LRRs), which participate in effector-triggered immune reaction (Cabot et al. 2019). Further, these domains represent chain pattern where histidine and cysteine residues coordinate with Zn ions to form secondary proteins, which then carry out specific biological functions (Peng et al. 2012; Zang et al. 2015; Guan et al. 2016; Noman et al. 2019). Understanding the function of Zn in planta, particularly when it interacts with histidine and cysteine in proteins, is crucial for applying molecular breeding strategies aimed at enhancing stress resistance and growth promotion. ZFPs are categorized into different groups depending upon the number, location and arrangement of C-H residues. These groups include CCCH, C2C2, C2HC, C2H2, C3HC4, C2HC5, C4HC3 C6, and C8 (Li et al. 2014; Seok et al. 2016; Xie et al. 2019). Zinc fingers perform twofold roles, binding to DNA as well as facilitating protein-protein associations (Noman et al. 2019; Tolosa and Zhang 2020). Apart from DNA, ZFPs interact with proteins, RNA, and lipids, playing a role in various cellular activities within plants (Guan et al. 2016). For instance, some ZFPs, like CCCH ZFPs, have motifs that enable direct binding to RNA, whereas most other ZFP families are characterized by their ability to bind DNA or proteins (Peng et al. 2012). The C2H2 zinc finger protein (ZFP) family is particularly large, comprising 189 members in rice, 176 in *Arabidopsis*, 124 associates in foxtail millet (Liu et al. 2015), 211 in maize, 122 in wheat, 321 in soybean, 301 in *Brassica*, and 129 in cucumber (Yin et al. 2020). Additionally, these ZFPs have significant functions in various stages of plant growth and well-being and in different environmental conditions (Noman et al. 2019). During stressful conditions, these proteins exhibit diverse roles, comprising RNA or DNA binding along with protein-protein interactions, which are crucial for stress and hormonal signalling in plants (Liu et al. 2015; Guan et al. 2016; Han et al. 2016; Yin et al. 2020).

2.7.2 Alcohol Dehydrogenase

Alcohol dehydrogenases (ADH) are Zn-metalloenzymes which belong to medium-chain dehydrogenase/reductase (MDR) family and often possess two Zn^{2+} ions per subunit. One Zn^{2+} ion is catalytic and located at active region, while the other is structural and situated at a region believed to influence subunit interactions (Auld and Bergman 2008). Alcohol dehydrogenases are NADP-reliant oxidoreductase enzymes involved in transferring alcoholic hydride ions to NAD^+, thereby facilitating the inter-convertible oxidation of alcohols into ketones or aldehydes (Kumar et al. 2016). Anaerobic conditions, due to the absence of oxygen, such as underwater logging alter chemical reactions leading to the synthesis of toxic compounds through microbial activity. Alcohol dehydrogenase metabolizes the alcohol produced under these conditions (Castillo-Gonzalez et al. 2018). In response to Zn deficiency, activity of ADH decreases. In rice, flooding leads to a doubling of ADH activity in the underground parts once there is enough zinc, but this increase does not occur when zinc is deficient. A lack of ADH activity is linked to damaged roots in submerged rice fields (Arshad and Frankenberger 2012). Zn-binding ADH (AdZADH2) was reported to be stimulated in *Arachis diogoi* in response to the inoculated *Phaeoisariopsis personata*. In tobacco leaves, AdZADH2 from peanut, driven by an estradiol-inducible promoter, induced hypersensitive reaction such as cell death along with increased induction of defense-linked genes, antioxidant enzymatic activities, and marker genes (Kumar et al. 2016). Similarly, binding of Zn to cinnamyl alcohol dehydrogenase (CAD) contributes to biosynthesis of lignin, enhancing shielding response by strengthening cell wall (Kumar et al. 2016). Wheat CAD (TaCAD12) plays a crucial function in resisting sharp eyespot disease produced by *Rhizoctonia cerealis* in wheat, by positively regulating the expression of specific defense genes as well as monolignol-biosynthesis-related (MBR) genes. MBR genes (CCR, CAD, and COMT) contribute to resistance by aiding in lignin formation through monolignols, as lignin deposition is believed to toughen the cell wall because of its great tolerance to microbe-induced degradation (Tronchet et al. 2010). This process acts as a means of defense against pathogen infection.

2.7.3 Metallothioneins

Metallothioneins (MTs), rich in cysteine proteins, bind to heavy metal ions through the thiol groups present in their cysteine residues (Hassinen et al. 2011) and play a significant role in combating heavy metal toxicity (Yang et al. 2015). Moreover, MTs function as antioxidants against toxic reactive oxygen as well as nitrogen species generated by pathogen attack. MT can check free radicals and bind with several heavy metals like Cd, Zn, Cu, etc. (Hassinen et al. 2011). Zinc plays a crucial role in functioning of MTs owing to their close association. Zinc stimulates increased expression of Zn-binding MTs via metal regulatory transcription factor 1 (MTF-1) that specifically responds to elevated levels of Zn (Dumanović et al. 2021; Ruttkay-Nedecky et al. 2013). During *Magnaporthe grisea* infection, MTs-1 and four pathogenesis-linked genes were strongly induced in rice to provide defense against the pathogen (Kim et al. 2001). Although MTs-1 expression was also observed in healthy plants, its expression was significantly higher during infection, indicating a defensive response (Kim et al. 2001). Similarly, in *Fagus grandifolia*, the MT gene was found to encode mRNA for synthesis of MT-like proteins, whose activation was linked to resistance against beech bark infection (Chen et al. 2021). The researchers advocated that these MT proteins might play an important function in defending against beech-scale insect, the vector of the disease.

2.7.4 Antioxidant Enzymes

Superoxide dismutase (SOD) is an antioxidant enzyme located in all oxygen-respiring cells and intra-cellular compartments that are vulnerable to oxidative damage. It is involved in the conversion of superoxide anions into hydrogen peroxide. This hydrogen peroxide is then broken down by the enzymes glutathione peroxidase (GPX), ascorbate peroxidase (APX), and catalase (CAT) (Dumanović et al. 2021). Superoxide dismutase (SOD) is a crucial antioxidant defense enzyme that combats superoxide anions, a type of ROS in plant cells exposed to oxygen. Additional antioxidant enzymes include catalase (CAT), glutathione peroxidase (GPX), ascorbate peroxidase (APX), glutathione reductase (GR). monodehydroascorbate reductase (MDHAR), and dehydroascorbate reductase (DHAR). Further, non-enzymatic antioxidant compounds like glutathione (GSH), ascorbate (AsA), tocopherols, carotenoids, phenolics, and proline also check and neutralize ROS (Sharma et al. 2012). The activities of CAT, GPX, and APX indirectly rely on the presence of Zn (Cakmak 2000). ROS act as signalling molecules and as damaging agents, depending on the equilibrium between their production and scavenging (Sharma et al. 2012). Cells must manage ROS levels, which increase during pathogenic infection or abiotic stresses (Wongpia and Lomthaisong 2010). Excessive accumulation of ROS is detrimental for cells as it causes oxidative damage (Subramanian et al. 2005). In enzymatic reactions, the elimination of ROS generated during stress is greatly helpful when they are catalyzed by SOD (Dumanović et al. 2021). SOD enzymes are categorized based on their responses to hydrogen peroxide and cyanide. CuZn-SOD is deactivated by hydrogen peroxide as well as cyanide, whereas Fe-SOD is repressed by hydrogen peroxide only. Mn-SOD is not affected by either cyanide or hydrogen peroxide (Cakmak 2000). CuZn-SOD is the main prevalent form of SOD in higher plants, whereas Fe-SOD and Mn-SOD constitute a minute portion of the total SOD (Jackson et al. 1978). In plants, Zn is linked with copper at the functional region of CuZn-SOD, and this association is crucial for the enzyme's activity. CuZn-SOD significantly contributes to overall SOD activity in plants. SOD activity rises with higher Zn levels and declines under Zn-deficit conditions (Bharti et al. 2014; Castillo-Gonzalez et al. 2018). Besides Zn, CuZn-SOD activity decreases when plants face photo-oxidative environment, which results in the production of large amounts of H_2O_2 (Cakmak 2000). Zinc participates in regulating the generation of ROS by interfering with oxidation of NADPH and removing O_2 radicals. When Zn is deficient, there is an increase in O_2 radicals, resulting in accumulation of toxic ROS. These free radicals rupture the bonds present in phospholipids and polyunsaturated fatty acids of cellular membranes, resulting in enhanced membrane electrolyte leakage. Consequently, this heightened permeability causes carbohydrates and amino acids to leak from the underground and foliage parts of the plant, making it more susceptible to pathogenic infection (Castillo Gonzalez et al. 2018).

2.8 Phytohormones

Zinc, necessary for the production of tryptophan, is needed in auxin synthesis. Consequently, a lack of Zn leads to decreased auxin production in plants (Mousavi 2011). Additionally, participation of Zn in synthesis of auxins might involve zinc finger proteins (ZFPs), such as AtAZF2, which control auxin-mediated genes in plants exposed to stress (Yin et al. 2020). Salicylic acid (SA), abscisic acid (ABA), and jasmonic acid (JA) are crucial natural phytohormones that are involved in defense against external stress in plants (Yang et al. 2018). When tobacco leaves were exposed to ZnO NPs (100 μg/mL), the levels of ABA and SA rose significantly by 5.17 fold and 1.60 fold, respectively, relative to the controls (Cai et al. 2019). The upregulation of SA and ABA induced by ZnO NPs may be essential in enhancing plant tolerance by triggering defense-related genes and regulating hormonal responses (Cai et al. 2019). Contrastingly it has also been reported that ZnO NPs contribute to the enhancement of ethylene biosynthesis which may induce toxicity in rice seedlings (Khan et al. 2021).

3 Zinc in Humans (Including Animals)

Of the entire Zn present in the human body, bones and muscles have 85%, 11% is present in liver and skin, and the remaining is contained in several other tissues (Prasad 2009; Bhowmik et al. 2010). The largest amount of Zn is found in prostate and area around the eye. Zinc in an adult human body ranges from 2.3 to 1.4 g (Prasad 2009; Bhowmik et al. 2010). Recommended daily dosage of Zn for adult men is 11 mg, while for mature women 8 mg is required (Brown et al. 2001; Erdman et al. 2012). In accord with estimates, one-third of the global population is at threat in terms of Zn deficiencies. Due to their greater needs of Zn for the process of development, children (under the age of 5) are at more risk (Wessells and Brown 2012). An estimated half a million children die annually as a result of Zn paucity (Black et al. 2008; Krebs et al. 2014). Micronutrient deficiencies, particularly Zn deficiency, are a major source of economic loss in impoverished nations. Productivity of people declines as their expenses on health services grow, which has a direct influence on the GDP (Darnton-Hill et al. 2005; Stein 2014).

Because Zn is so vital for human health, even a small deficit can have fatal consequences. Humans who suffer from Zn deficiency experience loss of appetite, smell and taste, anorexia, and other warning signs. Their immune systems may also be impacted, leading to anaemia and arteriosclerosis. Zn deficiency leads to impaired homeostasis because of faulty platelet accumulation, reduction in number of T cells, and decreased response of T-lymphocytes to phytoestrogens (Keen and Gershwin 1990; Tapiero and Tew 2003). Actually, the sole naturally occurring lymphocyte mitogen is Zn (Keen and Gershwin 1990; Tapiero and Tew 2003). Because Zn serves as a cofactor in several enzymes, RNA and DNA, it is an essential nutrient that the living body needs for a number of functions (WHO 1996). Growth retardation, congenital defects, and pregnancy and birth problems have all been related to

Zn deficiency in humans. Zinc is imperative for gut immune system and for maintaining the structure and function of the gut. The jejunum's villi shrivel and flatten when there is a Zn shortage. Because of this alteration in morphology, Zn absorption from the surface is reduced, which is important for an organism's proper operation. In addition to its function as a neural modulator in excitatory syndesis of the brain, it is engaged in biochemical processes that influence an organism's immunological response (Sun et al. 2011). Applying Zn in reaction to stressful events is another option. Zinc is involved in gene expression and is vital for cellular integrity of the cell walls. It affects how quickly an organism ages (Sun et al. 2011).

3.1 Source, Absorption, and Distribution of Zinc

The primary source of Zn is food. Meat (pork, beef, chicken, and turkey), seafood (oysters), legumes, and cereals all have relatively high quantities of this mineral. In general, meals derived from animals are a better source of Zn than foods derived from plants. There are very few substances in animal-based diets that prevent the absorption of Zn. What matters most is that there is no phytate content at all (Brown et al. 2001). Bioavailability of Zn is decreased by phytic acid through the formation of insoluble complexes. Zn, cysteine, and histidine are better absorbed when specific amino acids are present (Brown et al. 2001). Animals absorb Zn mostly from the small intestine, where the proximal jejunum and distal duodenum are important organs. There are two ways through which Zn is absorbed: actively and passively. Certain transporters are responsible for carrying out active transport, and the efficiency of this process rises with dietary Zn intake. The efficiency of passive transport, which relies on a diffusion process, is correlated with the Zn levels in lumen of intestine (Cousins 1998; Krebs 2000). Zinc finger proteins, metallothioneins (MTs), and zinc transport proteins are examples of non-enzymatic Zn proteins. ZnT proteins and ZIP proteins are two protein families that are connected to transport of Zn. Zinc enters the absorptive cells of intestine via basolateral membrane and moves into the portal blood system during the cellular process of Zn absorption. Several proteins referred to as Zn transporters are used in this process (Krebs 2000). Zinc that has been absorbed is carried by plasma, predominantly attached to albumin (60–80%) and to a lesser extent on transferrin and α-2-macroglobulin, additionally being combined with free amino acids, particularly cysteine and histidine. Although it only makes up roughly 0.1% of the body's entire Zn content, Zn linked with plasma proteins is the most readily accessible and extensively available form of Zn in living body. Zinc is fivefold more abundant in blood compared to plasma (Cousins et al. 2006). Zn-superoxide dismutase and carbonic anhydrase in erythrocytes contain 80% of Zn (Cousins et al. 2006).

Role of Zn in various regions of human body has been described in the following sub-sections and represented in Fig. 2:

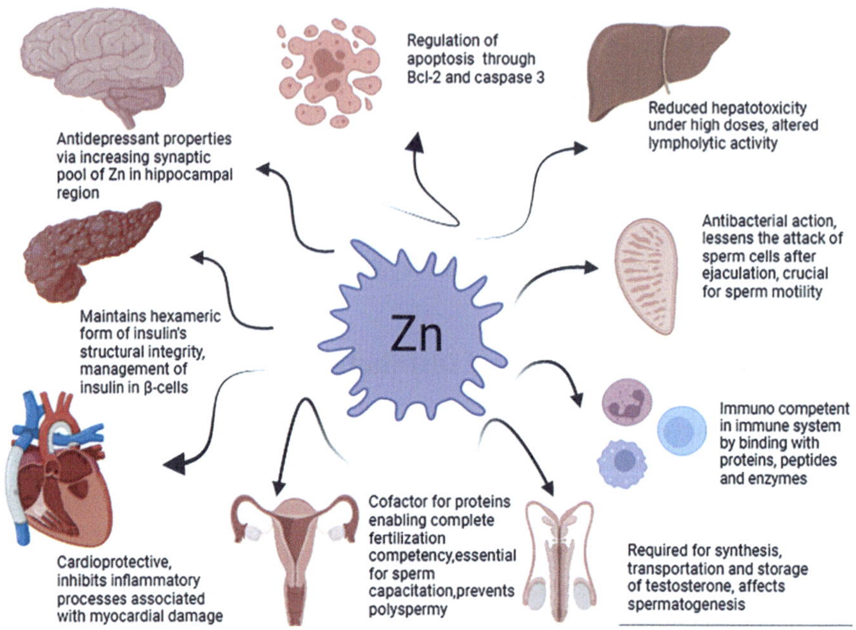

Fig. 2 Physiological significance of zinc (Zn) in humans as well as animals

3.2 Brain and Nervous System

Impact of Zn on the brain was investigated by Nowak et al. (2005) using a suitable mouse model. Zinc raises the amount Zn in hippocampal region and, like antidepressants, stimulates neurotrophin formed by the brain for gene expression (Nowak et al. 2005). Vesicular Zn co-localizes with glutamate in a part of glutamatergic nerve endings across the anterior brain in the mammalian brain. Additionally, Zn^{2+} has been shown to possess antidepressant properties (Holst et al. 2007). These properties are linked to the function of G protein-linked receptor 39 (GPR39), a Zn^{2+} receptor that is also a strong activator of GPR39 activity through Gα(s), Gα(q), and Gα (12/13) pathways (Popovics and Stewart 2011). It is considered that Zn plays a crucial part in neuronal function and signalling (Frederickson et al. 2005). Zn^{2+} transporters contribute in neurulation, growth of nervous system, and neural cell differentiation (Chowanadisai et al. 2013). Zinc transporters such as ZnT3 are believed to control presynaptic Erk1/2 signaling and hippocampal-dependent memories (Sindreu et al. 2011), indicating the function of Zn in maintaining memory. Few reports have revealed a positive interaction of Zn level with one or more attributes of cognitive abilities. Further, imbalance in Zn status results in brain-related diseases such as Parkinson's and Alzheimer's disease (Adani et al. 2020).

3.3 Cardiovascular System

Cardiomyocyte ionic currents were shown to alter in response to Zn, indicating that both external and intracellular Zn levels rise in cardiomyocytes under pathological circumstances and could change the L-type Ca^{2+} channel currents (I Ca); however not all of these effects occur in the same sequence or with the same intensity (Turan 2003). Regarding Zn^{2+}cardioprotective role, Zn functions by inhibiting the inflammatory processes initiated during myocardial damage and also functions by wound healing and repairing by preserving the cardiac stem cells which are required for cardiac healing and play a protective role against acute redox stress in cardiomyocytes (Little et al. 2010). Several roles of Zn have also been reported in controlling arterial blood pressure (Tubek 2007).

3.4 Liver

Very high doses of Zn may lessen hepatotoxicity caused by oxidative breakdown and undesirable compounds being deposited in the liver (Stehbens 2003). According to Savarino et al. (2001), the levels of selenium and Zn in the body and bodily fluids could act as initial warning signs of the underlying illness. Several other roles of Zn are also reported in the endocrine system, for example, zinc's function in preserving thymic output in healthy older people (Mitchell et al. 2006). A decrease in Zn content is linked with impairment of T-cell activity. Zinc supplements are known to rise or fall in response to thymic hormones. Moreover, Zn alone has the power to alter lympholytic activity.

3.5 Pancreas

Out of all the cells in the body, pancreatic β cells possess the most Zn. Zn^{2+} contributes to the accurate management of insulin within the β cells, spanning from its storage to its actual use (Prost et al. 2004). According to certain theories, Zn^{2+} activation on both sides of the membrane regulates ATP-induced K^+ channel (KATP) opening in the pancreas, with possibly distinct regions of action on sulfonylurea receptor (SUR) subunit (Prost et al. 2004). Zinc participates in the production, secretion, and storage of insulin as well as in maintaining the hexameric form of insulin's structural integrity (Fukunaka and Fujitani 2018; Norouzi et al. 2017). Any reduction in Zn levels can exacerbate the issue, especially in type 2 diabetes, by impairing islet of Langerhans cell's capacity to generate and produce insulin (Chausmer 1998; Ohta et al. 2019). It has been demonstrated that mouse endocrine cells are unaffected by a single low-dose injection of Zn, while pancreatic exocrine cells are unaffected by Zn (Minami et al. 2001).

3.6 Immune System

Immune responses are believed to depend critically on Zn as it interacts and impacts immune system components, which rapidly grow and increase in number (Wellinghausen and Rink 1998). Because Zn binds to proteins, peptides, and enzymes with varying binding affinities, it is important for immunocompetence (Mocchegiani et al. 2000). Free Zn^{2+} ions appear to be biologically active, but Zn is delivered to cells attached to proteins, primarily α2-macroglobulin, albumin, and transferrin (Vallee and Falchuk 1993). Zinc controls the activity of α2-macroglobulin itself. Immune function is indirectly impacted by Zn, which modifies the configuration of α2-macroglobulin and improves its association with proteases and cytokines. Zinc insufficiency has been linked to immune function impairment, which could be the primary basis of secondary immunodeficiency conditions in human beings (Tapiero and Tew 2003).

Further, low levels of Zn reduce the vaccination response as well as perturb the working of native immune cells (Haase and Rink 2009). Decreased Zn status is also linked to greater IgE levels, a hypersensitivity-related antibody, indicating involvement of low Zn levels in hypersensitivity (Seo et al. 2017). The role of Zn in growth and functioning of T regulatory cells implicated in the avoidance of auto-immune disorders has also been reported (Hogstrand et al. 2009).

3.7 Spermatogenesis

Poor spermatogenesis can result from the Zn content being affected by modified expression of Zn transporters (Kerns et al. 2013). Ellis and Stanfield (2014) reported the requirement for Zn and its involvement in survival of germ cells earlier to maturity and in replacement by protamine throughout spermatogenesis. Due to its high concentration in spermatocytes and its occurrence in chromatin and nucleus, Zn participates in spermatogenesis as well as initial phases of growth of sperm cells (Zhao et al. 2013). Further, due to the regulatory role of Zn in spermatogonial proliferation and its ability to maintain germ cells unharmed during meiosis, Zn concentrations in the testes rise during the initial phases of spermatogenesis (Hai et al. 2014). At this point, Zn deficiency will cause spermatogonia to multiply unauthentically and germ cells to die without producing functionally mature sperm cells (Bradbury 2017). An elevated Zn concentration in the prostate gland suggests that Zn is involved in epididymal transit, stabilizing sperm cells either before or during ejaculation. Spermatogenesis is significantly impacted by zinc finger proteins (ZFPs) (Ishizuka et al. 2016). Zinc is an important element in ZFPs, a broad category of transcription factors (Ecco et al. 2017). These transcription factors are involved in a number of growth and development processes, like activating transcription and translational processes, binding DNA, and triggering cell death.

3.8 Prostate Gland

Although Zn is found all across the human body, it is a necessary component of the prostate where it is present in higher amounts than other soft tissues (Prashanth et al. 2015; Verze et al. 2016). Prostate gland is three times more Zn-rich (150 μg/g Zn) than any other soft tissue. Prostatic fluid, on the other hand, has about 500 μg Zn/mL (Leitzmann et al. 2003). During ejaculation, the prostate gland secretes elevated Zn into the seminal fluid, where it is crucial for sperm motility (Agarwal et al. 2014). The primary role of Zn in prostate gland is antimicrobial, i.e it lessens the harm to sperm cells after ejaculation. Numerous active bacteria are present in female upper reproductive canal, and when sperm enters, the bacteria may damage it (Rametse et al. 2014). Since Zn possesses antimicrobial attributes, it can protect the sperm cells from damage (Jeng et al. 2015).

3.9 Sex Hormones

Zinc is integral for the synthesis, transportation, and storage of the major sex hormones, comprising testosterone, which is thought to be a key regulator of spermatogenesis (Peterson et al. 2014; Teerds and Huhtaniemi 2015). Thus, monitoring Zn intake at the dietary level is crucial for figuring out testosterone synthesis. Zinc deficiency in the diet raises the amount of luteinizing hormone in the blood; nevertheless, serum and seminal plasma have low testosterone levels (Sengupta and Dutta 2018). Zinc deficiency is positively corelated with reduced or damaged Leydig cell population, alterations in Leydig cell apoptosis, proliferation and differentiation, and damage to the testes (Acharyya 2020). Testicular inflammation and oxidative damage to the Leydig cells have been reported to result from Zn deficiency (Maremanda et al. 2014). Common signs of low or insufficient Zn levels include impaired spermatogenesis, a drop in testosterone levels, impaired Leydig cells and luteinizing hormone (LH) receptors, and alterations in appearance of Leydig cells (Carruthers 2013).

3.10 Female Reproductive Tract

Zinc serves as a cofactor in several proteins of reproductive tract in females and induces them, permitting total fertilization ability (Yeste et al. 2016). This helps to decrease immune reactions against entry of sperm cells. Once they enter the female reproductive tract (Holt and Fazeli 2015), they signal that it is for reproduction and the process should not be interrupted (Pfaus et al. 2015). Zinc ions are essential for capacitation of sperms inside female reproductive tract as well as regulating other critical processes, resulting in successful fertilization (De Jonge 2017). Zinc efflux is necessary for Ca^{2+} influx, which is a necessary process for capacitation to occur. Zinc ion efflux dysfunction or deficiency will decrease capacitation (Que et al. 2019). Zinc also functions in regulating polyspermy prevention via process known as Zn spark in the female reproductive tract. Furthermore, Zn also removes the other sperms which bind onto the fertilized egg by forming a Zn shield around the fertilized egg (Hamad 2017).

3.11 Anti-apoptosis and Cancer

A growing tendency of apoptosis, or the death of cells and tissues, is linked to Zn deficiency (Roscioli et al. 2013). A Zn deficit causes apoptosis through the crucial genes and proteins caspase 3 and Bcl-2 (Thomas et al. 2013; Siddiqui et al. 2015). More investigation is required into the method by which Zn is delivered to vital targets and the labile Zn that shields cells from harm (Zhang et al. 2013). Numerous reports have revealed zinc's anti-apoptotic qualities; however the exact mechanism by which zinc prevents apoptosis is unclear and varies depending on the quantity of exposure. DNA fragmentation may trigger an endonuclease activity that is reliant on Ca^{2+} and Mg^{2+} and causes the Leydig cells to undergo apoptosis (Evgeni et al. 2014). Zinc has the ability to block these ions, which stops apoptosis and DNA breakage. The over-expression of Zn transporter, namely, ZnT1 resulting in weakened Zn homeostasis, has been reported in no less than five diverse types of cancer (Lehvy et al. 2019). Drop in serum Zn levels has been reported to enhance the risks associated with breast cancer along with its inception in women (Jouybari et al. 2019). Contribution of an anti-apoptotic kinase CK2 and Zn channel protein ZIP7 in cancer has been described (Hogstrand et al. 2009; Taylor et al. 2012). Similarly, the role of ZIP4 (SLC39A4) (a Zn transporter) in pancreatic cancer was documented (Li et al. 2007). Even though the molecular mechanism is not so far well known, Zn has been reported to defend against prostate cancer (Kolenko et al. 2013).

3.12 Miscellaneous Roles

Zinc is involved in several health diseases, such as hepatitis, cardiac health, pneumonia, and obesity (Barnett et al. 2010; Kaur et al. 2014). Low Zn levels have been considered as a risk for developing pneumonia in old aged persons (Barnett et al. 2010). Zinc interference has been reported to lessen the recurrent pneumonia and diarrhoea in kids, thereby decreasing the chances of child death (Hambidge 2006; Lassi et al. 2016). Zinc could be employed as an economical addition for anti-viral treatment in the case of hepatitis C virus (Read et al. 2018). On the contrary, few reports have suggested presence of a Zn-binding domain in hepatitis C virus replicase (Tellinghuisen et al. 2005). In recent times, the involvement of Zn against COVID-19 virus has also been established (Sethuram et al. 2021).

4 Conclusions

Undoubtedly Zn is an element of survival since it is involved in several essential roles in physiology of plants as well as humans. Its functions in basic processes for existence are becoming gradually more apparent, spanning from photosynthesis in plants to brain functioning in humans. The combined efforts of law-makers, agricultural experts, agronomists, doctors, and dieticians are necessary to deal with Zn shortage. Solving the Zn deficiency issue would rescue lives of millions of people and increase the global economy by improving crop efficiency and lowering medical expenses.

References

Acharyya S (2020) Inflammation and ageing: probable role in male infertility. Chem Biol Lett 7(99–112):80

Adani G, Filippini T, Michalke B, Vinceti M (2020) Selenium and other trace elements in the etiology of Parkinson's disease: a systematic review and meta-analysis of case-control studies. Neuroepidemiology 54:1–23

Agarwal A, Durairajanayagam D, Halabi J, Peng J, Vazquez-Levin M (2014) Proteomics, oxidative stress and male infertility. Reprod Biomed 29:32–58

Ahmad S, Mfarrej MFB, El-Esawi MA, Waseem M, Alatawi A, Nafees M et al (2022) Chromium-resistant Staphylococcus aureus alleviates chromium toxicity by developing synergistic relationships with zinc oxide nanoparticles in wheat. Ecotoxicol Environ Saf 230:113142

Akay A (2011) Effect of zinc fertilizer applications on yield and element contents of some registered chickpeas varieties. Afr J Biotechnol 10(60):12890–12896

Al Jabri H, Saleem MH, Rizwan M, Hussain I, Usman K, Alsafran M (2022) Zinc oxide nanoparticles and their biosynthesis: overview. Life 12(4):594

Al-Zahrani HS, Alharby HF, Hakeem KR, Rehman RU (2021) Exogenous application of zinc to mitigate the salt stress in *Vigna radiata* (L.) Wilczek – evaluation of physiological and biochemical processes. Plan Theory 10(5):1005

An L, Zhou Z, Sun L, Yan A, Xi W, Yu N et al (2012) A zinc finger protein gene ZFP5 integrates phytohormone signaling to control root hair development in *Arabidopsis*. Plant J 72(3):474–490

Andreini C, Banci L, Bertini I, Rosato A (2006) Zinc through the three domains of life. J Proteome Res 5(11):3173–3178

Arshad M, Frankenberger WT Jr (2012) Ethylene: agricultural sources and applications. Springer

Auld DS, Bergman T (2008) Medium-and short-chain dehydrogenase/reductase gene and protein families: The role of zinc for alcohol dehydrogenase structure and function. Cell Mol Life Sci 65:3961–3970

Awan ZA, Shoaib A, Khan KA (2019) Crosstalk of Zn in combination with other fertilizers underpins interactive effects and induces resistance in tomato plant against early blight disease. Plant Pathol J 35(4):330

Barker AV, Pilbeam DJ (eds) (2015) Handbook of plant nutrition. CRC Press

Barnett JB, Hamer DH, Meydani SN (2010) Low zinc status: a new risk factor for pneumonia in the elderly? Nutr Rev 68:30–37

Bhantana P, Rana MS, Sun XC, Moussa MG, Saleem MH, Syaifudin M et al (2021) Arbuscular mycorrhizal fungi and its major role in plant growth, zinc nutrition, phosphorous regulation and phytoremediation. Symbiosis 84:19–37

Bharti K, Pandey N, Shankhdhar D, Srivastava PC, Shankhdhar SC (2014) Effect of different zinc levels on activity of superoxide dismutases & acid phosphatases and organic acid exudation on wheat genotypes. Physiol Mol Biol Plants 20:41–48

Bhatla SC, Lal MA, Kathpalia R, Bhatla SC (2018) Plant mineral nutrition. In: Plant physiology, development and metabolism. Springer, Singapore, pp 37–81
Bhowmik D, Chiranjib K, Kumar S (2010) A potential medicinal importance of zinc in human health and chronic. Int J Pharm 1(1):05–11
Black RE, Allen LH, Bhutta ZA, Caulfield LE, De Onis M, Ezzati M, Maternal and Child Undernutrition Study Group (2008) Maternal and child undernutrition: global and regional exposures and health consequences. Lancet 371(9608):243–260
Borah L, Saikia J, Basumatary A (2021) Effect of foliar application of zinc on growth and yield of garden pea (*Pisum sativum L.*) in Assam condition. Int J Chem Stud 9(2):869–872
Bradbury NA (2017) All cells have a sex: studies of sex chromosome function at the cellular level. In: Principles of gender-specific medicine. Academic Press, Cambridge, pp 269–290
Broadley MR, White PJ, Hammond JP, Zelko I, Lux A (2007) Zinc in plants. New Phytol 173(4):677–702
Brown KH, Wuehler S, Peerson J (2001) The importance of zinc in human nutrition and estimation of the global prevalence of zinc deficiency. Food Nutr Bull 22:113–125
Cabot C, Martos S, Llugany M, Gallego B, Tolrà R, Poschenrieder C (2019) A role for zinc in plant defense against pathogens and herbivores. Front Plant Sci 10:1171
Cai L, Liu C, Fan G, Liu C, Sun X (2019) Preventing viral disease by ZnONPs through directly deactivating TMV and activating plant immunity in *Nicotiana benthamiana*. Environ Sci Nano 6(12):3653–3669
Cakmak I (2000) Tansley review no. 111. Possible roles of zinc in protecting plant cells from damage by reactive oxygen species. New Phytol 146(2):185–205
Carruthers M (2013) Testosterone deficiency syndrome: cellular and molecular mechanism of action. Curr Aging Sci 6(1):115–124
Casson S, Gray JE (2008) Influence of environmental factors on stomatal development. New Phytol 178(1):9–23
Castillo-González J, Ojeda-Barrios D, Hernández-Rodríguez A, González-Franco AC, Robles-Hernández L, López-Ochoa GR (2018) Zinc metalloenzymes in *plants*. Interciencia 43(4):242–248
Chausmer A (1998) Zinc, insulin and diabetes. J Am Coll Nutr 17:109–115
Chen X, Pei Z, Peng L, Qin Q, Duan Y, Liu H et al (2021) Genome-wide identification and functional characterization of CCHC-type zinc finger genes in *Ustilaginoidea virens*. J Fungi 7(11):947
Chowanadisai W, Graham DM, Keen CL, Rucker RB, Messerli MA (2013) Neurulation and neurite extension require the zinc transporter ZIP12 (slc39a12). Proc Natl Acad Sci USA 110:9903–9908
Ciftci-Yilmaz S, Mittler R (2008) The zinc finger network of plants. Cell Mol Life Sci 65:1150–1160
Costa MI, Sarmento-Ribeiro AB, Gonçalves AC (2023) Zinc: from biological functions to therapeutic potential. Int J Mol Sci 24(5):4822
Cousins RJ (1998) A role of zinc in the regulation of gene expression. Proc Nutr Soc 57(2):307–311
Cousins RB, Liuzzi JP, Lichten LA (2006) Mammalian zinc transport, trafficking and signals. J Biol Chem 281:24085–24089
Couto EAA, Dias-Arieira CR, Kath J, Homiak JA, Puerari HH (2016) Boron and zinc inhibit embryonic development, hatching and reproduction of *Meloidogyne incognita*. Acta Agric Scand Sect B Soil Plant Sci 66(4):346–352
Darnton-Hill I, Webb P, Harvey PW, Hunt JM, Dalmiya N, Chopra M, De Benoist B (2005) Micronutrient deficiencies and gender: social and economic costs. Am J Clin Nutr 81(5):1198S–1205S
De Jonge C (2017) Biological basis for human capacitation – revisited. Hum Reprod Update 23:289–299
Dordas C (2008) Role of nutrients in controlling plant diseases in sustainable agriculture. A review. Agron Sustain Dev 28:33–46
Dröge-Laser W, Snoek BL, Snel B, Weiste C (2018) The *Arabidopsis* bZIP transcription factor family – an update. Curr Opin Plant Biol 45:36–49

Dumanović J, Nepovimova E, Natić M, Kuča K, Jaćević V (2021) The significance of reactive oxygen species and antioxidant defense system in plants: a concise overview. Front Plant Sci 11:552969

Dutta S, Ghosh PP, Ghorai AK, De Roy M, Das S (2017) Micronutrients and plant disease suppression. Fertil Environ News 3(2):2–6

Ecco G, Imbeault M, Trono DKRAB (2017) Zinc finger proteins. Development 144:2719–2729

Eisvand HR, Kamaei H, Nazarian F (2018) Chlorophyll fluorescence, yield and yield components of bread wheat affected by phosphate bio-fertilizer, zinc and boron under late-season heat stress. Photosynthetica 56:1287–1296

El-Fawy MM, El-Said MAA (2018) Effect of foliar application of some zinc and phosphorus sources on controlling *Helminthosporium* leaf spot disease and production of sesame. J Plant Prot Pathol 9(3):201–207

Ellis RE, Stanfield GM (2014) The regulation of spermatogenesis and sperm function in nematodes. Semin Cell Dev Biol 29:17–30

Elsby LM, Roberts SG (2008) Interaction of the TFIIB zinc ribbon with RNA polymerase II

Englbrecht CC, Schoof H, Böhm S (2004) Conservation, diversification and expansion of C2H2 zinc finger proteins in the *Arabidopsis thaliana* genome. BMC Genomics 5:1–17

Erdman JW Jr, Macdonald IA, Zeisel SH (2012) Present knowledge in nutrition. Wiley

Escobedo Monge MF, Barrado E, Alonso Vicente C, Redondo del Río MP, Marugán M, de Miguelsanz J (2019) Zinc nutritional status in patients with cystic fibrosis. Nutrients 11(1):150

Escudero-Almanza DJ, Ojeda-Barrios DL, Hernández-Rodríguez OA, Chávez ES, Ruíz-Anchondo T, Sida-Arreola JP (2012) Carbonic anhydrase and zinc in plant physiology. Chil J Agric Res 72(1):140

Evgeni E, Charalabopoulos K, Asimakopoulos B (2014) Human sperm DNA fragmentation and its correlation with conventional semen parameters. J Reprod Infertil 15:2

Fones H, Preston GM (2013) The impact of transition metals on bacterial plant disease. FEMS Microbiol Rev 37(4):495–519

Frederickson CJ, Koh J-Y, Bush AI (2005) The neurobiology of zinc in health and disease. Nat Rev Neurosci 6:449–462

Fukunaka A, Fujitani Y (2018) Role of zinc homeostasis in the pathogenesis of diabetes and obesity. Int J Mol Sci 19:476

Gan Y, Kumimoto R, Liu C, Ratcliffe O, Yu H, Broun P (2006) GLABROUS INFLORESCENCE STEMS modulates the regulation by gibberellins of epidermal differentiation and shoot maturation in *Arabidopsis*. Plant Cell 18(6):1383–1395

Gan Y, Liu C, Yu H, Broun P (2007) Integration of cytokinin and gibberellin signalling by *Arabidopsis* transcription factors GIS, ZFP8 and GIS2 in the regulation of epidermal cell fate

Gatlin DM III, Wilson RP (1983) Dietary zinc requirement of fingerling channel catfish. J Nutr 113(3):630–635

Gondal et al (2021) A detailed review study of Zn involvement in animal, plants and humans. Ind J Pure App Biosci 9(2):262–271. ISSN: 2582-2845

Grotz N, Guerinot ML (2002) Limiting nutrients: an old problem with new solutions? Curr Opin Plant Biol 5(2):158–163

Guan QJ, Ma HY, Wang ZJ, Wang ZY, Bu QY, Liu SK (2016) A rice LSD1-like-type ZFP gene OsLOL5 enhances saline-alkaline tolerance in transgenic *Arabidopsis thaliana,* yeast and rice. BMC Genomics 17:1–11

Haase H, Rink L (2009) The immune system and the impact of zinc during aging. Immun Ageing 6:9

Hai Y, Hou J, Liu Y, Liu Y, Yang H, Li Z, He Z (2014) The roles and regulation of Sertoli cells in fate determinations of spermatogonial stem cells and spermatogenesis. Semin Cell Dev Biol 29:66–75

Hamad AM (2017) Molecular and physical interactions of human sperm with female tract secretions. Ph.D. thesis, University of Birmingham, Birmingham

Hambidge KM (2006) Zinc and pneumonia. Am J Clin Nutr 83:991–992

Han YC, Fu CC, Kuang JF, Chen JY, Lu WJ (2016) Two banana fruit ripening-related C2H2 zinc finger proteins are transcriptional repressors of ethylene biosynthetic genes. Postharvest Biol Technol 116:8–15

Hassan MU, Aamer M, Chattha MU, Haiying T, Shahzad B, Barbanti L et al (2020) The critical role of zinc in plants facing the drought stress. Agriculture 10:396

Hassanpouraghdam MB, Mehrabani LV, Tzortzakis N (2020) Foliar application of nano-zinc and iron affects physiological attributes of *Rosmarinus officinalis* and quietens NaCl salinity depression. J Soil Sci Plant Nutr 20:335–345

Hassinen VH, Tervahauta AI, Schat H, Kärenlampi SO (2011) Plant metallothioneins–metal chelators with ROS scavenging activity? Plant Biol 13(2):225–232

Helfenstein J, Pawlowski ML, Hill CB, Stewart J, Lagos-Kutz D, Bowen CR et al (2015) Zinc deficiency alters soybean susceptibility to pathogens and pests. J Plant Nutr Soil Sci 178(6):896–903

Hogstrand C, Kille P, Nicholson RI, Taylor KM (2009) Zinc transporters and cancer: a potential role for ZIP7 as a hub for tyrosine kinase activation. Trends Mol Med 15:101–111

Holst B, Egerod KL, Schild E, Vickers SP, Cheetham S, Gerlach LO, Storjohann L, Stidsen CE, Jones R, BeckSickinger AG, Schwartz TW (2007) GPR39 signaling is stimulated by zinc ions but not by obestatin. Endocrinology 148:13–20

Holt WV, Fazeli A (2015) Do sperm possess a molecular passport? Mechanistic insights into sperm selection in the female reproductive tract. MHR Basic Sci Reprod Med 21:491–501

Hussein MM, Abou-Baker NH (2018) The contribution of nano-zinc to alleviate salinity stress on cotton plants. R Soc Open Sci 5(8):171809

Hussein MM, Alva AK (2014) Effects of zinc and ascorbic acid application on the growth and photosynthetic pigments of millet plants grown under different salinity. Agric Sci 5(13):1253

Ibrahim HA, Ibrahim M, Bondok A (2016) Improving growth, yield and resistance to viral diseases of potato plants through modifying some metabolites using Zinc Sulphate and Jasmonic Acid. J Hortic Sci Ornam Plants 8:161–172

Ishizuka M, Ohtsuka E, Inoue A, Odaka M, Ohshima H, Tamura N, Yoshida K, Sako N, Baba T, Kashiwabara S et al (2016) Abnormal spermatogenesis and male infertility in testicular zinc finger protein Zfp318-knockout mice. Develop Growth Differ 58:600–608

Jackson C, Dench J, Moore AL, Halliwell B, Foyer CH, Hall DO (1978) Subcellular localisation and identification of superoxide dismutase in the leaves of higher plants. Eur J Biochem 91(2)

Jamsheer M, Kumar M (2021) Transcription factors as zinc sensors in plants. Trends Plant Sci 26(8):761–763

Jeng HA, Huang YL, Pan CH, Diawara N (2015) Role of low exposure to metals as male reproductive toxicants. Int J Environ Health Res 25:405–417

Jouybari L, Kiani F, Akbari A, Sanagoo A, Sayehmiri F, Aaseth J, Chartrand MS, Sayehmiri K, Chirumbolo S, Bjørklund G (2019) A meta-analysis of zinc levels in breast cancer. J Trace Elem Med Biol 56:90–99

Kalsoom H, Ali S, Sahi GM, Habib A, Zeshan MA, Anjum R, ..., Abdullah A (2019) Differential response of micronutrients and novel insecticides to reduce cotton leaf curl virus disease and its vector in *Gossypium hirsutum* varieties

Kareem HA, Hassan MU, Zain M, Irshad A, Shakoor N, Saleem S et al (2022) Nanosized zinc oxide (n-ZnO) particles pretreatment to alfalfa seedlings alleviate heat-induced morpho-physiological and ultrastructural damages. Environ Pollut 303:119069

Kaur H, Garg N (2021) Zinc toxicity in plants: a review. Planta 253(6):129

Kaur K, Gupta R, Saraf SA, Saraf SK (2014) Zinc: the metal of life. Comp Rev Food Sci Food Saf 13:358–376

Keen CL, Gershwin ME (1990) Zinc deficiency and immune function. Annu Rev Nutr 10:415–431

Kerns K, Zigo M, Drobnis EZ, Sutovsky M, Sutovsky P (2013) Zinc ion flux during mammalian sperm capacitation. Nat Commun 9:1–10

Khan AR, Azhar W, Wu J, Ulhassan Z, Salam A, Zaidi SHR et al (2021) Ethylene participates in zinc oxide nanoparticles induced biochemical, molecular and ultrastructural changes in rice seedlings. Ecotoxicol Environ Saf 226:112844

Khan ST, Malik A, Alwarthan A, Shaik MR (2022) The enormity of the zinc deficiency problem and available solutions; an overview. Arab J Chem 15(3):103668

Khoshgoftarmanesh AH, Kabiri S, Shariatmadari H, Sharifnabi B, Schulin R (2010) Zinc nutrition effect on the tolerance of wheat genotypes to *Fusarium* root-rot disease in a solution culture experiment. Soil Sci Plant Nutr 56(2):234–243

Kim S, Ahn IP, Lee YH (2001) Analysis of genes expressed during rice-*Magnaporthe grisea* interactions. Mol Plant-Microbe Interact 14(11):1340–1346

Kinraide TB, Poschenrieder C, Kopittke PM (2011) The standard electrode potential (Eθ) predicts the prooxidant activity and the acute toxicity of metal ions. J Inorg Biochem 105(11):1438–1445

Kirkby EA, Römheld V (2004) Micronutrients in plant physiology: functions, uptake and mobility

Kolenko V, Teper E, Kutikov A, Uzzo R (2013) Zinc and zinc transporters in prostate carcinogenesis. Nat Rev Urol 10:219–226

Krebs NF (2000) Overview of zinc absorption and excretion in the human gastrointestinal tract. J Nutr 130:1374–1377

Krebs NF, Miller LV, Michael Hambidge K (2014) Zinc deficiency in infants and children: a review of its complex and synergistic interactions. Paediatr Int Child Health 34(4):279–288

Kumar D, Rampuria S, Singh NK, Kirti PB (2016) A novel zinc-binding alcohol dehydrogenase 2 from *Arachis diogoi,* expressed in resistance responses against late leaf spot pathogen, induces cell death when transexpressed in tobacco. FEBS Open Bio 6(3):200–210

Laity JH, Lee BM, Wright PE (2001) Zinc finger proteins: new insights into structural and functional diversity. Curr Opin Struct Biol 11(1):39–46

Lassi ZS, Moin A, Bhutta ZA (2016) Zinc supplementation for the prevention of pneumonia in children aged 2 months to 59 months. Cochrane Database Syst Rev 12:Cd005978

Lehvy AI, Horev G, Golan Y, Glaser F, Shammai Y, Assaraf YG (2019) Alterations in ZnT1 expression and function lead to impaired intracellular zinc homeostasis in cancer. Cell Death Discov 5:144

Leitzmann MF, Stampfer MJ, Wu K, Colditz GA, Willett WC, Giovannucci EL (2003) Zinc supplement use and risk of prostate cancer. J Natl Cancer Inst 95:1004–1007

Li M, Zhang Y, Liu Z, Bharadwaj U, Wang H, Wang X, Zhang S, Liuzzi JP, Chang S-M, Cousins RJ, Fisher WE, Brunicardi FC, Logsdon CD, Chen C, Yao Q (2007) Aberrant expression of zinc transporter ZIP4 (SLC39A4) significantly contributes to human pancreatic cancer pathogenesis and progression. Proc Natl Acad Sci 104:18636–18641

Li XJ, Guo CJ, Lu WJ, Duan WW, Miao Z, Gu JT, Xiao K (2014) Expression pattern analysis of zinc finger protein genes in wheat (*Triticum aestivum L.*) under phosphorus deprivation. J Integr Agric 13(8):1621–1633

Li Z, Fan Y, Gao L, Cao X, Ye J, Li G (2016) The dual roles of zinc sulfate in mitigating peach gummosis. Plant Dis 100(2):345–351

Lin JS, Li X, Luo Z, Mysore KS, Wen J, Xie F (2018) NIN interacts with NLPs to mediate nitrate inhibition of nodulation in *Medicago truncatula*. Nat Plants 4(11):942–952

Lin J, Bjørk PK, Kolte MV, Poulsen E, Dedic E, Drace T et al (2024) Zinc mediates control of nitrogen fixation via transcription factor filamentation. Nature:1–6

Little PJ, Bhattacharya R, Moreyra AE, Korichneva IL (2010) zinc and cardiovascular disease. Nutrition 26:1050–1057

Liu Q, Wang Z, Xu X, Zhang H, Li C (2015) Genome-wide analysis of C2H2 zinc-finger family transcription factors and their responses to abiotic stresses in poplar (*Populus trichocarpa).* PLoS One 10(8):e0134753

Machado PP, Steiner F, Zuffo AM, Machado RA (2018) Could the supply of boron and zinc improve resistance of potato to early blight? Potato Res 61:169–182

Maier AT, Stehling-Sun S, Offenburger SL, Lohmann JU (2011) The bZIP transcription factor PERIANTHIA: a multifunctional hub for meristem control. Front Plant Sci 2:79

Maremanda KP, Khan S, Jena G (2014) Zinc protects cyclophosphamide-induced testicular damage in rat: Involvement of metallothionein, tesmin and Nrf2. Biochem Biophys Res Commun 445:591–596

Martos S, Gallego B, Cabot C, Llugany M, Barceló J, Poschenrieder C (2016) Zinc triggers signaling mechanisms and defense responses promoting resistance to *Alternaria brassicicola* in *Arabidopsis thaliana*. Plant Sci 249:13–24

Mendoza-Cózatl DG, Rangel-González E, Moreno-Sánchez R (2006) Simultaneous Cd^{2+}, Zn^{2+}, and Pb^{2+} uptake and accumulation by photosynthetic *Euglena gracilis*. Arch Environ Contam Toxicol 51:521–528

Minami T, Shimane M, Tanaka H, Namikawa K, Ichida S (2001) Pancreatic exocrine damage induced by subcutaneous injection of a low dosage of zinc. Biol Trace Elem Res 84:169–179

Mitchell W, Meng I, Nicholson SA, Aspinall R (2006) Thymic output, ageing and zinc. Biogerontology 7:461–470

Mocchegiani E, Muzzioli M, Giacconi R (2000) Zinc and immunoresistance to infections in ageing: new biological tools. Trends Pharmacol Sci 21:205–208

Mousavi SR (2011) Zinc in crop production and interaction with phosphorus. Aust J Basic Appl Sci 5(9):1503–1509

Natasha N, Shahid M, Bibi I, Iqbal J, Khalid S, Murtaza B et al (2022) Zinc in soil-plant-human system: a data-analysis review. Sci Total Environ 808:152024

Noman A, Aqeel M, Khalid N, Islam W, Sanaullah T, Anwar M et al (2019) Zinc finger protein transcription factors: integrated line of action for plant antimicrobial activity. Microb Pathog 132:141–149

Norouzi S, Adulcikas J, Sohal SS, Myers S (2017) Zinc transporters and insulin resistance: therapeutic implications for type 2 diabetes and metabolic disease. J Biomed Sci 24:87

Nowak G, Szewczyk B, Pilc A (2005) Zinc and depression: an update. Pharmacol Rep 57:713–718

Ohta S, Ikemoto T, Wada Y, Saito Y, Yamada S, Imura S, Morine Y, Shimada M (2019) A change in the zinc ion concentration reflects the maturation of insulin-producing cells generated from adipose-derived mesenchymal stem cells. Sci Rep 9:18731

Okamoto S, Ohnishi E, Sato S, Takahashi H, Nakazono M, Tabata S, Kawaguchi M (2009) Nod factor/nitrate-induced CLE genes that drive HAR1-mediated systemic regulation of nodulation. Plant Cell Physiol 50(1):67–77

Pan L, Zhao L, Jiang W, Wang M, Chen X, Shen X, ..., Mao Z (2023) Effect of Zinc Oxide Nanoparticles on the Growth of *Malus hupehensis Rehd.* Soil-plant-microbe interactions: An innovative approach towards improving soil health and plant growth, 16648714

Pandey N, Gupta B (2012) Improving seed yield of black gram (*Vigna mungo L.* var. DPU-88-31) through foliar fertilization of zinc during the reproductive phase. J Plant Nutr 35(11):1683–1692

Patel K, Kumar A, Durani S (2007) Analysis of the structural consensus of the zinc coordination centers of metalloprotein structures. Biochim Biophys Acta 1774(10):1247–1253

Peck AW, McDonald GK (2010) Adequate zinc nutrition alleviates the adverse effects of heat stress in bread wheat. Plant Soil 337:355–374

Peng X, Zhao Y, Cao J, Zhang W, Jiang H, Li X et al (2012) CCCH-type zinc finger family in maize: genome-wide identification, classification and expression profiling under abscisic acid and drought treatments. PLoS One 7(7):e40120

Peterson MP, Rosvall KA, Taylor CA, Lopez JA, Choi JH, Ziegenfus C, Tang H, Colbourne JK, Ketterson ED, Tang H et al (2014) Potential for sexual conflict assessed via testosterone-mediated transcriptional changes in liver and muscle of a songbird. J Exp Biol 217:507–517

Pfaus JG, Jones SL, Flanagan-Cato LM, Blaustein JD (2015) Female sexual behavior. In: Physiology of reproduction. Elsevier. New York, pp 2287–2370, 108

Popovics P, Stewart AJ (2011) GPR39: a Zn^{2+} -activated G protein-coupled receptor that regulates pancreatic, gastrointestinal and neuronal functions. Cell Mol Life Sci 68:85–95

Prasad AS (2009) Zinc: role in immunity, oxidative stress and chronic inflammation. Curr Opin Clin Nutr Metab Care 12:646–652

Prasad AS, Halsted JA, Nadimi M (1961) Syndrome of iron deficiency anemia, hepatosplenomegaly, hypogonadism, dwarfism, and geophagia. Am J Med 31:532–546

Prashanth L, Kattapagari KK, Chitturi RT, Baddam VRR, Prasad LK (2015) A review on role of essential trace elements in health and disease. J Dr NTR Univ Health Sci 4:75–85. 53

Prost AL, Bloc A, Hussy N, Derand R, Vivaudou M (2004) Zinc is both an intracellular and extracellular regulator of KATP channel function. J Physiol 559:157–167

Quaglia M, Bocchini M, Orfei B, D'Amato R, Famiani F, Moretti C, Buonaurio R (2021) Zinc phosphate protects tomato plants against *Pseudomonas syringae pv.* tomato. J Plant Dis Prot 128:989–998

Quaglia M, Troni E, D'amato R, Ederli L (2022) Effect of zinc imbalance and salicylic acid co-supply on *Arabidopsis* response to fungal pathogens with different lifestyles. Plant Biol 24(1):30–40

Que EL, Bleher R, Duncan FE, Kong BY, Gleber SC, Vogt S, Chen S, Garwin SA, Bayer AR, Dravid VP et al (2019) Quantitative mapping of zinc fluxes in the mammalian egg reveals the origin of fertilization-induced zinc sparks. Nat Chem 7:130–139

Rametse CL, Olivier AJ, Masson L, Barnabas S, McKinnon LR, Ngcapu S, Liebenberg LJ, Jaumdally SZ, Gray CM, Jaspan HB et al (2014) Role of semen in altering the balance between inflammation and tolerance in the female genital tract: Does it contribute to HIV risk? Viral Immunol 27(200–206):3

Raulin J (1869) Chemical studies on vegetation. Ann Sci Nat 11:93–99

Read SA, Parnell G, Booth D, Douglas MW, George J, Ahlenstiel G (2018) The antiviral role of zinc and metallothioneins in hepatitis C infection. J Viral Hepat 25(5):491–501

Roscioli E, Hamon R, Lester S, Murgia C, Grant J, Zalewski P (2013) Zinc-rich inhibitor of apoptosis proteins (IAPs) as regulatory factors in the epithelium of normal and inflamed airways. Biometals 26:205–227

Running MP, Meyerowitz EM (1996) Mutations in the PERIANTHIA gene of *Arabidopsis* specifically alter floral organ number and initiation pattern. Development 122(4):1261–1269

Ruttkay-Nedecky B, Nejdl L, Gumulec J, Zitka O, Masarik M, Eckschlager T et al (2013) The role of metallothionein in oxidative stress. Int J Mol Sci 14(3):6044–6066

Savarino L, Granchi D, Ciapetti G, Cenni E, Ravaglia G, Forti P, Maioli F, Mattioli R (2001) Serum concentrations of zinc and selenium in elderly people: results in healthy nonagenarians/centenarians. Exp Geront 36(2):327–339

Savi GD, Bortoluzzi AJ, Scussel VM (2013) Antifungal properties of Zinc-compounds against toxigenic fungi and mycotoxin. Int J Food Sci Technol 48(9):1834–1840

Sengupta P, Dutta S (2018) Thyroid disorders and semen quality. Biomed Pharmacol J 11:1–10

Seo H-M, Kim YH, Lee JH, Kim JS, Park YM, Lee JY (2017) Serum zinc status and its association with allergic sensitization: the Fifth Korea National Health and Nutrition Examination Survey. Sci Rep 7:12637

Seok HY, Woo DH, Park HY, Lee SY, Tran HT, Lee EH et al (2016) AtC3H17, a non-tandem CCCH zinc finger protein, functions as a nuclear transcriptional activator and has pleiotropic effects on vegetative development, flowering and seed development in *Arabidopsis*. Plant Cell Physiol 57(3):603–615

Sethuram R, Bai D, Abu-Soud HM (2021) Potential role of zinc in the COVID-19 disease process and its probable impact on reproduction. Reprod Sci

Sharma P, Jha AB, Dubey RS, Pessarakli M (2012) Reactive oxygen species, oxidative damage, and antioxidative defense mechanism in plants under stressful conditions. J Bot 2012(1):217037

Shaukat SS, Siddiqui IA (2003) Zinc improves biocontrol of Meloidogyne javanica by the antagonistic rhizobia. Pak J Biol Sci 6(6):575–579

Shoaib A, Akhtar M, Javaid A, Ali H, Nisar Z, Javed S (2021) Antifungal potential of zinc against leaf spot disease in chili pepper caused by *Alternaria alternata*. Physiol Mol Biol Plants 27(6):1361–1376

Siddiqui WA, Ahad A, Ahsan H (2015) The mystery of BCL2 family: Bcl-2 proteins and apoptosis: an update. Arch Toxicol 89:289–317

Siddiqui ZA, Khan A, Khan MR, Abd-Allah EF (2018) Effects of zinc oxide nanoparticles (ZnO NPs) and some plant pathogens on the growth and nodulation of lentil (*Lens culinaris Medik.*). Acta Phytopathol Entomol Hung 53(2):195–211

Simoglou KB, Dordas C (2006) Effect of foliar applied boron, manganese and zinc on tan spot in winter durum wheat. Crop Prot 25(7):657–663

Sinclair SA, Krämer U (2012) The zinc homeostasis network of land plants. Biochim Biophys Acta 1823(9):1553–1567
Sindreu C, Palmiter RD, Storm DR (2011) Zinc transporter ZnT-3 regulates presynaptic Erk1/2 signaling and hippocampus-dependent memory. Proc Natl Acad Sci USA 108:3366–3370
Singh AK, Meena MK, Upadhyaya A (2012) Effect of sulphur and zinc on rice performance and nutrient dynamics in plants and soil of Indo Gangetic plains. J Agric Sci 4(11):162
Skrajnowska D, Bobrowska-Korczak B (2019) Role of zinc in immune system and anti-cancer defense mechanisms. Nutrients 11(10):2273
Sofo A, Moreira I, Gattullo CE, Martins LL, Mourato M (2018) Antioxidant responses of edible and model plant species subjected to subtoxic zinc concentrations. J Trace Elem Med Biol 49:261–268
Sofy AR, Sofy MR, Hmed AA, Dawoud RA, Alnaggar AEAM, Soliman AM, El-Dougdoug NK (2021) Ameliorating the adverse effects of tomato mosaic tobamovirus infecting tomato plants in Egypt by boosting immunity in tomato plants using zinc oxide nanoparticles. Molecules 26(5):1337
Stehbens WE (2003) Oxidative stress, toxic hepatitis, and antioxidants with particular emphasis on zinc. Exp Mol Pathol 75(3):265–276
Stein AJ (2014) Rethinking the measurement of undernutrition in a broader health context: Should we look at possible causes or actual effects? Glob Food Sec 3(3–4):193–199
Sturikova H, Krystofova O, Huska D, Adam V (2018) Zinc, zinc nanoparticles and plants. J Hazard Mater 349:101–110
Subramanian B, Bansal VK, Kav NN (2005) Proteome-level investigation of *Brassica carinata*-derived resistance to *Leptosphaeria maculans*. J Agric Food Chem 53(2):313–324
Sun JY, Wang JF, Zi NT, Jing MY, Weng XY (2011) Effects of zinc supplementation and deficiency on bone metabolism and related gene expression in rat. Biol Trace Elem Res 143:394–402
Taheri N, Abad HHS, Yousefi K, Mousavi SR (2011) Effect of organic manure with phosphorus and zinc on yield of seed potato. Aust J Basic Appl Sci 5(8):775–780
Tapiero H, Tew KD (2003) Trace elements in human physiology and pathology: zinc and metallothioneins. Biomed Pharmacother 57(9):399–411
Tavallali V, Rahemi M, Eshghi S, Kholdebarin B, Ramezanian A (2010) Zinc alleviates salt stress and increases antioxidant enzyme activity in the leaves of pistachio (*Pistacia vera L. 'Badami'*) seedlings. Turk J Agric For 34(4):349–359
Taylor KM, Hiscox S, Nicholson RI, Hogstrand C, Kille P (2012) Protein kinase CK2 triggers cytosolic zinc signaling pathways by phosphorylation of zinc channel. Sci Signal ZIP7. 5:ra11-ra11
Teerds KJ, Huhtaniemi IT (2015) Morphological and functional maturation of Leydig cells: from rodent models to primates. Hum Reprod Update 21:310–328
Tellinghuisen TL, Marcotrigiano J, Rice CM (2005) Structure of the zinc-binding domain of an essential component of the hepatitis C virus replicase. Nature 435:374–379
Thomas S, Quinn BA, Das SK, Dash R, Emdad L, Dasgupta S, Wang X-Y, Dent P, Reed JC, Pellecchia M et al (2013) Targeting the Bcl-2 family for cancer therapy. Expert Opin Ther Targets 17:61–75
Thounaojam TC, Panda P, Choudhury S, Patra HK, Panda SK (2014) Zinc ameliorates copper-induced oxidative stress in developing rice (*Oryza sativa L.*) seedlings. Protoplasma 251:61–69
Tiwari S (2005) An inventory and assessment of invasive alien plant species of Nepal. IUCN Nepal
Todd WR, Elvehjem CA, Hart EB (1933) Zinc in the nutrition of the rat. Am J Physiol Leg Content 107(1):146–156
Tolay I (2021) The impact of different Zinc (Zn) levels on growth and nutrient uptake of Basil (*Ocimum basilicum L.*) grown under salinity stress. PLoS One 16(2):e0246493
Tolosa LN, Zhang Z (2020) The role of major transcription factors in Solanaceous food crops under different stress conditions: current and future perspectives. Plan Theory 9(1):56
Tomaž Š, Gruden K, Coll A (2022) TGA transcription factors – structural characteristics as basis for functional variability. Front Plant Sci 13:935819
Toor MD, Adnan M, Rehman FU, Tahir R, Saeed MS, Khan AU, Pareek V (2021) Nutrients and their importance in agriculture crop production; A review. Indian J Appl Pure Biosci 9(1):1–6

Trepreau J, Girard E, Maillard AP, de Rosny E, Petit-Haertlein I, Kahn R, Covès J (2011) Structural basis for metal sensing by CnrX. J Mol Biol 408(4):766–779

Tronchet M, Balagué C, Kroj T, Jouanin L, Roby D (2010) Cinnamyl alcohol dehydrogenases-C and D, key enzymes in lignin biosynthesis, play an essential role in disease resistance in *Arabidopsis*. Mol Plant Pathol 11(1):83–92

Tubek S (2007) Selected zinc metabolism parameters in women with arterial hypotension. Biol Trace Elem Res 116:73–79

Turan B (2003) Zinc-induced changes in ionic currents of cardiomyocytes. Biol Trace Elem Res 94:49–60

Uematsu K, Suzuki N, Iwamae T, Inui M, Yukawa H (2012) Increased fructose 1, 6-bisphosphate aldolase in plastids enhances growth and photosynthesis of tobacco plants. J Exp Bot 63(8):3001–3009

ul Hassan Z, Ali S, Rizwan M, Hussain A, Akbar Z, Rasool N, Abbas F (2017) Role of zinc in alleviating heavy metal stress. In: Essential plant nutrients: uptake, use efficiency, and management, pp 351–366

Vallee BL, Falchuk KF (1993) The biochemical basis of zinc physiology. Physiol Rev 73:79–118

Verze P, Cai T, Lorenzetti S (2016) The role of the prostate in male fertility, health and disease. Nat Rev Urol 13:379–386

Wadhwa N, Joshi UN, Mehta N (2014) Zinc induced enzymatic defense mechanisms in *Rhizoctonia* root rot infected clusterbean seedlings. J Bot 2014(1):735760

Wang Y, Li X, Fan B, Zhu C, Chen Z (2021) Regulation and function of defense-related callose deposition in plants. Int J Mol Sci 22(5):2393

Watanabe T, Kiron V, Satoh S (1997) Trace minerals in fish nutrition. Aquaculture 151(1–4):185–207

Weisany W, Sohrabi Y, Heidari G, Siosemardeh A, Ghassemi-Golezani K (2012) Changes in antioxidant enzymes activity and plant performance by salinity stress and zinc application in soybean (*'Glycine max' L.*). Plant Omics 5(2):60–67

Wellinghausen N, Rink L (1998) The significance of zinc for leukocyte biology. J Leukoc Biol 64:571–577

Wessells KR, Brown KH (2012) Estimating the global prevalence of zinc deficiency: results based on zinc availability in national food supplies and the prevalence of stunting. PLoS One 7(11):e50568

White PJ, Brown PH (2010 Jun) Plant nutrition for sustainable development and global health. Ann Bot 105(7):1073–1080

WHO (1996) Trace elements in human nutrition and health. Geneva, pp 72–104

Wongpia A, Lomthaisong K (2010) Changes in the 2DE protein profiles of chilli pepper (*Capsicum annuum*) leaves in response to *Fusarium oxysporum* infection. ScienceAsia 36(4):259–270

Xie M, Sun J, Gong D, Kong Y (2019) The roles of Arabidopsis C1-2i subclass of C2H2-type zinc-finger transcription factors. Genes 10(9):653

Yang M, Zhang F, Wang F, Dong Z, Cao Q, Chen M (2015) Characterization of a type 1 metallothionein gene from the stresses-tolerant plant *Ziziphus jujuba*. Int J Mol Sci 16(8):16750–16762

Yang T, Lv R, Li J, Lin H, Xi D (2018) Phytochrome A and B negatively regulate salt stress tolerance of *Nicotiana tabacum* via ABA–jasmonic acid synergistic cross-talk. Plant Cell Physiol 59(11):2381–2393

Yeste M, Jones C, Amdani SN, Patel S (2016) Coward, K. Oocyte activation deficiency: A role for an oocyte contribution? Hum Reprod Update 22:23–47

Yin J, Wang L, Zhao J, Li Y, Huang R, Jiang X et al (2020) Genome-wide characterization of the C2H2 zinc-finger genes in *Cucumis sativus* and functional analyses of four CsZFPs in response to stresses. BMC Plant Biol 20:1–22

Zaheer IE, Ali S, Saleem MH, Arslan Ashraf M, Ali Q, Abbas Z et al (2020) Zinc-lysine supplementation mitigates oxidative stress in rapeseed (*Brassica napus L.*) by preventing phytotoxicity of chromium, when irrigated with tannery wastewater. Plan Theory 9(9):1145

Zaheer IE, Ali S, Saleem MH, Yousaf HS, Malik A, Abbas Z et al (2022) Combined application of zinc and iron-lysine and its effects on morpho-physiological traits, antioxidant capacity and chromium uptake in rapeseed (*Brassica napus L.*). PLoS One 17(1):e0262140

Zang D, Wang C, Ji X, Wang Y (2015) *Tamarix hispida* zinc finger protein ThZFP1 participates in salt and osmotic stress tolerance by increasing proline content and SOD and POD activities. Plant Sci 235:111–121

Zhang X, Liang D, Guo B, Deng W, Chi ZH, Cai Y, Wang L, Ma J (2013) Zinc transporter 5 and zinc transporter 7 induced by high glucose protects peritoneal mesothelial cells from undergoing apoptosis. Cell Signal 25:999–1010

Zhang Z, Chen Y, Li B, Chen T, Tian S (2020) Reactive oxygen species: A generalist in regulating development and pathogenicity of phytopathogenic fungi. Comput Struct Biotechnol J 18:3344–3349

Zhao Y, Zhao H, Zhai X, Dai J, Jiang X, Wang G, Li W, Cai L (2013) Effects of Zn deficiency, antioxidants, and low-dose radiation on diabetic oxidative damage and cell death in the testis. Toxicol Mech Methods 23:42–47

Zhao X, Song B, Riaz M, Li M, Lal MK, Adil MF et al (2024) Foliar zinc spraying improves assimilative capacity of sugar beet leaves by promoting magnesium and calcium uptake and enhancing photochemical performance. Plant Physiol Biochem 206:108277

Zhou Z, An L, Sun L, Zhu S, Xi W, Broun P et al (2011) Zinc finger protein5 is required for the control of trichome initiation by acting upstream of zinc finger protein8 in *Arabidopsis*. Plant Physiol 157(2):673–682

Zhu XG, De Sturler E, Long SP (2007) Optimizing the distribution of resources between enzymes of carbon metabolism can dramatically increase photosynthetic rate: a numerical simulation using an evolutionary algorithm. Plant Physiol 145(2):513–526

The Mechanisms of Uptake, Translocation, and Distribution of Zinc in Plants

Rachel Carmelita Mathias, Pavana Shree, M. Annapoorneshwari, P. M. Krupa, J. S. Rajatha Sriram, and Smitha Hegde

Abstract

Heavy metals are one of the major environmental pollutants introduced into various ecosystems by rapid industrial development globally. Heavy metals have detrimental impacts on human health and are the root cause of many major illnesses. The zinc content in the soil is affected by the complex process of global warming. Plants absorb zinc either as a bivalent cation or by building a complex with organic ligands in the xylem of their roots, which is then transferred to the plants' shoots. Zinc is a necessary component that triggers the activation of several enzymes necessary for protein synthesis, growth, and the creation of hormones that promote development in plants, animal, and humans. Nevertheless, increased zinc accumulation is harmful since it causes growth retardation, decreased biomass output, and delays in several metabolic processes, including photosynthesis, transpiration, and nitrogen metabolism. In fact, plants may even die because of elevated Zn concentrations. Certain plant species have the capacity to accumulate elevated levels of zinc in their tissues without displaying any harmful effects. These species that are resistant of zinc can be used in phytoremediation. This chapter focuses on advancements in the understanding of micronutrient absorption and translocation and distribution in plants. The current study also aims to assess zinc availability in the environment, plant uptake and molecular mechanisms of translocation, impact on physiological and biochemical status, hyperaccumulation of zinc, and strategies for enhancing Zn uptake in plants.

R. C. Mathias · P. Shree · M. Annapoorneshwari · P. M. Krupa · J. S. Rajatha Sriram · S. Hegde (✉)
Nitte University Centre for Science Education and Research, Nitte (Deemed to be University), Mangaluru, Karnataka, India
e-mail: smitha.hegde@tlc.edu.in

S. Srivastava, H. Kaur (eds.), *Zinc in Soil-Plant Continuum*,
https://doi.org/10.1007/978-981-96-4253-3_3

Keywords
Zinc deficiency · Zinc content · Zinc uptake · Hyperaccumulator · Protein transporters

Abbreviations

FRD	Ferric reductase defective 3
HMA	Heavy metal ATPase
IRT	Iron-regulated transporter
MTP	Metal tolerance proteins
NRAMP	Natural resistance-associated macrophage protein
YSL	Yellow stripe-like (YSL) family
ZIF	Zinc-induced facilitator
ZIP	ZRT-IRT-like protein
ZNT	Zinc transporter

1 Introduction

Environmental pollution is a major source of health risk and a major cause of various diseases worldwide; it has emerged as a serious public health concern (Briggs 2003). The earth's crust naturally contains heavy metals, but uncontrolled human activity has significantly changed the geological cycles and biological balance of these elements. Heavy metal pollution of the environment has increased due to industrial expansion, and it has been demonstrated that large concentrations of heavy metals are found in industrial locations (Sumiahadi and Acar 2018). This leads to the buildup of metals in plant portions. The harmful impact that various heavy metals have on human health varies. Human health can suffer when exposed to heavy metals including zinc, nickel, copper, lead, and cadmium for an extended period. Since heavy metals cannot be removed by deterioration, they constitute a greater worry than other environmental pollutants. Certain techniques are required to extract the metals from contaminated sites in the remediation process of heavy metal-affected soils, groundwater, and surface water.

Zinc (Zn) is a highly versatile catalytic cofactor in proteins due to its unique chemical characteristics when compared to other first-row transition metals like iron (Fe) and copper (Cu). Multiple coordination geometries can be accommodated by Zn^{2+} in binding sites, but it is primarily found in a distorted tetrahedral conformation where it interacts with side chains of histidine, cysteine, glutamate, or aspartate that donate nitrogen (N), oxygen (O), or sulfur (S) (Lee and Lim 2008; Habashi 2013).

Usually industrial waste water and solids, unregulated e-waste processing, mining, and smelting processes add zinc to soils (Huang et al. 2013; Ha et al. 2014; Wu et al. 2018; Kang et al. 2019; Zhong et al. 2020). Zinc is a necessary element, but excessive amounts of it can be hazardous to plants and lower their biomass and production (Broadley et al. 2007; Broadley 2011; Sinclair and Krämer 2012). Furthermore, the human body reacts adversely to excessive zinc in the form of nausea, vomiting, headaches, diarrhea, and appetite loss. Also, excessive use of agricultural pesticides might potentially be a factor in the rise in soil zinc concentrations (Fosmire 1990; Luo et al. 2009; Zhao et al. 2010; Liu et al. 2013).

Zinc is an essential structural element found in numerous proteins. Zn-binding protein domains, of which the Zn-finger domain proteins are the biggest family, are particularly well-suited for nucleotide-binding and protein-protein interactions due to their redox stability (Laity et al. 2001). However, it is challenging to examine Zn proteins using conventional spectroscopic techniques due to the redox stability of zinc and the colorlessness of Zn complexes under biological conditions (Penner-Hahn 2005). As a result, direct structural evidence is lacking for most Zn-binding proteins, which have instead been found by bioinformatics techniques (Andreini et al. 2009). A recent review provides additional details on the zinc proteome and a summary of zinc homeostasis (Clemens 2022).

Zinc deficiency can have serious physiological and biochemical effects on plants when the amount of bioavailable zinc in the soil is below ideal values. There are many biotechnological ramifications to the molecular understanding of plant metal accumulation, some of which may not be fully understood in the long run. Inappropriate binding to intracellular ligands and competition with other metal ions for enzyme active sites or transporters are two potential causes of zinc toxicity, while the exact mechanisms are yet unknown.

The roles of nicotianamine (NA) and the phytosiderophores of the mugineic acid (MAs) family in plants' iron (Fe) absorption and homeostasis have been extensively studied (Bashir et al. 2010; Mori 1999; Takagi 1976). According to Suzuki et al. (2006, 2008), Takagi (1976), and Welch and Shuman (1995) and other studies, MAs are involved in Zn absorption and translocation. While MAs are only generated in graminaceous plants, NA is found in all higher plants (Marschner 1995). Methionine (Met) is used to produce MAs, while L-Met is the starting point for the biosynthesis of both NA and 2′-deoxymugineic acid (DMA) (Mori and Nishizawa 1987). Trimerization of S-adenosyl Met to NA is catalyzed by NA synthase (NAS) (Higuchi et al. 1994, 1999). Several genes have been identified and demonstrated to be involved in metal absorption and transport in plants (Eide et al. 1996; Korshunova et al. 1999; Connolly et al. 2002). These individuals belong to the iron (Fe)-regulated transporter-like protein (ZIP) gene family and the Zn-regulated transporter gene family (Guerinot 2000).

Plants that have excess zinc produce extremely reactive oxygen free radicals, which have an impact on the amount of chlorophyll, the integrity of membranes, plant development, water relations, and the activity of nitrogen metabolism-related enzymes (Cuypers et al. 2001; Vaillant et al. 2005). Additionally, it interferes with plants' ability to absorb nutrients and inhibits the activity of several enzymes, including nitrogenase (Chaney 1993; Kaya et al. 2000). How readily available zinc is to plants depends on a variety of factors, including plant age, weather, soil pH, plant species, and how zinc interacts with other metals. The present state of knowledge about the harmful effects of zinc on different physiological and metabolic systems in plants, the phytotoxic effects on plants that target different physiological and metabolic pathways, biochemical parameters, genotoxicity, and the usage of certain plants for Zn accumulation is summarized in this chapter.

2 Biochemical and Physiological Importance of Zinc (Zn)

Zinc (Zn) is an essential micronutrient for plants and is required in minute quantities; it plays a critical role in plant growth and physiology. This essential micronutrient acts as a key factor for many enzymes, impacting a wide range of biochemical and physiological processes crucial for plant health and productivity. Zinc acts as a cofactor for over 300 enzymes. These enzymes drive various metabolic pathways, including carbohydrate metabolism, protein synthesis, and gene regulation. Without sufficient zinc, these processes slow down, hindering overall plant health (Chaney 1993).

Zinc is a key component in Zn-metalloenzymes and stabilizes DNA-binding domains of regulatory proteins or transcription factors, known as "Zn-finger." These "zinc fingers" fold sub-domains of regulatory proteins, enabling them to recognize and bind to specific DNA sequences, inducing gene expression (Hasanuzzaman et al. 2018). Zinc is crucial for chlorophyll synthesis, the pigment responsible for capturing sunlight in photosynthesis (Henriques et al. 2012). Zinc contributes to the structural integrity of ribosomes (Prask and Plocke 1971).

Zinc is crucial for the synthesis of plant growth hormones, such as auxins, and is involved in the regulation of various developmental processes including stem elongation and leaf expansion (Henriques et al. 2012). Additionally, zinc is a structural element in the superoxide dismutase enzyme, which is important for protecting cells from oxidative damage (Singh and Singh 2011). Zinc plays a regulatory role in plant metabolism, influencing physiological processes like plant growth, reproduction, and seed quality (Pandey et al. 2013).

In conclusion, zinc is indispensable for plant health, affecting a multitude of physiological and biochemical pathways. Its role extends from basic cellular functions to complex stress responses, highlighting the importance of maintaining adequate zinc levels for optimal plant growth and development.

3 Soil Concentrations of Zinc and Its Availability to Plants

There are five stable isotope forms of zinc found in nature: ^{67}Zn (4.90%), ^{68}Zn (18.75%), ^{66}Zn (27.90%), ^{70}Zn (0.62%), and ^{64}Zn (48.63%) (Broadley et al. 2007). It has been shown that plants exhibit both light and heavy isotopic enrichment of zinc in their roots and shoots, respectively (Caldelas et al. 2010). High binding affinities of Zn^{2+} ions are found for nitrogen (N), oxygen (O), or sulfur (S) of amino acid residues in proteins and enzymes. The most common amino acid residue is that of histidine, which is followed by cysteine, glutamine/asparagine, O of aspartate/

glutamate, and carbonyl O of peptide bonds (Leuci et al. 2020). Zinc is an essential micronutrient that participates in structural and/or catalytic roles in several processes, including protein synthesis, cell division, and expansion (Jain et al. 2013). It is essential for the structure of chromatin, gene expression and regulation, protein, lipid, carbohydrate, and nucleic acid metabolism, as well as photosynthetic carbon metabolisms (Gai et al. 2017; Noulas et al. 2018).

Zinc (Zn) is an essential component for plants, which lack indicates low crop yields and concentrations of this element in staple food like cereals, which may encourage detrimental effects on human health when diets require this type of food. Zinc helps strengthen the human immune system and is essential for the basic operations of all living organisms (Recena et al. 2021). Many countries with Mediterranean climates have alkaline-calcareous soils, which are an important kind of agricultural soil and are often used for cereal production (Noulas et al. 2018).

A zinc deficit is associated with decreased grain quality for human consumption as well as crop productivity. The deficiency of micronutrients is an extended nutritional imbalance constraining crop production in many agricultural lands in the world. Around one-third of cultivated soils have low Zn availability levels for suitable crop production. Zinc has a significant impact on the growth of the human immune system and brain, as well as on enzymatic processes and plant metabolism (Suganya et al. 2020).

Zn deficiency causes yield decreases, particularly in cereals, which are staple food in many regions of the world, frequently without clear deficiency symptoms (Moreno-Lora and Delgado 2020). The sandy, calcareous, saline, and wetland soils that are susceptible to zinc deficiency are compacted and rich in organic matter with high amounts of nitrogen and phosphate. The levels of zinc in uncontaminated, unfertilized soil range from 10 to 300 mg/kg (Noulas et al. 2018). The potential availability of zinc to plants is influenced by the dynamics of zinc in the soil, which is largely determined by carbonates and iron oxides (Recena et al. 2021). The roots of plants absorb zinc from the soil solution, mainly as Zn^{2+} ions or complexes with organic acid chelates, and translocate it through the xylem into the portion of the plant that is above ground (Sturikova et al. 2018).

Comprehensive knowledge of plant root nutrient absorption capability, especially of plant membrane transporters involved in phosphorous (P) and Zn transport, is necessary to efficiently control the amounts of soil Zn and that are available. Furthermore, it is critical to know about the variables that influence the regulation of the P and Zn transport genes, such as relationships with arbuscular mycorrhizal fungi (AMF), which can improve plant nutrient uptake (Nguyen et al. 2019). Arbuscular mycorrhizal (AM) symbiosis is the relationship that occurs between the roots of plants and a certain fungus called glomeromycetes. This mutualistic relationship is widespread and affects most plant species (Malbreil et al. 2014).

Over 80% of land plants have required symbionts called arbuscular mycorrhizal (AM) fungus colonizing their roots, which enhances the nutrition, growth, and resistance to disease of the plants (Bago et al. 1998). There might be processes at work that prevent the host plant from absorbing heavy metals, which lowers the concentration of heavy metals in the plant tissues (Li and Christie 2001).

As the soil pH increases within 6.5–7.5, zinc is lesser toxic to plants. The content of zinc in soils worldwide varies within a broad range from 35 mg to 12,400 mg kg–1 of soil. These components could function as building blocks. Zinc levels in plants range from 10 to 86 mg kg–1 dm (Wyszkowska et al. 2013). Heavy metal-contaminated soil may cause plants to absorb more of these substances, which are subsequently passed on to plants at later stages of the food system (Nadgórska-Socha et al. 2015). There are two distinct processes by which plant roots absorb trace elements: the passive system, which involves diffusion, and the active mechanism, which uses metabolic energy to accelerate against the gradient of concentrations (Rajkumar and Freitas 2008). The ability of different plant species or even cultivars to absorb these elements and tolerate high concentrations of zinc may vary (Graham et al. 1992).

Plant growth is slowed down by zinc contamination of the soil, and when it builds up in crops, it can reach levels that are harmful to the living organism. Zinc's phytotoxic effect is mostly determined by the type of crop and stage of development at which it is growing, as well as soil characteristics such as soil response, organic matter concentration, and soil absorption complex (Zalewska 2012). Enzymatic activity is a strong predictor of soil health because it accurately captures alterations in the cycling of organic matter brought on by heavy metals (Wyszkowska et al. 2016). The disruption of physiological processes is brought on by abnormalities in the uptake of micro- and macro-elements required for plants to function properly the process that promotes phytotoxicity because of excessive metals in the soil (Wyszkowska et al. 2013). Several plant species (e.g., sunflower) are used in phytoremediation to eliminate, stabilize, or degrade metals and pollutants from soil and groundwater (Soares et al. 2021).

The objective of phytoremediation techniques is to restore severely damaged ecosystems to a level of contamination that is considered tolerable (Benshoff et al. 2003). According to the findings, the zinc is essential for all living organisms, which lack in concentration indicates deficiency in living organism. Zn concentration in the grain is inherently very low, particularly when grown on Zn-deficient soils. Zn is essential for the development of immune system and brain function in humans and plays an important role in enzymatic reactions and metabolic activities in plant system. And the Zn in soil solution is absorbed by the plant roots and spreads into other parts of the plants.

Arbuscular mycorrhizal fungi (AMF) can improve plant nutrient uptake from soil. The fungi prevent the plants from absorbing heavy metals, which decrease the concentration of Zn in plants tissue. The physicochemical properties that indicate more concentrations of zinc have an impact on microbial growth and the activity of soil enzymes. Some plant biosorbents are used in phytoremediation technique which eliminate the contaminated heavy metals in soil.

4 Prevalence and Causes for Zinc Deficiencies in Plants

The prevalence of zinc (Zn) deficiency in plants is a significant global issue, affecting both crop yield and quality. This deficiency is widespread and has been recorded in many parts of the world, leading to poor yield levels and reduced quality of produce, which in turn can cause malnutrition in animals and humans.

The prevalence of Zinc deficiencies is attributed to various soil factors like the following:

Low soil zinc levels: Soils naturally vary in their zinc content, and regions with inherently low zinc levels can lead to deficiencies in plants grown in those areas (Sharma et al. 2013; De Groote et al. 2021).

pH conditions: Zinc availability to plants is reduced in high pH soils, as zinc tends to form insoluble compounds under alkaline conditions, making it less accessible to plant roots. Rapid changes in soil pH levels, often caused by factors like acid rain or excessive liming, can impact zinc solubility and availability to plants (Adamczyk-Szabela and Wolf 2022).

Excessive phosphorus: High levels of phosphorus in the soil can interfere with zinc uptake by plants, as these two nutrients can exhibit antagonistic interactions, with phosphorus inhibiting zinc absorption (Recena et al. 2021).

High levels of other elements: Other elements such as Na, Ca, Mg, and bicarbonates which can interfere with zinc availability (Recena et al. 2021).

Low levels of organic matter: Sustainable improvement to soil health depends heavily on the management of organic matter and micronutrients. Zinc (Zn) micronutrient availability in plants is typically lowered by low soil organic matter levels. Crop development and productivity are severely reduced when there is zinc deficiency (Dawar et al. 2022).

Soil compaction: Compacted soils can restrict root growth and nutrient uptake, including zinc, leading to deficiencies in plants even when zinc levels in the soil are sufficient (Parlak and Parlak 2011). There are several soil types that are susceptible to zinc deficiency, including sandy-textured soils in various climatic zones, tropical regions with heavily weathered soils, and semi-arid locations with calcareous soils (Noulas et al. 2018).

Imbalanced fertilization: Inadequate zinc application or improper ratios of zinc to other nutrients in fertilizers can result in zinc deficiencies in plants. This results in restricted growth of plants with symptoms of chlorosis and small leaves, spikelet sterility, and many other plant abnormalities (Sharma et al. 2013).

Waterlogging: Waterlogged or poorly drained soils can create anaerobic conditions that reduce zinc availability to plants, contributing to deficiencies (Tuiwong et al. 2021).

Climate factors: Environmental conditions such as temperature, rainfall patterns, and humidity can influence zinc availability in the soil and its uptake by plants, affecting the prevalence of deficiencies (Suganya et al. 2020). Certain crops, including maize, wheat, and rice, are particularly susceptible to Zn deficiency, which can lead to poor plant health and reduced productivity (Alloway 2009; Noulas et al. 2018; Singh and Singh 2017; Yoshida and Tanaka 1969).

Interestingly, while soil properties and environmental factors are significant contributors to Zn deficiency, plant adaptations and microbial interactions in the rhizosphere also affect Zn uptake (Pantigoso et al. 2022). Plants can undergo morphological and biochemical changes to cope with nutrient deficiencies, and microorganisms can enhance nutrient availability and translocation (Francis et al. 2023). Moreover, the relationship between Zn and other nutrients, such as phosphorus, is complex and can impact Zn availability to plants (Recena et al. 2021).

Zn deficiency in plants is prevalent due to a combination of soil characteristics, environmental conditions, and biological interactions. Addressing these deficiencies requires a comprehensive understanding of the factors affecting Zn availability and the strategies plants employ to adapt to nutrient scarcity.

5 Uptake of Zinc by Roots

In addition to the amount of uptake of zinc, like any other mineral nutrient, factors such as concentration in the soil solution and transportation to the root surface are crucial. Major elements also include root growth and surface area, which establish the distances between the "sink" (the root surface) and the "source" (quantity, such as exchangeable percentage). Because most of the zinc is incorporated into the soil matrix, it is typically found in relatively small concentrations in soil solutions, which also limits its mass flow to the roots (Marschner 1993). The rhizosphere is where metals and plants first come into contact. Zinc is primarily found in soil as crystalline iron-magnesium minerals like zinc sulfide (ZnS). Mostly as Zn^{2+}, $ZnOH^{+}$, and $ZnCl^{+}$, zinc is adsorbed in variable forms onto clay surfaces and organic materials (Moreira et al. 2018). Zinc is mainly taken up by roots as Zn^{2+} ions, apart from high pH, where it is taken up as ZnOH (Cui Li 2018).

6 Radial Transport of Zinc in Roots

Radial transport is the movement of water and mineral nutrients in the direction of the stele. Root cells absorb water and micronutrients, including zinc, from the soil and move them radially into the xylem, where they are then delivered to the stem and leaves (Li et al. 2015). The pericycle, endodermis, cortex, and epidermis are among the tissues that these materials must travel through (Hanikenne et al. 2008). As these tissue layers moved in the direction of the stellar region in the center, their diameter reduces. Large accumulations of ions (metal cations) in the pericycle help to load the xylem because the radial pattern of transport speeds up water flow in its journey to the xylem and concentration increases quicker at smaller radii for the same flux (lesser solvent volume) (Claus et al. 2013). The intracellular zinc concentration of root tissues is ultimately determined by the strict regulation of specialized transporter proteins' expression and activity (rate of zinc transport per unit time), notwithstanding their presence on the cell membranes of corresponding tissues. Zinc compartmentation in roots, translocation, usage in shoots, sink growth stage, etc. are some of the factors that affect the xylem loading of zinc in roots, which in turn causes variations in symplastic zinc levels.

There are two primary pathways that Zn^{2+} (mineral ions) use on their way to the xylem: (1) the symplastic pathway and (2) the apoplastic pathway. In root tissues, cytoplasmic bridges, or plasmodesmata, in the cell wall link the cytoplasms of neighboring cells, creating a symplastic continuum devoid of membrane barriers. Additionally, the apoplast, which plays a crucial role in root Zn transport, allows water and mineral ions to pass across intercellular gaps and cell walls (Steudle 1994). However, the apoplastic route is disrupted in endodermal regions because suberin lamellae are deposited in the Casparian strip. Both water and charged ions are very impermeable to suberin. To continue their journey through the symplastic route, mineral ions and water molecules must first cross the membrane before reaching the casparian strip (Hong-Qiang and Yu-Ling 2005). In areas where the casparian band is not fully developed, such as the vicinity of the root apex and the region of lateral root initiation, as shown by White et al. (2002) and Ranathunge et al. (2005), some Zn^{2+} ions are also transported extracellularly to the stellar apoplast. The net zinc fluxes in the shoot are influenced by both apoplastic and symplastic fluxes. Compared to symplastic transport, apoplastic zinc transport is less selective as it includes zinc entering the cytosol from the cell wall plasma membrane contact. Symplastic transport, on the other hand, regulates the amount and selectivity of nutrient delivery. Zinc is mostly absorbed by plant roots as hydrated zinc and Zn^{2+} free ions, which are subsequently distributed throughout the plant in a complex way (Kisko et al. 2015). The symplast pathway allows zinc to be transported from root epidermal cells to cortical cells and ultimately to xylem, where it is injected into the stellar apoplast (Caldelas and Weiss 2017). In regions around the root apex and lateral root initiation zones, for example, where the endodermal casparian band is not fully established, zinc can be transferred extracellularly to the stellar apoplast (White et al. 2002). Together, the apoplastic and symplastic routes allow Zn to flow to the aerial portions (Noulas et al. 2018). With the exception of Poaceae, zinc moves through plasma membranes in dicots and monocots in the direction of a greater negative electrical potential along cation channels.

The non-protein amino acids known as "phytosiderophores (PS/MAs)" or "phytometallophores" in Poaceae form complexes with zinc and transfer it to the root's outer surface (Suzuki et al. 2006). For Poaceae to acquire Fe, increased uptake of Fe (III) through PS secretion is well documented; however, this has not been proven decisively for Zn acquisition so far. The research suggesting the usage of Zn-PS complexes as the primary absorption mechanism in Poaceae members is not entirely consistent. For instance, Zn-deficient barley received more Zn (II)-DMA (deoxymugineic acid) from the soil than Zn^{2+} ions, according to Suzuki et al. (2006). In another study, Suzuki et al. (2008) found that while Zn-DMA complexes were favored over Zn^{2+} for internal transport and seed loading in rice, Zn^{2+} was preferred for root uptake.

In contrast, the findings of Arnold et al. (2010), Widodo et al. (2010) and Ptashnyk et al. (2011) offered compelling proof that zinc is absorbed in large quantities in rice as Zn-DMA complexes. These discrepancies in the literature imply that, in graminaceous plants experiencing a Zn deficit, Zn may absorb through the same mechanism as Fe. Concentration-dependent absorption of free zinc ions through a carrier-mediated mechanism is thought to be saturable. Hacisalihoglu et al. (2001) conducted uptake studies that demonstrated the presence of two unique low- and high-affinity transport channels for Zn uptake in wheat plants.

Zinc can be absorbed through the xylem of zinc hyperaccumulator plants (*Noccaea caerulescens* and *Arabidopsis halleri*), where it combines with a variety of molecules to produce complexes that include amino acids like histidine and organic acids like citrate and malate (Cornu et al. 2015). Most of the poisonous zinc in vacuoles, such as those found in *N. caerulescens*, is chelated with the help of MTs and phytate (Monsant et al. 2011). Zinc's root-to-shoot transfer is restricted in hyperaccumulator plants by its seclusion in the vacuoles of the root cells; this process may be caused by zinc's unloading in the shoots and sequestration in the vacuoles of the leaf cells (Verbruggen et al. 2009). Roots of *Brassica napus* have been found to have Zn-rhamnolipid complexes (Stacey et al. 2008).

While Zn-NA is the main compound present in *N. caerulescens* mesophyll, Zn-malate and Zn-citrate complexes predominate in the leaf epidermis (Schneider et al. 2013). Zinc was mostly bound to chelator NA in phloem sap, but it was also present in rice xylem as free ions and to some amount in bound form (Yoneyama et al. 2015). Zn attached to oxygen/nitrogen-ligands/chelators was the predominant type of zinc collected in *Zygophyllum fabago*'s vascular bundle and spongy mesophyll (Lefèvre et al. 2014). Zinc is mostly loaded as Zn^{2+} ions in the xylem of strategy I plants (non-graminaceous), with organic acids accounting for the remaining percentage (Lu et al. 2014). Zn-DMA complexes have been found in wheat root extracts (Xuan et al. 2006) and in japonica rice shoots in plants of the Poaceae family (strategic II) (Tsednee et al. 2016). This suggests that Zn is delivered in xylem as Zn-MA complexes.

Studies have shown that Zn is distributed differently in hyperaccumulators and non-accumulators. Zinc was mostly restricted to the root meristematic area in cowpea (Kopittke et al. 2011; Wang et al. 2013). In non-accumulators like tobacco (*Nicotiana tabacum L.*) (Straczek et al. 2008; Sarret et al. 2009) and *Arabidopsis lyrate* (Sarret et al. 2009), zinc accumulates at the base of trichomes. Weremczuk et al. (2017) discovered elevated zinc concentrations in groups of tobacco "Zn-accumulating mesophyll cells," which underwent programmed cell death and eventually formed necrotic lesions. Zinc was found to be restricted in mesophyll cells and the base of trichomes in hyperaccumulators such as *A. halleri*, according to Mishra et al. (2017), while Zn was stored in the epidermis of leaves in *Nymphaea caerulea* (Wang et al. 2013) scenes. Zn also collected in the stem and leaf epidermis of *Sedum alfredii* (Tian et al. 2009). An exterior barrier preventing zinc from entering the protoplasm is created by the cell wall's widening and increased cation exchange capacity (CEC) (Liu et al. 2014). Zinc has been shown to attach to proteins found in plant cell walls as well as to the carboxyl and hydroxyl groups of pectin, hemicellulose, and cellulose hydroxyl groups (Sarret et al. 2009). Zn^{2+} poisoning changed *Triticum aestivum* root meristematic cells by increasing the number of vacuoles, causing granular materials to accumulate within the vacuoles and thickening the cell walls (Glińska et al. 2016). These changes suggest that excess Zn speeds up the process of cell wall component synthesis, enhancing apoplast's capacity to sequester the excess metal in areas where it poses less of a threat. Furthermore, an increase in the quantity of vacuoles containing granular deposits implies that plants are trapping extra zinc inside the vacuoles to avoid metal toxicity. The outcomes of research have also demonstrated the transfer of zinc through phloem. The xylem-mediated transport of 65Zn to the aerial portions of white lupin and its subsequent allocation from mature to immature leaves through the phloem were demonstrated by Page et al. (2006). Crop plants exhibit significant variances in the symplastic Zn allocation. Variations in the Zn competency manifestation were shown by Erenoglu et al. (2002) to be unrelated to the Zn reallocation in wheat plants.

On the other hand, Ozturk et al. (2006) found higher zinc concentrations in wheat grains after applying zinc topically, suggesting strong plant zinc mobility. Transport systems of the xylem and phloem interact to control the movement of zinc to grains that are maturing. In this case, it was thought that zinc would flow continuously via xylem from the rice grain's stem to its vascular bundle during grain filling (Krishnan and Dayanandan 2003). Moreover, there is a clear correlation between the concentrations of Zn added to plants and the Zn contents of grains. Most significantly, understanding Zn mobility in phloem is essential for controlling Zn feeding as well as determining how various plant species may express signs of possible Zn poisoning.

7 Xylem and Phloem Transport

Zinc moves through symplastic and apoplastic routes before entering the xylem to continue to the phloem. Zn gets loaded into the xylem using HMA (heavy metal ATPase) pumps. After being absorbed from the soil, some zinc may sequester in the vacuole on its way to the xylem. The vascular pathway of zinc translocation into developing seed begins when zinc exits from the xylem and travels through transport proteins to the living xylem parenchyma cells in the leaf symplast. From there, it travels to the leaf apoplast and finally to phloem (Rehman et al. 2018).

However, zinc ions must traverse the Casparian strip barrier in the root endodermis before reaching the xylem. Zinc penetrates live cells in the pericycle and xylem parenchyma upon crossing this barrier. Xylem loading, a critical stage in nutrient transfer, presents another obstacle to zinc transit. Here, the constant activity of H + -ATPase in xylem parenchyma cells leads to membrane hyperpolarization, which restricts the efflux of positive ions from the cytosol. Consequently, zinc ions are actively loaded into the apoplastic xylem from xylem parenchyma/pericycle cells, ensuring continued transport within the plant (Sondergaard et al. 2004).

Specific transporters (HMA family or P1B-type ATPase) found on pericycle cells and xylem parenchyma promote the active efflux of symplastic Zn^{2+}. Increased plasma membrane H-ATPase expression and activity in xylem parenchyma contribute to acidic xylem sap (pH 5–6.5). Zn^{2+}, Ni^{2+}, Cu^{2+}, Fe^{2+}, and other metal ions are primarily carried in xylem sap as metal complexes with asparagines, histidine, organic acids, and nicotianamine. Nicotianamine (NA) is a common chelator of many metal cations in plants. Research on the tomato chloronerva mutant indicates that this unique amino acid controls the metal distribution between the apoplastic and symplastic compartments (Pich and Scholz 1991; Takahashi et al. 2003).

There have also been reports of nutrient absorption being activated at the xylem unloading location due to transpirational pull dilution of xylem sap caused by a constant water supply (Sondergaard et al. 2004). At these transport interfaces, specialized cells are present to improve solute symplastic transfer. This is mainly the responsibility of transfer and vascular-associated cells, which are members of the companion cell complex of the phloem's sieve element. Transfer cells are considered the most specialized cell type for membrane transport. The transport of nutrients in and out of the xylem is facilitated by the more significant number of mitochondria, an abundance of H^+-ATPases, and multiple plasma membrane invasions of these cells toward the tracheary elements.

However, a unique kind of xylem parenchyma cell known as the vessel-related cell is connected to nearby xylem vessels through several huge pits (Sondergaard et al. 2004). Short- and long-distance routes mediate Zn's subsequent transport to different plant organs and developing sinks once it enters the phloem. Furthermore, because chelating solutes (peptides, organic acids, etc.) are more concentrated in phloem sap than in xylem, Zn is more mobile than in xylem. Zinc is believed to be transported in phloem tissues as ions or complexes such as zinc-nicotinamide, zinc-malate, and zinc-histidine. The xylem is crucial for delivering nutrients to different organs despite reduced solute content. Phloem primarily nourishes developing grains, tubers, and other young sink tissues.

8 Transporter Proteins and Metabolites Facilitating Zinc Transport

Plants uptake Zinc as Zn^{2+}. The charged zinc ions do not enter the lipid bilayer of the plant cell. It first enters the root's cell wall free space via diffusion from soil. Zn^{2+} transports through the root cortex via apoplast and symplast.

Metal chelating compounds called siderophores enhance the uptake of metal ions from soil. Siderophores that help in zinc uptake are called zincophores. Deoxymugineic acids are one such zincophore produced by graminaceous plants. Plants secrete DMA to increase the zinc uptake by roots. At a $pH > 6.5$, DMA chelates with zinc ion, and at a $pH < 4.7$ the DMA-Zn complex dissociates. This means that DMA can obtain zinc from alkaline soils and that the DMA-Zn complex is stable at the root interface (Weiss et al. 2021).

The transporters involved in zinc transport include the ZIP-IRT family of transporters, natural resistance-associated macrophage protein (NRAMP), heavy metal

ATPase (HMA), cation diffusion facilitator (CDF), yellow stripe-like (YSL), and ATP-binding cassette (ABC) transporters.

NRAMP belongs to the conserved metal transport family in natural evolution involved in the transport, uptake, and intercellular stability of metal ions such as Fe, Zn, and Mn. In *Areca catechu* the expression of the genes AcNRAMP1, AcNRAMP2, AcNRAMP3, AcNRAMP6, AcNRAMP8, AcNRAMP11, and AcNRAMP12 increased in zinc-deficient plants. AcNRAMP2, AcNRAMP4, and AcNRAMP12 were closely related with areca's roots' tolerance under zinc deficiency stress (Zhou et al. 2023).

The ZIP family of proteins is actively involved in the zinc uptake, distribution, and transport across the plant. *Arabidopsis thaliana* has about 15 ZIP transporters. ZIP transporters transport a wide range of metals that include iron, zinc, cadmium, and manganese. ZIP transporters are found in various cell organelles that help in maintaining the zinc homeostasis in the cell. Since charged zinc ions cannot diffuse through the lipid bilayer, the ZIP proteins help the transportation of zinc across the plasma membrane. Identification and characterization of ZIPs are majorly done in model crops such as *Arabidopsis* and rice (Ajeesh Krishna et al. 2020).

HMAs (heavy metal associated) are P1B-type ATPase involved in metal ion transportation. AtHMA1 located on the chloroplast envelope enables detoxification of zinc. AtHMA2 does the translocation of zinc and cadmium ions through coordination with conserved residues of cysteine and glutamine. HMA4 transporters are involved in root to shoot translocation of zinc. HMA combines ATP hydrolysis with metal ion transport across membranes (Nazmul Hasan et al. 2022).

Tonoplast localized metal tolerance proteins (MTP) of cation diffusion facilitator (CDF) family are involved in sequestering zinc into the root cell resulting in zinc immobilization in roots and reducing the root to shoot zinc translocation. They are involved in the zinc efflux from the cytosol, either into the cellular components or outside the cell (Wang et al. 2022).

Lazarus1 (LAZ1) is a transmembrane protein belonging to the DUF300 protein family involved in organic solute transporters in vertebrates. Heterologous expression of ZmLAZ1–4 proteins in zinc-sensitive mutant, *Arabidopsis*, and maize showed that these proteins accumulated zinc ions. These transmembrane proteins were localized on plasma membrane, tonoplast, and chloroplast to maintain zinc homeostasis. They transported zinc under the negative regulation of ZmBES1/BZR1–11 transcription factor (Liu et al. 2022). Brassinosteroid Insensitive 1-Ems-Suppressor 1 (BES1) And Brassinazole Resistant 1 (BZR1) are involved in the signaling of brassinosteroids. Brassinosteroids are growth hormones that play a role in cell elongation, vascular differentiation, light-dependent architecture of plant architecture, pollen fertility, and senescence (Otani et al. 2022).

Yellow strip-like (YSL) assists in the translocation of zinc via xylem and phloem vessels. YSL can also translocate zinc in leaf and seed tissues with the assistance of nicotianamine-zinc (NA-Zn) complexes. They maintain zinc homeostasis in *Arabidopsis*, rice, and barley (Z. Song et al. 2024). Nicotianamine (NA) is a non-proteinogenic amino acid found in the root and shoot of meristematic tissues. They form hexadentate complexes with metal ions such as Fe, Ni Cu, Zn, and Mn. NA is

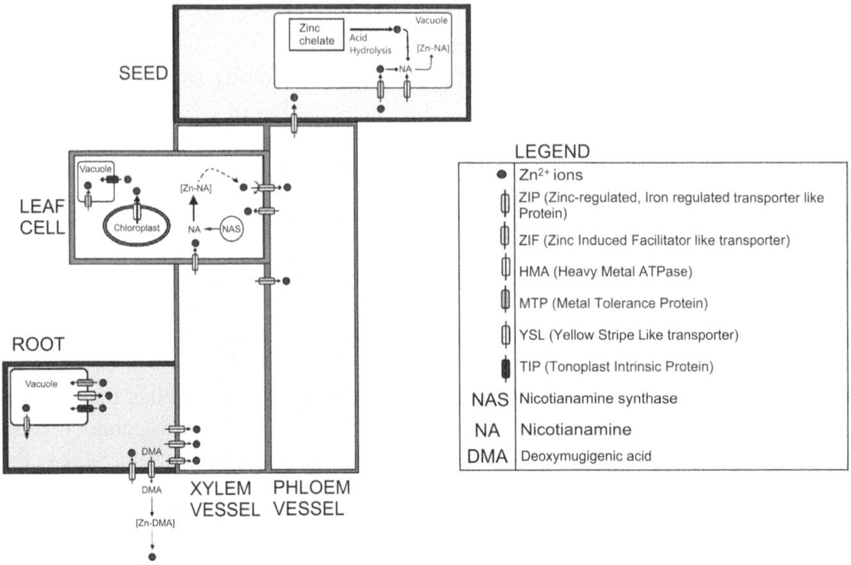

Fig. 1 Transporters and metabolites involved in the zinc uptake, transport, and sequestration in plants

a symplastic chelator that delivers metals to the transporters (Seregin and Kozhevnikova 2023).

Major facilitator superfamily (MFS) is a secondary active membrane transporter found in eukaryotes. Zinc-induced facilitators (ZIF) are a type of MFS transporters involved in the basal tolerance of zinc, and its expression is upregulated under excess zinc. Overexpression of ZIF1 increases zinc retention in roots due to the increase in concentration of endogenous nicotianamine in root vacuoles (Lee et al. 2021) (Fig. 1).

9 Intracellular Mechanisms for Detoxification and Sequestration of Zinc

Both zinc hyperaccumulators and non-hyperaccumulators store excess zinc in the stem/leaves' trichomes. ZIP (zinc-regulated, iron-regulated transporter-like proteins) transporters are located in some cell organelles and are involved in zinc homeostasis and adaptation of the plant toward high or low zinc content present in the soil. Nicotianamine synthase (NAS) gene is responsible for the chelation of different minerals including zinc (Lee et al. 2023).

Zinc is sequestered in the shoots, chelated to organic acids. It is translocated from the roots in chelated form via Xylem. Zn reaches the shoot, where it is stored bound to chelators (organic acids). A small portion of Zn can also be found as free ions or bound to the cell wall or histidine. In *Noccaea caerulescens* shoots, concentrations of malate, oxalate, succinate, and citrate are detected only after high Zn exposure; 38%, 9%, 16%, and 12% of the total Zn were associated with citrate, oxalate, His, and the cell wall, respectively, while 26% remained as free Zn^{2+} ions in the shoots *of N. caerulescens* (Zhao et al. 2006). Expression of *MsZIP2* is positively correlated with the increase in foliar zinc which is related to the detoxification of zinc in xylem parenchyma cells. Increase in the expression of *MsNAS1* in shoot indicated the chelation of excessive zinc in the shoot enabling the sequestration of zinc in vacuoles or redistribution of zinc back to the root through the phloem (Cardini et al. 2021).

The MTP (metal tolerance protein) family helps in zinc sequestration in cell organelles such as vacuoles. The effect of the MTP family was observed in *Malus halliana.* MTP2, MTP11, and MTPC2-like genes were negatively correlated to zinc concentration in roots, and they were assisted in zinc detoxification and sequestration. MTP1 was organ specific and showed more expression in shoots and leaves compared to the roots (Cheng et al. 2020). Along with MTP, cation diffusion facilitator (CDF) sequesters zinc in the vacuoles of the root cells. In contrast to MTP and CDF, the heavy metal ATPase (HMA) family of proteins facilitates loading zinc into the root xylem and enables root-to-shoot translocation. Tonoplast intrinsic proteins (TIPs) are involved in retaining excess zinc in roots, facilitating transport to shoot and accumulating zinc concentration in trichomes (Wang et al. 2022).

Moreover, Zn transport in the roots is facilitated by NcZNT 2 and 5 (Lin et al. 2016) which are the orthologs of *Arabidopsis thaliana* iron-regulated transporter 3 (IRT3) and ZIP5. Furthermore, following exposure to zinc, IRT3 was found in the plasma membrane of *Arabidopsis halleri* cells in the roots. Interestingly, Zn-hyperaccumulating species like *N. caerulescens* and *A. halleri* have high levels of IRT3 expression in their roots. Additional research by Shanmugam et al. (2013) demonstrated that Zn transport involves IRT1, which is found in the plasma membrane. This transporter's subcellular position in the plasma membrane was also noted by Lin et al. (2016) indicating that it is crucial for the movement of these ions into the cellular cytoplasm. *A. halleri* and *N. caerulescens* showed a considerable induction of ZIP19 and ZIP23 expression (Lin et al. 2016).

Excess zinc is transported into root vessels and xylem vessels once it reaches the endoderm of hyperaccumulator plants. Low molecular weight ligands will chelate zinc in the xylem parenchyma to stop surrounding cell walls from retaining it. For example, significant secretion of the non-proteinogenic amino acid nicotianamine (NA) coincides with Zn translocation in *A. halleri.* Three molecules of S-adenosylmethionine are converted into one NA molecule by the enzyme nicotianamine synthase (NAS). The expression of the NAS1, NAS2, and NAS3 genes distinguished hyperaccumulator plants from non-hyperaccumulator plants. Furthermore, expression investigations have demonstrated that NAS2 regulates NA synthesis in *N. caerulescens*, whereas NAS2 and NAS3 regulate it in *A. halleri*. Zn translocation is reduced when NA production in the root is inhibited (Lin et al. 2016).

A few ZIP family genes—likely IRT3 and ZIP4—are also implicated in root-to-shoot translocation. *A. thaliana* overexpressing AtIRT3 has been demonstrated to enhance zinc accumulation in the shoots, indicating that this gene may be involved in loading zinc for long-distance transport. Yellow stripe-like (YSL) genes are among the other genes implicated in Zn transfer. Zn is loaded into and unloaded from xylem by YSL transporters. YSL proteins are involved in the transport of Fe, Mn, Zn, Cu, Ni, and Cd. Their primary function is to facilitate the absorption of metals into cells by complexing them with non-proteinogenic amino acids. Like YSL, the MATE (multidrug and toxin efflux) transporter FRD3 (ferric reductase deficient 3) has also been called critical (Pineau et al. 2012). Afterward, chelated Zn is stored in different cells/tissues depending on the plant species. For example, *A. halleri* promotes the accumulation of Zn at the base of trichomes and in mesophyll cells, whereas *N. caerulescens* stores metals in the leaves' epidermis. In *Sedum alfredii*, Zn is at the level of the epidermis of the stems and leaves. In contrast, mesophyll cells accumulate small amounts of Zn (Yang et al. 2018).

Once in the cytosol, Zn is stored in the vacuoles as a strategy of hyperaccumulator species to prevent Zn damage of vital organelles, such as chloroplast. Sitko et al. (2017) showed that metallicolous populations of *A. halleri* were characterized by a high chlorophyll content compared to non-tolerant plants. This finding can be explained by the inhibition of Zn accumulation in chloroplasts in this species. It was confirmed that HMA1, a member of the P1B-type ATPase superfamily, was identified and localized on the chloroplast envelope in *A. thaliana* (Kim et al. 2009). Furthermore, *Arabidopsis* HMA1 knockout plants were more sensitive to Zn and accumulated a high amount of Zn in the chloroplast compared to wild-type plants. These authors demonstrated that HMA1 is associated with Zn^{2+} transport from the chloroplasts to the cytoplasm under conditions of high levels of zinc, avoiding damage to the organelle. However, the implication of HMA1 in Zn transport in hyperaccumulator species has not been confirmed yet (Kim et al. 2009). We can thus suggest the possible involvement of HMA1 in the inhibition of Zn accumulation in the chloroplasts of hyperaccumulator species. It was confirmed that members of the MTP/ABCC (metal tolerance proteins/ATP binding cassette C), belonging to the superfamily of the cation diffusion facilitator (CDF) proteins (Dräger et al. 2004), are involved in Zn loading in the vacuoles. In the leaves, the vacuolar protein MTP1 plays a key role in Zn sequestration and detoxification. It was reported that MTP1 contributes to Zn accumulation in the shoots of *A. halleri*. Likewise, in *N. goesingense*, TgMTP1 protein is accumulated in high levels in the vacuolar membrane in the shoot, which confirms its implication in Zn tolerance and accumulation enhancement in this species (Gustin et al. 2009). Next to MTP1, upregulation by high Zn of MTP8 and MTP11 was also confirmed in the shoots of the Zn hyperaccumulators *N. caerulescens* and *A. halleri*. Like MTP1, ZNT5 is also an important player in the sequestration of Zn into the epidermal storage cells. Zinc transportation in plant generally takes place via symplastic or apoplastic pathway (Hu et al. 2009). Table 1 shows the function of the transporters and their role in transportation.

Heavy metal ATPase "HMA" family proteins are involved in the transfer of zinc from the roots to the shoots. Heavy metal ATPase 2 (HMA2) and 4 (HMA4) in A. *halleri* play a major role in Zn hypertolerance and hyperaccumulation (Hanikenne et al. 2008). With a plasma membrane localization in the root pericycle cells, they promote Zn^{2+} efflux from these cells and are associated with its loading in the root xylem.

Table 1 Zinc transporters and their function in the transportation

Zinc transporter	Location	Function and transport location	References
IRT 1 ZIP 19 ZIP 4	Plasma membrane	Zinc absorption to epidermis cell	Lin et al. (2016), Yang et al. (2018)
ZIF 2	Tonoplast	Zinc sequestration to vacuoles from cortex	Remy et al. (2014)
ZIP 23 ZIP 19 ZIP 5 IRT 3	Plasma membrane	Transportation from cortex to endodermis	Van De Mortel et al. (2006), Lin et al. (2016)
ZIP 4 ZNT 1	Plasma membrane	Zinc from apoplastic route enters to epidermis	Van De Mortel et al. (2006)
YSL FRD 3 HMA 4	Vascular tissues	Transportation in xylem	Curie et al. (2009), Van De Mortel et al. (2006), Lin et al. (2016)
MTP 1 MTP8 MTP 11 NRAMP 4 HMA 3 HMA 4	Vacuolar membrane	Sequestration in vacuole	Ishida et al. (2018), Hanikenne and Nouet (2011)

In *A. thaliana*, there is another important efflux transporter: plant cadmium resistance 2 (PCR2). PCR2 functions as a Zn efflux transporter, which contributes to Zn translocation and detoxification. However, its implication in Zn hyperaccumulation is not understood.

After Zn loading in the xylem using different transporters from three major families HMA, ZIP, and YLS, Zn reaches the leaves, where it is mainly bound to organic acids, such as malate and citrate, and then sequestered in vacuoles (Song et al. 2010).

Phytohormones also regulate zinc content. Phytohormones such as brassinosteroids (BR) and gibberellic acid (GA) are shown to reduce the accumulation of heavy metals such as Cd, Pb, Cu, Cr, and zinc. Both brassinosteroids and gibberellic acid reduced the excessive accumulation of zinc in the roots and shoots of *Medicago sativa* L. BR and GA assisted in the upregulation of MsMTP which helped in the detoxification of zinc (Ren et al. 2023). Autophagy is an intracellular mechanism by which nutrient recycling occurs under carbon or nitrogen starvation. Inducing autophagy in zinc-deficient plants improved the zinc uptake by recycling zinc ions from proteins and organelles (Shinozaki et al. 2020).

10 Zinc Accumulation in Plants' Organ and Tissue Level

Zinc (Zn), being the second most common transition metal in living things due to its structural, catalytic, and activating properties, has a significant impact on plant development, reproduction, and signaling. Additionally, zinc is a cofactor for numerous enzymes, including carbonic anhydrase, Zn-superoxide dismutase, and carboxypeptidase. The average Zn concentration needed for healthy growth in most crops is around 15–20 mg kg −1 dry weight. Zn can be harmful to humans, animals, and plants at quantities higher than this. Zinc toxicity in plants rises as it becomes more bioavailable (Adamczyk-Szabela et al. 2015). The latter is dependent on various elements that either limit or increase its bioavailability, including pH, root exudate, microbial populations, and soil organic matter. There are certain plants known to accumulate high amount of metals such as nickel (Ni), manganese (Mg), and arsenic (As) in its parts to more than 1% of their dry weight (Shinozaki et al. 2020); they are known as hyperaccumulators. Generally, they can accumulate abnormal amount in the place with low contact with plant parts. Zn, being one among the toxic heavy metal, is seen to cause stress in plants; low level of zinc leads to greater degree of cell vacuolation and increase in the compartmentalization potential reducing the cytotoxic effects of the metal and plants chelating with malate. The appearance of dense globule in the vacuole was considered as detoxification mechanism in pigeon pea genotypes, and root cells of zinc-tolerant *Deschampsia cespitosa* can pump zinc actively into their vacuoles and store it. There are concentrations which would severely inhibit the plant (Sresty and Madhava Rao 1999). *Phragmites australis*, a common reed when exposed to 2 mM $ZnCl_2$ for 7 days, showed highest accumulation in roots than in shoots. The study was done using X-ray microanalysis to check concentration of zinc in compartments. The concentration gradient showed a sequence like intercellular space > cell wall > vacuole > cytoplasm, which highlighted that Zn was immobilized in the apoplast or stored into the vacuolar lumen. Also this study shows the accumulation was 10 times more in roots than in leaves and in stem was two third of that in leaves (Jiang and Wang 2008). There was a study for zinc and nickel hyperaccumulators which revealed that *Arabidopsis halleri* and *Noccaea caerulescens* have the ability to accumulate extremely high concentrations of Zn of up to 13,620 and 43,710 ppm of Zn, respectively, when growing in Zn-enriched metalliferous soils (Peer et al. 2006). In a study by Lin et al. (2012), aboveground parts of *Corydalis davidii* grown in contaminated soil have shown zinc in phytoextracted sample. Tolerance for zinc is highest tolerance as it can grow normally at 800 mg ZnL^{-1}. The concentrations of SOD, CAT, POD, and MDA in *C. davidii* increased significantly.

The rhizosphere is where metals and plants first come into contact (Moreira et al. 2018). Zinc is primarily found in soil as crystalline iron-magnesium minerals like zinc sulfide (ZnS). Zinc is mostly adsorbed onto clay surfaces and organic matter in variable forms, primarily as Zn^{2+}, $ZnOH^+$, and $ZnCl^+$. With the exception of high pH, where zinc is taken as ZnOH (Chahal et al. 2005), roots typically absorb zinc as Zn^{2+} ions. Numerous transporters involved in Zn radial transport to the stele in roots have been found, and they belong to the zinc-regulated transporter/iron-regulated transporter-like proteins (ZRT-IRT-like protein or ZIP family). NcZNT1, a significant factor in Zn transport throughout the endodermal cell, was identified in *N. caerulescens*. In *A. thaliana*, constitutive overexpression of the metal transporter NcZNT1, which is situated in the plasma membrane, improved the tolerance to excessive Zn exposure and raised the accumulation of Zn compared to wild-type plants.

11 Plant Factors Influencing Zinc Uptake

The plant uptake of the heavy metal depends upon certain factors such as physiological factors such as root intrusion, water, ion flux, and the relationship to the kinetics of metal solubilization in soil; biological parameters, including the kinetics of membrane transport, ion interactions, and microflora in rhizosphere; and other non-edaphic factors and metabolic fate of absorbed ions and availability of plants to adapt metabolically changing metal stress in the environment (Cataldo and Wildung 1978).

At an acidic pH, zinc is more bioavailable than at an alkaline pH. At pH 7.7 or lower, Zn^{2+} is the more advantageous form of zinc; yet, $ZnOH^{+}$ is present in pH 7.7–9.1. The neutral form Zn(OH)2 predominates above it. Adamczyk-Szabela et al. (2015) have shown that a decrease in the intra- and intermolecular hydrogen bonds in humic acid molecules can account for an increase in zinc's bioavailability at high pH values which suggests that slight variations can have an impact on solubility. The primary source for the zinc is physical and chemical rock weathering. Most of zinc which is present in soil is bioavailable to plant that can be because of certain factors and zinc which is present as insoluble or in adsorbed exchangeable form; soluble form is freely available to plant. Root activity can make the exchangeable form available for the uptake to some extent via the ion exchange and release of organic acid, and 90% of the zinc present in soil generally is not available to plants (Gupta et al. 2016).

12 Detection of Zinc

Light microscope (LM) with histochemical method, scanning electron microscope combined with energy dispersive spectrometry (SEM-EDS), and transmission electron microscope (TEM) are used to determine the distribution of Zn and Cd in *P. griffithii* at tissue and cellular levels. SEM-EDS analysis of roots revealed that xylem parenchyma cells had the highest concentration of zinc, whereas results from epidermal cells suggested that zinc was deposited in the endodermis, cortex, epidermis, and certain xylem parenchyma cells (Hu et al. 2009).

13 Chemical Forms of Accumulated Zinc

Chemical form of the accumulated zinc in plants can differ. In the case of *Arabidopsis halleri*, studies have shown that zinc can be sequestered in various forms in the plant tissues. Zinc phosphate precipitate was observed in the root surface; a transient presence of zinc silicate was also seen in the cytoplasm before being translocated and stored in vacuoles in undetermined forms (Sarret et al. 2002). The globular deposits from *D. cespitosa* were identified as Zn-phytate crystals through X-ray microanalysis (Sresty and Madhava Rao 1999). Zinc accumulation was checked in *Silene vulgaris*, where the leaves had high concentrations of citrate and oxalate. In leaves the malate concentration was high but not in the roots, so further work is required to verify their functionality (Harmens et al. 1994).

By combining chemical analyses, synchrotron-based X-ray microfluorescence, muEXAFS, and extended X-ray absorption spectroscopy (EXAFS), the molecular forms of zinc (Zn) in the Zn-tolerant and hyperaccumulator *Arabidopsis halleri* and in the non-tolerant and nonaccumulator *Arabidopsis lyrata subsp.* petraea were ascertained. Hydroponically produced plants with different Zn concentrations were gathered, as were examples of *A. halleri* growing in a contaminated natural habitat. In *A. halleri*, zinc speciation occurred regardless of the plants' origin (contaminated or not) and zinc exposure. Zn was primarily octahedrally coordinated and complexed to malate in the aerial sections. Zn concentrations in the bases of the trichomes revealed the presence of a secondary organic species (Sarret et al. 2002).

14 Unique Features in Zinc Hyperaccumulators

Zinc hyperaccumulators are plants that can accumulate more than 3000 ppm of zinc. To date, there are about 28 plant species that are zinc hyperaccumulators. Zinc hyperaccumulators include plants belonging to the family *Brassicaceae*. Hyperaccumulators have root exudates to improve the bioavailability. Hyperaccumulator uses the genes MTP and HMA4 for the phenotypes connected to zinc hyperaccumulation (Ricachenevsky et al. 2021).

There are two types of hyperaccumulation: active and passive hyperaccumulation. Hyper-accumulating plants increase zinc concentration by increasing root-to-shoot translocation and subsequently sequester zinc in the vacuole of the leaf cell without showing toxic symptoms.

Hyperaccumulators have a different metal partitioning among the roots and shoots. Non-hyperaccumulators have a root shoot metal uptake ratio less than 1, while the ratio in hyperaccumulators is above 1. Hyperaccumulator plants also have a bioaccumulation factor above 1. Bioaccumulation factor is the ratio of metal concentration in the plant versus soil metal concentration (Siyar et al. 2022).

Zinc hyperaccumulators accumulate zinc in the base of leaf trichome cells. Zinc accumulation in trichomes changes depending on the zinc concentration in the leaves (Ricachenevsky et al. 2021). Zincophilic root foraging is one mechanism by which hyperaccumulators increase root proliferation to increase the zinc uptake. *N. rotundifolia*, a zinc hyperaccumulator, showed a larger root mass in soils that had heterogenous metal distribution. The roots avoided the metal-rich patches in soil, which meant avoidance was one of the tolerance mechanisms used by the hyperaccumulator (Golestanifard et al. 2020). Roots of zinc hyperaccumulators contain histidine which promotes the loading of zinc in xylem vessels. L-Histidine provided exogenously in *Noccaea caerulescens* reduced zinc concentration in its roots. Histidine also improved the translocation factor of zinc (Kozhevnikova et al. 2021).

Zinc hyperaccumulating plants have special characteristics that allow them to flourish in surroundings high in zinc and store large amounts of zinc in their tissues. Some of the major key points for the Zinc accumulation are as follows:

1. *High zinc accumulation*: Zinc hyperaccumulators have the ability to accumulate zinc at abnormally high quantities, frequently more than 1% of their dry weight, for instance, while growing in zinc-enriched soils, *Arabidopsis halleri* and *Noccaea caerulescens* can accumulate up to 13,620 and 43,710 ppm of zinc, respectively, in their aerial parts. It has ecological significance as it plays a crucial role in phytoremediation (Gupta et al. 2016). Agromining is the technique of extracting precious metals from plant biomass grown on soil rich in metals. The high concentrations of zinc in hyperaccumulator plants can be utilized for this purpose.
2. *Specific genetic traits*: Studies have identified specific genes and genetic variations associated with zinc hyperaccumulation in plants. For instance, research has revealed divergent orthologous genes involved in various processes such as transcription, translation, protein metabolism, stress response, and signal transduction in hyperaccumulating plants. When the genomic variation of the hyperaccumulating ecotype (HE) and non-hyperaccumulating ecotype (NHE) of *Sedum alfredii* was compared using the random amplified polymorphic DNA (RAPD) method, it was discovered that the Zn/Cd hyperaccumulation phenotype was associated with compounds that contained SH groups (Chao et al. 2008). Simple sequence repeats (SSRs) and single nucleotide polymorphisms (SNPs) of *S. alfredii* are molecular markers that are useful for genome mapping in order to detect hyperaccumulation and hypertolerance features (Yang et al. 2017).

   By cross-referencing tolerant and sensitive species, the genetic basis of hyperaccumulation features can be revealed. There are three quantitative trait loci (QTLs) associated with zinc accumulation in *Noccaea caerulescens*, according to a study (González-Alcaraz and van Gestel 2015). *Arabidopsis lyrata* ssp. *petraea*, the non-tolerant relative of *A. halleri* ssp., and its first-generation backcross progeny were analyzed using QTL mapping of zinc tolerance. Zn tolerance was enhanced by *A. halleri* alleles in all three of the QTLs that the authors discovered. The enhanced Zn tolerance that this species has undergone positive selection for accounts for this outcome. Variations among populations continue to be a crucial area of research to understand whether there is an adaptation or a decline in this ability (Willems et al. 2007).
3. *Physiological adaptation techniques*: To withstand and hyperaccumulate zinc in their aboveground components, zinc hyperaccumulators have evolved a variety of physiological adaptation techniques. The stimulation of particular genes and processes linked to zinc absorption, transport, and sequestration may be one of these tactics (Tian et al. 2009). Plants that hyperaccumulate metals demonstrate effective metal import, vacuole loading, and controlled zinc redistribution within

their tissues. Numerous transporter families, including MTP, ZIP, YSL, and HMA, are essential for zinc tolerance and transport (Kim et al. 2009).

15 Effect of Other Mineral Nutrients on Zinc Uptake and Utilization

Zinc uptake by plants depends on the mineral content in soil. Calcareous soil (soil with pH > 7) containing organic matter causes zinc deficiency in soil due to the presence of bicarbonate ions. Zinc forms insoluble complexes with carbonates and bicarbonates present in soil. Highly acidic soils will have less zinc content due to soil leaching (Hacisalihoglu 2020).

Phosphorus displays antagonist behavior against zinc when it comes to nutrient uptake and plant development. This is also known as "P-induced Zn deficiency." Phosphorus-zinc antagonism depends on the reactive minerals of soil that interact with zinc and phosphorus. Calcareous soils are less sensitive to Zn-P antagonism compared to non-calcareous soil (Sacristán et al. 2019). Phosphorus and zinc joint fertilization did increase the zinc content, increased the translocation of both the nutrients, and enhanced zinc content in the seeds of maize (Sánchez-Rodríguez et al. 2021).

Application of potassium decreased the Zn-P antagonism. Zinc and potassium showed synergetic relations in soil and plants by influencing the uptake of both the nutrients in wheat (Bashir et al. 2023). Zinc competes against transitional earth metals such as cadmium and copper. ZIP (zinc-regulated/iron-regulated transporter protein) transporters transport zinc ($^{2+}$) along with transitional metal cations such as iron (Fe^{2+}), manganese (Mn^{2+}), cadmium (Cd^{2+}), cobalt (Co^{2+}), copper (Cu^{2+}), and nickel (Ni^{2+}). Majority of ZIP are transmembrane channels with histidyl-rich metal-binding sites (Ajeesh Krishna et al. 2020).

Zinc and cadmium have similar physical and chemical properties. Zinc-cadmium interaction occurs from the root uptake, translocation in the xylem from root to shoot, re-translocation via the phloem vessel, and the accumulation in edible parts of the plant. Zn-Cd antagonism is dependent on the ratio of Zn:Cd and pH of the soil. Rice grown in more acidic soils were more vulnerable to increased cadmium uptake. Zinc fertilization inhibited cadmium uptake in both rice and wheat (Yang et al. 2020).

Nitrogen has a synergistic effect on zinc uptake. Zinc accumulation in shoots increased with the application of nitrogen fertilizers when zinc supply was not limiting (Xue et al. 2021). Sulfur increased zinc content in rice under low and adequate zinc content in soil. This is due to the zinc and sulfur reaction at cellular levels. Thiols (compounds containing -SH) are zinc binding compounds in cellular systems (Khampuang et al. 2023).

16 Strategies for Enhancing Zinc Uptake in Plant Tissues

Some of the techniques employed to enhance zinc uptake include application of organic ligand or chelator, application of mycorrhizal fungi, zinc solubilizing bacteria, and nanotechnology. Zinc nanofertilizers can be provided via foliar application or are placed in the roots of the plants. Nanotechnology can be employed to enhance zinc uptake by roots. Nanoparticles act as transport systems and can transport mineral nutrients to the target site as nanofertilizers (Khampuang et al. 2023). Nanofertilizers such as zinc oxide nanoparticles enhance zinc uptake and were also shown to decrease the Cd concentration in wheat (Rizwan et al. 2019). Zinc nanoparticles can be more efficient than the conventional zinc fertilizers available. Foliar application of zinc oxide nanoparticles showed higher yield increment over control that were exposed to zinc fertilizers and were comparatively more efficient in nutrient uptake (Ahmed et al. 2023).

Ligands bind to zinc stabilizing them and making it more accessible for the plant. These ligands include ethylene diamine tetra-acetic acid (EDTA), hydroxyethylene diamine triacetic acid (HEEDTA), and diethylene triamine penta-acetic acid (DTPA). Organic ligands are more efficient and are relatively environment friendly compared to the conventional ligands. [S,S]-EDDS a biodegradable ligand was used as a soil solubilizer to increase the uptake of Fe, Mn, Zn, and Cu in nutrient-deficient soil (López-Rayo et al. 2023).

Biofertilizers are another strategy by which zinc uptake can be increased. Mycorrhizae have symbiotic relation to the plant roots. Arbuscular mycorrhizal fungi (AMF) is one the most prominent of the fungal types. AMF inoculation in maize increased the zinc uptake by the roots and showed higher accumulation in shoot, leaves, and seed compared to the non-inoculated plants (Saboor et al. 2021).

Zinc-solubilizing rhizobacteria enhance the bioavailability of zinc to plants. Zinc-solubilizing bacteria produce organic acids that will acidify the soil environment and solubilize the zinc present in soil. Some bacteria also produce siderophores that chelate metal nutrients such as zinc (Upadhayay et al. 2021). Rhizobacteria isolated from zinc mines and sodic environments were assessed for its effectiveness in zinc uptake in rice. *Bacillus paramycoides* strain 1 (BP1) improved the zinc uptake by over 17% compared to the control. BS-1 also showed a 22–24% increase in water-soluble plus exchangeable fraction of zinc in alkaline soil (Singh et al. 2023).

Genetic modification of plants is another method of biofortification that can alter the zinc uptake by plants. Transgenic *Arabidopsis thaliana* and *Populous alba* were overexpressed with the gene ZRC1 extracted from *Saccharomyces cerevisiae*. ZRC1 is responsible for uptake of zinc into the vacuole. Both these plants showed high zinc tolerance, and *Populous alba* had higher accumulation of zinc in shoot and leaves of the plant and had lower zinc content in the root compared to wild type control (DalCorso et al. 2021).

17 Conclusion

Zinc is a vital nutrient that plants need for a variety of metabolic processes. On the other hand, elevated zinc levels in plants impact a range of physiological and biochemical aspects. Understanding the soil's ability to bind zinc and the bioavailable forms of the metal is essential for assessing the possible danger of zinc transmission throughout the soil-plant-human chain. The plant system's mechanism for absorbing zinc is strictly regulated at every level of hierarchy. At the organ level, shoots—leaves, stems, flowers, and fruits—absorb zinc more so than roots. Regardless of the root or shoot, vascular tissues and cortical parenchyma have the highest tissue-level zinc absorption. Plants use a variety of transporters to move zinc and maintain equilibrium. Proteomic and genomic techniques have undoubtedly made it easier to comprehend the roles of zinc transporters in the current omic era, but there are still a lot of unanswered questions. Membrane protein expression may help to clarify the interface between transporter proteins and other elements involved in cellular zinc trafficking, which could lead to a better knowledge of plant mechanisms pertaining to zinc transport. Moreover, the identification of secondary genes and messengers that control zinc uptake, accumulation, and translocation may be a valuable means of gaining deeper understanding of the several mechanisms involved in plant zinc absorption. Research on antinutritional compounds (e.g., phytic acid) that decrease the bioavailability of micronutrients (e.g., zinc) can potentially be a promising strategy to fill the target tissue/organ with micronutrients. In conclusion, achieving comprehensive understanding of zinc signalling would be crucial for achieving sustainable agricultural expansion worldwide and would have significant scientific implications.

References

Adamczyk-Szabela D, Wolf WM (2022) The impact of soil pH on heavy metals uptake and photosynthesis efficiency in Melissa officinalis, Taraxacum officinale, Ocimum basilicum. Molecules 27(15). https://doi.org/10.3390/molecules27154671

Adamczyk-Szabela D, Markiewicz J, Wolf WM (2015) Heavy metal uptake by herbs. IV. Influence of soil pH on the content of heavy metals in Valeriana officinalis L. Water Air Soil Pollut 226(4):106. https://doi.org/10.1007/s11270-015-2360-3

Ahmed R, Uddin MK, Quddus MA, Samad MYA, Hossain MAM, Haque ANA (2023) Impact of foliar application of zinc and zinc oxide nanoparticles on growth, yield, nutrient uptake and quality of tomato. Horticulturae 9(2). https://doi.org/10.3390/horticulturae9020162

Ajeesh Krishna TP, Maharajan T, Victor Roch G, Ignacimuthu S, Antony Ceasar S (2020) Structure, function, regulation and phylogenetic relationship of ZIP family transporters of plants. Front Plant Sci 11(May):1–18. https://doi.org/10.3389/fpls.2020.00662

Alloway BJ (2009) Soil factors associated with zinc deficiency in crops and humans. Environ Geochem Health 31(5):537–548. https://doi.org/10.1007/s10653-009-9255-4

Andreini C, Bertini I, Rosato A (2009) Metalloproteomes: a bioinformatic approach. Acc Chem Res 42(10):1471–1479. https://doi.org/10.1021/ar900015x

Arnold T, Kirk GJD, Wissuwa M, Frei M, Zhao FJ, Mason TFD, Weiss DJ (2010) Evidence for the mechanisms of zinc uptake by rice using isotope fractionation. Plant Cell Environ 33(3):370–381. https://doi.org/10.1111/j.1365-3040.2009.02085.x

Bago B, Azcón-Aguilar C, Goulet A, Piché Y (1998) Branched absorbing structures (BAS): a feature of the extraradical mycelium of symbiotic arbuscular mycorrhizal fungi. New Phytol 139(2):375–388. https://doi.org/10.1046/j.1469-8137.1998.00199.x

Bashir K, Ishimaru Y, Nishizawa NK (2010) Iron uptake and loading into rice grains. Rice 3(2–3):122–130. https://doi.org/10.1007/s12284-010-9042-y

Bashir A, Khan QU, Alem A, Hendi AA, Zaman U, Khan SU, Rehman K et al (2023) Zinc and potassium fertilizer synergizes plant nutrient availability and affects growth, yield, and quality of wheat genotypes. Plan Theory 12(12). https://doi.org/10.3390/plants12122241

Benshoff JJ, Harrawood LK, Koch DS (2003) Phytoremediation of potentially toxic elements: role, status and concerns. J Rehabil 69(2):43–48

Briggs D (2003) Environmental pollution and the global burden of disease. Br Med Bull 68:1–24. https://doi.org/10.1093/bmb/ldg019

Broadley MR (2011) Function of nutrients: micronutrients. In: Marschner's mineral nutrition of higher plants, 3rd edn. https://doi.org/10.1016/B978-0-12-384905-2.00007-8

Broadley MR, White PJ, Hammond JP, Zelko I, Lux A (2007) Zinc in plants: Tansley review. New Phytol 173(4):677–702. https://doi.org/10.1111/j.1469-8137.2007.01996.x

Caldelas C, Weiss DJ (2017) Zinc homeostasis and isotopic fractionation in plants: a review. Plant Soil 411(1–2):17–46. https://doi.org/10.1007/s11104-016-3146-0

Caldelas C, Dong S, Araus JL, Jakob Weiss D (2010) Zinc isotopic fractionation in Phragmites australis in response to toxic levels of zinc. J Exp Bot 62(6):2169–2178. https://doi.org/10.1093/jxb/erq414

Cardini A, Pellegrino E, White PJ, Mazzolai B, Mascherpa MC, Ercoli L (2021) Transcriptional regulation of genes involved in zinc uptake, sequestration and redistribution following foliar zinc application to Medicago sativa. Plan Theory 10(3):1–19. https://doi.org/10.3390/plants10030476

Cataldo DA, Wildung RE (1978) Soil and plant factors influencing the accumulation of heavy metals by plants. Environ Health Perspect 27(December):149–159. https://doi.org/10.1289/ehp.7827149

Chahal DS, Sharma BD, Singh PK (2005) Distribution of forms of zinc and their association with soil properties and uptake in different soil orders in semi-arid soils of Punjab, India. Commun Soil Sci Plant Anal 36(19–20):2857–2874. https://doi.org/10.1080/00103620500306031

Chaney RL (1993) Zinc Phytotoxicity. Soil Sci 147(6):390–397. https://doi.org/10.1097/00010694-198906000-00002

Chao YE, Zhang M, Tian SK, Lu LL, Yang XE (2008) Differential generation of hydrogen peroxide upon exposure to zinc and cadmium in the hyperaccumulating plant species (sedum alfredii Hance). J Zhejiang Univ Sci B 9(3):243–249. https://doi.org/10.1631/jzus.B0710624

Cheng L, Zhao T, Wang H, Zhu Y, Hu Y, Wu Y, Jia X, Zhu Z, Wang Y, xiu. (2020) Correlation of the expression of metal tolerance protein genes with zinc accumulation and distribution in Malus halliana seedlings under zinc toxicity stress. Arid Land Res Manag 34(3):336–352. https://doi.org/10.1080/15324982.2019.1703060

Claus J, Bohmann A, Chavarría-Krauser A (2013) Zinc uptake and radial transport in roots of Arabidopsis thaliana: a modelling approach to understand accumulation. Ann Bot 112(2):369–380. https://doi.org/10.1093/aob/mcs263

Clemens S (2022) The cell biology of zinc. J Exp Bot 73(6):1688–1698. https://doi.org/10.1093/jxb/erab481

Connolly EL, Fett JP, Guerinot ML (2002) Expression of the IRT1 metal transporter is controlled by metals at the levels of transcript and protein accumulation. Plant Cell 14(6):1347–1357. https://doi.org/10.1105/tpc.001263

Cornu JY, Deinlein U, Höreth S, Braun M, Schmidt H, Weber M, Persson DP, Husted S, Schjoerring JK, Clemens S (2015) Contrasting effects of nicotianamine synthase knockdown on zinc and nickel tolerance and accumulation in the zinc/cadmium hyperaccumulator Arabidopsis halleri. New Phytol 206(2):738–750. https://doi.org/10.1111/nph.13237

Cui Li F (2018) Absorption and translocation of Zn Foliar. https://espace.library.uq.edu.au/data/UQ_6dedde7/s4344450_phd_final_thesis.pdf?Expires=1557654115andSignature=W-

QUk3U7Xa0~rD5tgHj8iH8hNtsqypfljFc0u7JbWSaLlFsap31C6YFDfDPfXoqz6lNeU4hHlTaaplRZPkMrur2W9FV3tnmfgzbexdbGXprfqE1ncEtrFOaD39~bnVsr8WWt23TjqAMarNCOUH2w

Curie C, Cassin G, Couch D, Divol F, Higuchi K, Le Jean M, Misson J, Schikora A, Czernic P, Mari S (2009) Metal movement within the plant: contribution of nicotianamine and yellow stripe 1-like transporters. Ann Bot 103(1):1–11. https://doi.org/10.1093/aob/mcn207

Cuypers A, Vangronsveld J, Clijsters H (2001) The redox status of plant cells (AsA and GSH) is sensitive to zinc imposed oxidative stress in roots and primary leaves of Phaseolus vulgaris. Plant Physiol Biochem 39(7–8):657–664. https://doi.org/10.1016/S0981-9428(01)01276-1

DalCorso G, Martini F, Fasani E, Manara A, Visioli G, Furini A (2021) Enhancement of Zn tolerance and accumulation in plants mediated by the expression of Saccharomyces cerevisiae vacuolar transporter ZRC1. Planta 253(6):1–17. https://doi.org/10.1007/s00425-021-03634-z

Dawar K, Ali W, Bibi H, Mian IA, Ahmad MA, Hussain MB, Ali M, Ali S, Fahad S, Ur Rehman S, Datta R, Syed A, Danish S (2022) Effect of different levels of zinc and compost on yield and yield components of wheat. Agronomy 12(7):1–15. https://doi.org/10.3390/agronomy12071562

De Groote H, Tessema M, Gameda S, Gunaratna NS (2021) Soil zinc, serum zinc, and the potential for agronomic biofortification to reduce human zinc deficiency in Ethiopia. Sci Rep 11(1):1–11. https://doi.org/10.1038/s41598-021-88304-6

Dräger DB, Desbrosses-Fonrouge AG, Krach C, Chardonnens AN, Meyer RC, Saumitou-Laprade P, Krämer U (2004) Two genes encoding Arabidopsis halleri MTP1 metal transport proteins co-segregate with zinc tolerance and account for high MTP1 transcript levels. Plant J 39(3):425–439. https://doi.org/10.1111/j.1365-313X.2004.02143.x

Eide D, Broderius M, Fett J, Guerinot ML (1996) A novel iron-regulated metal transporter from plants identified by functional expression in yeast. Proc Natl Acad Sci USA 93(11):5624–5628. https://doi.org/10.1073/pnas.93.11.5624

Erenoglu B, Nikolic M, Römheld V, Cakmak I (2002) Uptake and transport of foliar applied zinc (65Zn) in bread and durum wheat cultivars differing in zinc efficiency. Plant Soil 241(2):251–257. https://doi.org/10.1023/A:1016148925918

Fosmire G (1990) Zinc toxicity. Am J Clin Nutr 51(2):225–227. https://doi.org/10.1093/AJCN/51.2.225

Francis B, Aravindakumar CT, Brewer PB, Simon S (2023) Plant nutrient stress adaptation: A prospect for fertilizer limited agriculture. Environ Exp Bot 213:105431. https://doi.org/10.1016/J.ENVEXPBOT.2023.105431

Gai APC, dos Santos DS, Vieira EA (2017) Effects of zinc excess on antioxidant metabolism, mineral content and initial growth of Handroanthus impetiginosus (Mart. ex DC.) Mattos and Tabebuia roseoalba (Ridl.) Sandwith. Environ Exp Bot 144:88–99. https://doi.org/10.1016/j.envexpbot.2017.09.006

Glińska S, Gapińska M, Michlewska S, Skiba E, Kubicki J (2016) Analysis of Triticum aestivum seedling response to the excess of zinc. Protoplasma 253(2):367–377. https://doi.org/10.1007/s00709-015-0816-3

Golestanifard A, Puschenreiter M, Aryan A, Santner J, Wenzel WW (2020) Metal accumulation and rhizosphere characteristics of Noccaea rotundifolia ssp. cepaeifolia. Environ Pollut 266:115088. https://doi.org/10.1016/j.envpol.2020.115088

González-Alcaraz MN, van Gestel CAM (2015) Climate change effects on enchytraeid performance in metal-polluted soils explained from changes in metal bioavailability and bioaccumulation. Environ Res 142:177–184. https://doi.org/10.1016/j.envres.2015.06.027

Graham RD, Ascher JS, Hynes SC (1992) Selecting zinc-efficient cereal genotypes for soils of low zinc status. Plant Soil 146(1–2):241–250. https://doi.org/10.1007/BF00012018

Guerinot ML (2000) The ZIP family of metal transporters. Biochim Biophys Acta Biomembr 1465(1–2):190–198. https://doi.org/10.1016/S0005-2736(00)00138-3

Gupta N, Ram H, Kumar B (2016) Mechanism of zinc absorption in plants: uptake, transport, translocation and accumulation. Rev Environ Sci Biotechnol 15(1):89–109. https://doi.org/10.1007/s11157-016-9390-1

Gustin JL, Loureiro ME, Kim D, Na G, Tikhonova M, Salt DE (2009) MTP1-dependent Zn sequestration into shoot vacuoles suggests dual roles in Zn tolerance and accumulation in Zn-hyperaccumulating plants. Plant J 57(6):1116–1127. https://doi.org/10.1111/j.1365-313X.2008.03754.x

Ha H, Olson JR, Bian L, Rogerson PA (2014) Analysis of heavy metal sources in soil using kriging interpolation on principal components. Environ Sci Technol 48(9):4999–5007. https://doi.org/10.1021/es405083f

Habashi F (2013) Lead, physical and chemical properties. In: Kretsinger RH, Uversky VN, Permyakov EA (eds) Encyclopedia of metalloproteins. Springer, New York. https://doi.org/10.1007/978-1-4614-1533-6

Hacisalihoglu G (2020) Zinc (Zn): the last nutrient in the alphabet and shedding light on Zn efficiency for the future of crop production under suboptimal Zn. Plan Theory 9(11):1–9. https://doi.org/10.3390/plants9111471

Hacisalihoglu G, Hart JJ, Kochian LV (2001) High- and low-affinity zinc transport systems and their possible role in zinc efficiency in bread wheat. Plant Physiol 125(1):456–463. https://doi.org/10.1104/pp.125.1.456

Hanikenne M, Nouet C (2011) Metal hyperaccumulation and hypertolerance: a model for plant evolutionary genomics. Curr Opin Plant Biol 14(3):252–259. https://doi.org/10.1016/j.pbi.2011.04.003

Hanikenne M, Talke IN, Haydon MJ, Lanz C, Nolte A, Motte P, Kroymann J, Weigel D, Krämer U (2008) Evolution of metal hyperaccumulation required cis-regulatory changes and triplication of HMA4. Nature 453(7193):391–395. https://doi.org/10.1038/nature06877

Harmens H, Koevoets PL, Verkleij JA, Ernst WH (1994) The role of low molecular weight organic acids in the mechanism of increased zinc tolerance in Silene vulgaris (Moench) Garcke. New Phytol 126(4):615–621. https://doi.org/10.1111/j.1469-8137.1994.tb02956.x

Hasanuzzaman M, Fujita M, Oku H, Nahar K, Hawrylak-Nowak B (2018) Plant nutrients and abiotic stress tolerance. In: Plant nutrients and abiotic stress tolerance. https://doi.org/10.1007/978-981-10-9044-8

Henriques AR, Chalfun-Junior A, Aarts M (2012) Strategies to increase zinc deficiency tolerance and homeostasis in plants. Braz J Plant Physiol 24(1):3–8. https://doi.org/10.1590/S1677-04202012000100002

Higuchi K, Kanazawa K, Nishizawa NK, Chino M, Mori S (1994) Purification and characterization of nicotianamine synthase from Fe-deficient barley roots. Plant Soil 165(2):173–179. https://doi.org/10.1007/BF00008059

Higuchi K, Suzuki K, Nakanishi H, Yamaguchi H, Nishizawa NK, Mori S (1999) Cloning of nicotianamine synthase genes, novel genes involved in the biosynthesis of phytosiderophores. Plant Physiol 119(2):471–479. https://doi.org/10.1104/pp.119.2.471

Hong-Qiang Y, Yu-Ling J (2005) Uptake and transport of calcium in plants. J Plant Physiol Mol Biol 31(3):227–234. https://doi.org/10.13592/j.cnki.ppj.2015.0024

Hu PJ, Qiu RL, Senthilkumar P, Jiang D, Chen ZW, Tang YT, Liu FJ (2009) Tolerance, accumulation and distribution of zinc and cadmium in hyperaccumulator Potentilla griffithii. Environ Exp Bot 66(2):317–325. https://doi.org/10.1016/j.envexpbot.2009.02.014

Huang Q, Li M, Liu Y, Hu W, Liu M, Chen Z, Li F (2013) Using construction expansion regulation zones to manage urban growth in Hefei City, China. J Urban Plann Dev 139(1):62–69. https://doi.org/10.1061/(asce)up.1943-5444.0000135

Ishida JK, Caldas DGG, Oliveira LR, Frederici GC, Leite LMP, Mui TS (2018) Genome-wide characterization of the NRAMP gene family in Phaseolus vulgaris provides insights into functional implications during common bean development. Genet Mol Biol 41(4):820–833. https://doi.org/10.1590/1678-4685-gmb-2017-0272

Jain A, Sinilal B, Dhandapani G, Meagher RB, Sahi SV (2013) Effects of deficiency and excess of zinc on morphophysiological traits and spatiotemporal regulation of zinc-responsive genes reveal incidence of cross talk between micro- and macronutrients. Environ Sci Technol 47(10):5327–5335. https://doi.org/10.1021/es400113y

Jiang X, Wang C (2008) Zinc distribution and zinc-binding forms in Phragmites australis under zinc pollution. J Plant Physiol 165(7):697–704. https://doi.org/10.1016/j.jplph.2007.05.011

Kang MJ, Kwon YK, Yu S, Lee PK, Park HS, Song N (2019) Assessment of Zn pollution sources and apportionment in agricultural soils impacted by a Zn smelter in South Korea. J Hazard Mater 364:475–487. https://doi.org/10.1016/j.jhazmat.2018.10.046

Kaya C, Higgs D, Burton A (2000) Plant growth, phosphorus nutrition, and acid phosphatase enzyme activity in three tomato cultivars grown hydroponically at different zinc concentrations. J Plant Nutr 23(5):569–579. https://doi.org/10.1080/01904160009382041

Khampuang K, Chaiwong N, Yazici A, Demirer B, Cakmak I, Prom-U-Thai C (2023) Effect of sulfur fertilization on productivity and grain zinc yield of rice grown under low and adequate soil zinc applications. Rice Sci 30(6):632–640. https://doi.org/10.1016/j.rsci.2023.07.003

Kim YY, Choi H, Segami S, Cho HT, Martinoia E, Maeshima M, Lee Y (2009) AtHMA1 contributes to the detoxification of excess Zn(II) in Arabidopsis. Plant J 58(5):737–753. https://doi.org/10.1111/j.1365-313X.2009.03818.x

Kisko M, Bouain N, Rouached A, Choudhary SP, Rouached H (2015) Molecular mechanisms of phosphate and zinc signalling crosstalk in plants: phosphate and zinc loading into root xylem in Arabidopsis. Environ Exp Bot 114:57–64. https://doi.org/10.1016/j.envexpbot.2014.05.013

Kopittke PM, Menzies NW, de Jonge MD, Mckenna BA, Donner E, Webb RI, Paterson DJ, Howard DL, Ryan CG, Glover CJ, Scheckel KG, Lombi E (2011) In situ distribution and speciation of toxic copper, nickel, and zinc in hydrated roots of cowpea. Plant Physiol 156(2):663–673. https://doi.org/10.1104/pp.111.173716

Korshunova YO, Eide D, Clark WG, Guerinot ML, Pakrasi HB (1999) The IRT1 protein from Arabidopsis thaliana is a metal transporter with a broad substrate range. Plant Mol Biol 40(1):37–44. https://doi.org/10.1023/A:1026438615520

Kozhevnikova AD, Seregin IV, Zhukovskaya NV, Kartashov AV, Schat H (2021) Histidine-mediated nickel and zinc translocation in intact plants of the Hyperaccumulator Noccaea caerulescens. Russ J Plant Physiol 68(1):S37–S50. https://doi.org/10.1134/S1021443721070074

Krishnan S, Dayanandan P (2003) Structural and histochemical studies on grain-filling in the caryopsis of rice (Oryza sativa L.). J Biosci 28(4):455–469. https://doi.org/10.1007/BF02705120

Laity JH, Lee BM, Wright PE (2001) Zinc finger proteins: new insights into structural and functional diversity. Curr Opin Struct Biol 11(1):39–46. https://doi.org/10.1016/S0959-440X(00)00167-6

Lee Y, Lim C (2008) Physical basis of structural and catalytic Zn-binding sites in proteins. J Mol Biol 379(3):545–553. https://doi.org/10.1016/j.jmb.2008.04.004

Lee S, Ricachenevsky FK, Punshon T (2021) Functional overlap of two major facilitator superfamily transporter, ZIF1, and ZIFL1 in zinc and iron homeostasis. Biochem Biophys Res Commun 560:7–13. https://doi.org/10.1016/J.BBRC.2021.04.120

Lee S, Rahman MM, Nakanishi H, Nishizawa NK, An G, Nam HG, Jeon JS (2023) Concomitant activation of OsNAS2 and OsNAS3 contributes to the enhanced accumulation of iron and zinc in rice. Int J Mol Sci 24(7). https://doi.org/10.3390/ijms24076568

Lefèvre I, Vogel-Mikuš K, Jeromel L, Vavpetič P, Planchon S, Arčon I, Elteren JTV, Lepoint G, Gobert S, Renaut J, Pelicon P, Lutts S (2014) Differential cadmium and zinc distribution in relation to their physiological impact in the leaves of the accumulating Zygophyllum fabago L. Plant Cell Environ 37(6):1299–1320. https://doi.org/10.1111/pce.12234

Leuci R, Brunetti L, Laghezza A, Loiodice F, Tortorella P, Piemontese L (2020) Importance of biometals as targets in medicinal chemistry: an overview about the role of zinc (II) chelating agents. Appl Sci (Switzerland) 10(12). https://doi.org/10.3390/APP10124118

Li X, Christie P (2001) Changes in soil solution Zn and pH and uptake of Zn by arbuscular mycorrhizal red clover in Zn-contaminated soil. Chemosphere 42(2):201–207. https://doi.org/10.1016/S0045-6535(00)00126-0

Li J, Huang Y, Tan H, Yang X, Tian L, Luan S, Chen L, Li D (2015) An endoplasmic reticulum magnesium transporter is essential for pollen development in Arabidopsis. Plant Sci 231(7):212–220. https://doi.org/10.1016/j.plantsci.2014.12.008

Lin W, Xiao T, Wu Y, Ao Z, Ning Z (2012) Hyperaccumulation of zinc by Corydalis davidii in Zn-polluted soils. Chemosphere 86(8):837–842. https://doi.org/10.1016/j.chemosphere.2011.10.060

Lin YF, Hassan Z, Talukdar S, Schat H, Aarts MGM (2016) Expression of the Zntl zinc transporter from the metal hyperaccumulator Noccaea caerulescens confers enhanced zinc and cadmium tolerance and accumulation to Arabidopsis thaliana. PLoS One 11(3):1–30. https://doi.org/10.1371/journal.pone.0149750

Liu M, Li Y, Zhang W, Wang Y (2013) Assessment and spatial distribution of zinc pollution in agricultural soils of Chaoyang, China. Procedia Environ Sci 18:283–289. https://doi.org/10.1016/j.proenv.2013.04.037

Liu D, Chen J, Mahmood Q, Li S, Wu J, Ye Z, Peng D, Yan W, Lu K (2014) Effect of Zn toxicity on root morphology, ultrastructure, and the ability to accumulate Zn in Moso bamboo (Phyllostachys pubescens). Environ Sci Pollut Res 21(23):13615–13624. https://doi.org/10.1007/s11356-014-3271-3

Liu B, Yu H, Yang Q, Ding L, Sun F, Qu J, Feng W, Yang Q, Li W, Fu F (2022) Zinc transporter ZmLAZ1-4 modulates zinc homeostasis on plasma and vacuolar membrane in maize. Front Plant Sci 13(May). https://doi.org/10.3389/fpls.2022.881055

López-Rayo S, Valverde S, Lucena JJ (2023) [S,S]-EDDS ligand as a soil Solubilizer of Fe, Mn, Zn, and Cu to improve plant nutrition in deficient soils. J Agric Food Chem 71(25):9728–9737. https://doi.org/10.1021/acs.jafc.3c02057

Lu L, Liao X, Labavitch J, Yang X, Nelson E, Du Y, Brown PH, Tian S (2014) Speciation and localization of Zn in the hyperaccumulator sedum alfredii by extended X-ray absorption fine structure and micro-X-ray fluorescence. Plant Physiol Biochem 84:224–232. https://doi.org/10.1016/j.plaphy.2014.10.004

Luo L, Ma Y, Zhang S, Wei D, Zhu YG (2009) An inventory of trace element inputs to agricultural soils in China. J Environ Manag 90(8):2524–2530. https://doi.org/10.1016/j.jenvman.2009.01.011

Malbreil M, Tisserant E, Martin F, Roux C (2014) Genomics of arbuscular mycorrhizal fungi: out of the shadows. In: Advances in botanical research, vol 70, 1st edn. Elsevier Ltd. https://doi.org/10.1016/B978-0-12-397940-7.00009-4

Marschner H (1993) Zinc uptake from soils. In: Zinc in soils and plants, pp 59–77. https://doi.org/10.1007/978-94-011-0878-2_5

Marschner H (1995) Diagnosis of deficiency and toxicity of nutrients. In: Mineral nutrition of higher plants, 3rd edn, pp 299–312. https://doi.org/10.1016/B978-0-12-384905-2.00011-X

Mishra S, Mishra A, Küpper H (2017) Protein biochemistry and expression regulation of cadmium/zinc pumping ATPases in the hyperaccumulator plants Arabidopsis halleri and Noccaea caerulescens. Front Plant Sci 8(May):1–13. https://doi.org/10.3389/fpls.2017.00835

Monsant AC, Kappen P, Wang Y, Pigram PJ, Baker AJM, Tang C (2011) In vivo speciation of zinc in Noccaea caerulescens in response to nitrogen form and zinc exposure. Plant Soil 348(1–2):167–183. https://doi.org/10.1007/s11104-011-0887-7

Moreira A, Moraes LAC, dos Reis AR (2018) The molecular genetics of zinc uptake and utilization efficiency in crop plants. In: Plant micronutrient use efficiency: molecular and genomic perspectives in crop plants. Elsevier Inc. https://doi.org/10.1016/B978-0-12-812104-7.00006-X

Moreno-Lora A, Delgado A (2020) Factors determining Zn availability and uptake by plants in soils developed under Mediterranean climate. Geoderma 376(May):114509. https://doi.org/10.1016/j.geoderma.2020.114509

Mori S (1999) Iron acquisition Satoshi Mori. Curr Opin Plant Biol 2:250–253

Mori S, Nishizawa N (1987) Methionine as a dominant precursor of phytosiderophores in graminaceae plants. Plant Cell Physiol 28(6):1081–1092

Nadgórska-Socha A, Kandziora-Ciupa M, Ciepał R (2015) Element accumulation, distribution, and phytoremediation potential in selected metallophytes growing in a contaminated area. Environ Monit Assess 187(7):441. https://doi.org/10.1007/s10661-015-4680-6

Nazmul Hasan M, Islam S, Bhuiyan FH, Arefin S, Hoque H, Azad Jewel N, Ghosh A, Prodhan SH (2022) Genome wide analysis of the heavy-metal-associated (HMA) gene family in tomato

and expression profiles under different stresses. Gene 835:146664. https://doi.org/10.1016/J.GENE.2022.146664
Nguyen TD, Cavagnaro TR, Watts-Williams SJ (2019) The effects of soil phosphorus and zinc availability on plant responses to mycorrhizal fungi: a physiological and molecular assessment. Sci Rep 9(1):1–13. https://doi.org/10.1038/s41598-019-51369-5
Noulas C, Tziouvalekas M, Karyotis T (2018) Zinc in soils, water and food crops. J Trace Elem Med Biol 49:252–260. https://doi.org/10.1016/j.jtemb.2018.02.009
Otani Y, Kawanishi M, Kamimura M, Sasaki A, Nakamura Y, Nakamura T, Okamoto S (2022) Behavior and possible function of Arabidopsis BES1/BZR1 homolog 2 in brassinosteroid signaling. Plant Signal Behav 17(1). https://doi.org/10.1080/15592324.2022.2084277
Ozturk L, Yazici MA, Yucel C, Torun A, Cekic C, Bagci A, Ozkan H, Braun HJ, Sayers Z, Cakmak I (2006) Concentration and localization of zinc during seed development and germination in wheat. Physiol Plant 128(1):144–152. https://doi.org/10.1111/j.1399-3054.2006.00737.x
Page V, Weisskopf L, Feller U (2006) Heavy metals in white lupin: uptake, root-to-shoot transfer and redistribution within the plant. New Phytol 171(2):329–341. https://doi.org/10.1111/j.1469-8137.2006.01756.x
Pandey N, Gupta B, Pathak GC (2013) Foliar application of Zn at flowering stage improves plant's performance, yield and yield attributes of black gram. Indian J Exp Biol 51(7):548–555
Pantigoso HA, Newberger D, Vivanco JM (2022) The rhizosphere microbiome: plant–microbial interactions for resource acquisition. J Appl Microbiol 133(5):2864–2876. https://doi.org/10.1111/jam.15686
Parlak M, Parlak AÖ (2011) Effect of soil compaction on root growth and nutrient uptake of forage crops. J Food Agric Environ 9(3–4):275–278
Peer WA, Mahmoudian M, Freeman JL, Lahner B, Richards EL, Reeves RD, Murphy AS, Salt DE (2006) Assessment of plants from the Brassicaceae family as genetic models for the study of nickel and zinc hyperaccumulation. New Phytol 172(2):248–260. https://doi.org/10.1111/j.1469-8137.2006.01820.x
Penner-Hahn JE (2005) Characterization of "spectroscopically quiet" metals in biology. Coord Chem Rev 249(1–2):161–177. https://doi.org/10.1016/J.CCR.2004.03.011
Pich A, Scholz G (1991) Nicotianamine and the distribution of iron into Apoplast and Symplast of tomato (Lycopersicon esculentum mill.). J Exp Bot 187(1):1517–1523. https://doi.org/10.1007/bf00201622
Pineau C, Loubet S, Lefoulon C, Chalies C, Fizames C, Lacombe B, Ferrand M, Loudet O, Berthomieu P, Richard O (2012) Natural variation at the FRD3 MATE transporter locus reveals cross-talk between Fe homeostasis and Zn tolerance in Arabidopsis thaliana. PLoS Genet 8(12):e1003120. https://doi.org/10.1371/journal.pgen.1003120
Prask JA, Plocke DJ (1971) A role for zinc in the structural integrity of the cytoplasmic ribosomes of Euglena gracilis. Plant Physiol 48(2):150–155. https://doi.org/10.1104/pp.48.2.150
Ptashnyk M, Roose T, Jones DL, Kirk GJD (2011) Enhanced zinc uptake by rice through phytosiderophore secretion: a modelling study. Plant Cell Environ 34(12):2038–2046. https://doi.org/10.1111/j.1365-3040.2011.02401.x
Rajkumar M, Freitas H (2008) Influence of metal resistant-plant growth-promoting bacteria on the growth of Ricinus communis in soil contaminated with heavy metals. Chemosphere 71(5):834–842. https://doi.org/10.1016/j.chemosphere.2007.11.038
Ranathunge K, Steudle E, Lafitte R (2005) A new precipitation technique provides evidence for the permeability of Casparian bands to ions in young roots of corn (Zea mays L.) and rice (Oryza sativa L.). Plant Cell Environ 28(11):1450–1462. https://doi.org/10.1111/j.1365-3040.2005.01391.x
Recena R, García-López AM, Delgado A (2021) Zinc uptake by plants as affected by fertilization with Zn sulfate, phosphorus availability, and soil properties. Agronomy 11(2). https://doi.org/10.3390/agronomy11020390
Rehman A, Farooq M, Ozturk L, Asif M, Siddique KHM (2018) Zinc nutrition in wheat-based cropping systems. Plant Soil 422(1–2):283–315. https://doi.org/10.1007/s11104-017-3507-3

Remy E, Cabrito TR, Batista RA, Hussein MAM, Teixeira MC, Athanasiadis A, Sá-Correia I, Duque P (2014) Intron retention in the 5′UTR of the novel ZIF2 transporter enhances translation to promote zinc tolerance in Arabidopsis. PLoS Genet 10(5):15–19. https://doi.org/10.1371/journal.pgen.1004375

Ren Y, Li X, Liang J, Wang S, Wang Z, Chen H, Tang M (2023) Brassinosteroids and gibberellic acid actively regulate the zinc detoxification mechanism of Medicago sativa L. seedlings. BMC Plant Biol 23(1):1–13. https://doi.org/10.1186/s12870-023-04091-4

Ricachenevsky FK, Punshon T, Salt DE, Fett JP, Guerinot ML (2021) Arabidopsis thaliana zinc accumulation in leaf trichomes is correlated with zinc concentration in leaves. Sci Rep 11(1):1–10. https://doi.org/10.1038/s41598-021-84508-y

Rizwan M, Ali S, Ali B, Adrees M, Arshad M, Hussain A et al (2019) Zinc and iron oxide nanoparticles improved the plant growth and reduced the oxidative stress and cadmium concentration in wheat. Chemosphere 214:269–277. https://doi.org/10.1016/j.chemosphere.2018.09.120

Saboor A, Ali MA, Danish S, Ahmed N, Fahad S, Datta R, Ansari MJ, Nasif O, Rahman MH (2021) Effect of arbuscular mycorrhizal fungi on the physiological functioning of maize under zinc-deficient soils. Sci Rep 11(1):1–11. https://doi.org/10.1038/s41598-021-97742-1

Sacristán D, González-Guzmán A, Barrón V, Torrent J, Del Campillo MC (2019) Phosphorus-induced zinc deficiency in wheat pot-grown on noncalcareous and calcareous soils of different properties. Arch Agron Soil Sci 65(2):208–223. https://doi.org/10.1080/03650340.2018.1492714

Sánchez-Rodríguez AR, Rey MD, Nechate-Drif H, Castillejo MÁ, Jorrín-Novo JV, Torrent J, del Campillo MC, Sacristán D (2021) Combining P and Zn fertilization to enhance yield and grain quality in maize grown on Mediterranean soils. Sci Rep 11(1):1–14. https://doi.org/10.1038/s41598-021-86766-2

Sarret G, Saumitou-Laprade P, Bert V, Proux O, Hazemann JL, Traverse A, Marcus MA, Manceau A (2002) Forms of zinc accumulated in the hyperaccumulator Arabidopsis halleri. Plant Physiol 130(4):1815–1826. https://doi.org/10.1104/pp.007799

Sarret G, Willems G, Isaure MP, Marcus MA, Fakra SC, Frérot H, Pairis S, Geoffroy N, Manceau A, Saumitou-Laprade P (2009) Zinc distribution and speciation in Arabidopsis halleri × Arabidopsis lyrata progenies presenting various zinc accumulation capacities. New Phytol 184(3):581–595. https://doi.org/10.1111/j.1469-8137.2009.02996.x

Schneider T, Persson DP, Husted S, Schellenberg M, Gehrig P, Lee Y, Martinoia E, Schjoerring JK, Meyer S (2013) A proteomics approach to investigate the process of Zn hyperaccumulation in Noccaea caerulescens (J and C. Presl) F.K. Meyer. Plant J 73(1):131–142. https://doi.org/10.1111/tpj.12022

Seregin IV, Kozhevnikova AD (2023) Nicotianamine: a key player in metal homeostasis and hyperaccumulation in plants. Int J Mol Sci 24(13). https://doi.org/10.3390/ijms241310822

Shanmugam V, Lo JC, Yeh KC (2013) Control of Zn uptake in Arabidopsis halleri: a balance between Zn and Fe. Front Plant Sci 4(JUL):1–6. https://doi.org/10.3389/fpls.2013.00281

Sharma A, Patni B, Shankhdhar D, Shankhdhar SC (2013) Zinc - an indispensable micronutrient. Physiol Mol Biol Plants 19(1):11–20. https://doi.org/10.1007/s12298-012-0139-1

Shinozaki D, Merkulova EA, Naya L, Horie T, Kanno Y, Seo M, Ohsumi Y, Masclaux-Daubresse C, Yoshimoto K (2020) Autophagy increases zinc bioavailability to avoid light-mediated reactive oxygen species production under zinc deficiency. Plant Physiol 182(3):1284–1296. https://doi.org/10.1104/PP.19.01522

Sinclair SA, Krämer U (2012) The zinc homeostasis network of land plants. Biochim Biophys Acta, Mol Cell Res 1823(9):1553–1567. https://doi.org/10.1016/j.bbamcr.2012.05.016

Singh B, Singh BK (2011) Effect of reduced seed and applied zinc on zinc efficiency of wheat genotypes under zinc deficiency in nutrient solution culture. J Plant Nutr 34(3):449–464. https://doi.org/10.1080/01904167.2011.536884

Singh S, Singh V (2017) Productivity, quality and nutrients uptake of maize (Zea mays) as affected by sources and levels of zinc. Ann Plant Soil Res 19(1):95–99

Singh A, Mishra S, Choudhary M, Chandra P, Rai AK, Yadav RK, Chander Sharma P (2023) Rhizobacteria improve rice zinc nutrition in deficient soils. Rhizosphere 25:100646. https://doi.org/10.1016/J.RHISPH.2022.100646

Sitko K, Rusinowski S, Kalaji HM, Szopiński M, Małkowski E (2017) Photosynthetic efficiency as bioindicator of environmental pressure in A. Halleri. Plant Physiol 175(1):290–302. https://doi.org/10.1104/pp.17.00212

Siyar R, Doulati Ardejani F, Norouzi P, Maghsoudy S, Yavarzadeh M, Taherdangkoo R, Butscher C (2022) Phytoremediation potential of native hyperaccumulator plants growing on heavy metal-contaminated soil of Khatunabad copper smelter and refinery, Iran. Water (Switzerland) 14(22). https://doi.org/10.3390/w14223597

Soares E, Hamid A, Mangkoedihardjo S (2021) Phytoremediation of zinc polluted soil using sunflower (Helianthus annuus l.). J Phytology 13:9–12. https://doi.org/10.25081/jp.2021.v13.6872

Sondergaard TE, Schulz A, Palmgren MG (2004) Energization of transport processes in plants. Roles of the plasma membrane H+-ATPase. Plant Physiol 136(1):2475–2482. https://doi.org/10.1104/pp.104.048231

Song WY, Choi KS, Kim DY, Geisler M, Park J, Vincenzetti V, Schellenberg M, Kim SH, Lim YP, Noh EW, Lee Y, Martinoiaa E (2010) Arabidopsis PCR2 is a zinc exporter involved in both zinc extrusion and long-distance zinc transport. Plant Cell 22(7):2237–2252. https://doi.org/10.1105/tpc.109.070185

Song Z, Li S, Li Y, Zhou X, Liu X, Yang W, Chen R (2024) Identification and characterization of yellow stripe-like genes in maize suggest their roles in the uptake and transport of zinc and iron. BMC Plant Biol 24(1):1–17. https://doi.org/10.1186/s12870-023-04691-0

Sresty TVS, Madhava Rao KV (1999) Ultrastructural alterations in response to zinc and nickel stress in the root cells of pigeon pea. Environ Exp Bot 41(1):3–13. https://doi.org/10.1016/S0098-8472(98)00034-3

Stacey SP, McLaughlin MJ, Çakmak I, Hettiarachchi GM, Scheckel KG, Karkkainen M (2008) Root uptake of lipophilic zinc-rhamnolipid complexes. J Agric Food Chem 56(6):2112–2117. https://doi.org/10.1021/jf0729311

Steudle E (1994) Water transport across roots. Plant Soil 167(1):79–90. https://doi.org/10.1007/BF01587602

Straczek A, Sarret G, Manceau A, Philippe Hinsinger N, Geoffroy BJ (2008) Zinc distribution and speciation in roots of various genotypes of tobacco exposed to Zn. Environ Exp Bot 63:80–90

Sturikova H, Krystofova O, Huska D, Adam V (2018) Zinc, zinc nanoparticles and plants. J Hazard Mater 349:101–110. https://doi.org/10.1016/j.jhazmat.2018.01.040

Suganya A, Saravanan A, Manivannan N (2020) Role of zinc nutrition for increasing zinc availability, uptake, yield, and quality of maize (Zea mays L.) grains: an overview. Commun Soil Sci Plant Anal 51(15):2001–2021. https://doi.org/10.1080/00103624.2020.1820030

Sumiahadi A, Acar R (2018) A review of phytoremediation technology: heavy metals uptake by plants. IOP Conf Ser Earth Environ Sci 142(1):9. https://doi.org/10.1088/1755-1315/142/1/012023

Suzuki M, Takahashi M, Tsukamoto T, Watanabe S, Matsuhashi S, Yazaki J, Kishimoto N, Kikuchi S, Nakanishi H, Mori S, Nishizawa NK (2006) Biosynthesis and secretion of mugineic acid family phytosiderophores in zinc-deficient barley. Plant J 48(1):85–97. https://doi.org/10.1111/j.1365-313X.2006.02853.x

Suzuki M, Tsukamoto T, Inoue H, Watanabe S, Matsuhashi S, Takahashi M, Nakanishi H, Mori S, Nishizawa NK (2008) Deoxymugineic acid increases Zn translocation in Zn-deficient rice plants. Plant Mol Biol 66(6):609–617. https://doi.org/10.1007/s11103-008-9292-x

Takagi SI (1976) Naturally occurring iron. Chelating compounds in oat- and rice·rootwashings 1. Activity measurement and preliminary characterization. Soil Sci Plant Nutr 22(4):423–433. https://doi.org/10.1080/00380768.1976.10433004

Takahashi M, Terada Y, Nakai I, Nakanishi H, Yoshimura E, Mori S, Nishizawa NK (2003) Role of nicotianamine in the intracellular delivery of metals and plant reproductive development. Plant Cell 15(6):1263–1280. https://doi.org/10.1105/tpc.010256

Tian SK, Lu LL, Yang XE, Labavitch JM, Huang YY, Brown P (2009) Stem and leaf sequestration of zinc at the cellular level in the hyperaccumulator sedum alfredii. New Phytol 182(1):116–126. https://doi.org/10.1111/j.1469-8137.2008.02740.x

Tsednee M, Huang YC, Chen YR, Yeh KC (2016) Identification of metal species by ESI-MS/MS through release of free metals from the corresponding metal-ligand complexes. Sci Rep 6(May):1–13. https://doi.org/10.1038/srep26785

Tuiwong P, Lordkaew S, Prom-U-thai C (2021) Improving grain zinc concentration in wetland and upland rice varieties grown under waterlogged and well-drained soils by applying zinc fertilizer. Agronomy 11(3). https://doi.org/10.3390/agronomy11030554

Upadhayay VK, Singh AV, Khan A (2021) Cross talk between zinc-solubilizing bacteria and plants: a short tale of bacterial-assisted zinc biofortification. Front Soil Sci 1(January):1–16. https://doi.org/10.3389/fsoil.2021.788170

Vaillant N, Monnet F, Hitmi A, Sallanon H, Coudret A (2005) Comparative study of responses in four Datura species to a zinc stress. Chemosphere 59(7):1005–1013. https://doi.org/10.1016/J.CHEMOSPHERE.2004.11.030

Van De Mortel JE, Villanueva LA, Schat H, Kwekkeboom J, Coughlan S, Moerland PD, Van Themaat EVL, Koornneef M, Aarts MGM (2006) Large expression differences in genes for iron and zinc homeostasis, stress response, and lignin biosynthesis distinguish roots of Arabidopsis thaliana and the related metal hyperaccumulator Thlaspi caerulescens. Plant Physiol 142(3):1127–1147. https://doi.org/10.1104/pp.106.082073

Verbruggen N, Hermans C, Schat H (2009) Molecular mechanisms of heavy metal hyperaccumulation in plants. In: Phytoremediation of environmental pollutants, pp 99–116. https://doi.org/10.4324/9781315161549

Wang P, Menzies NW, Lombi E, McKenna BA, de Jonge MD, Donner E, Blamey FPC, Ryan CG, Paterson DJ, Howard DL, James SA, Kopittke PM (2013) Quantitative determination of metal and metalloid spatial distribution in hydrated and fresh roots of cowpea using synchrotron-based X-ray fluorescence microscopy. Sci Total Environ 463–464:131–139. https://doi.org/10.1016/j.scitotenv.2013.05.091

Wang Y, Kang Y, Yu W, Lyi SM, Choi HW, Xiao E, Li L, Klessig DF, Liu J (2022) AtTIP2;2 facilitates resistance to zinc toxicity via promoting zinc immobilization in the root and limiting root-to-shoot zinc translocation in Arabidopsis thaliana. Ecotoxicol Environ Saf 233(November 2021):113333. https://doi.org/10.1016/j.ecoenv.2022.113333

Weiss D, Northover G, Hanif M, García-España E, Vilar R, Arnold T, Markovic T, Wissuwa M, Delgado E (2021) Isotope fractionation of zinc in the paddy rice soil-water environment and the role of 2'deoxymugineic acid (DMA) as zincophore under Zn limiting conditions. Chem Geol 577(October 2020):120271. https://doi.org/10.1016/j.chemgeo.2021.120271

Welch RM, Shuman L (1995) Micronutrient nutrition of plants. Crit Rev Plant Sci 14(1):49–82

Weremczuk A, Ruszczyńska A, Bulska E, Antosiewicz DM (2017) NO-dependent programmed cell death is involved in the formation of Zn-related lesions in tobacco leaves. Metallomics 9(7):924–935. https://doi.org/10.1039/c7mt00076f

White PJ, Whiting SN, Alan JM, Baker MRB (2002) Does zinc move apoplastically to the xylem in roots of Thlaspi caerulescens? New Phytol 153(2):199–201. https://doi.org/10.1046/j.0028-646X.2001.00339.x

Widodo JA, Broadley MR, Rose T, Frei M, Pariasca-Tanaka J, Yoshihashi T, Thomson M, Hammond JP, Aprile A, Close TJ, Ismail AM, Wissuwa M (2010) Response to zinc deficiency of two rice lines with contrasting tolerance is determined by root growth maintenance and organic acid exudation rates, and not by zinc-transporter activity. New Phytol 186(2):400–414. https://doi.org/10.1111/j.1469-8137.2009.03177.x

Willems G, Dräger DB, Courbot M, Godé C, Verbruggen N, Saumitou-Laprade P (2007) The genetic basis of zinc tolerance in the metallophyte Arabidopsis halleri ssp. halleri (Brassicaceae): an analysis of quantitative trait loci. Genetics 176(1):659–674. https://doi.org/10.1534/genetics.106.064485

Wu W, Wu P, Yang F, Sun D et al (2018) Assessment of heavy metal pollution and human health risks in urban soils around an electronics manufacturing facility. Sci Total Environ 630:53–61. https://doi.org/10.1016/j.scitotenv.2018.02.183

Wyszkowska J, Borowik A, Kucharski M, Kucharski J (2013) Effect of cadmium, copper and zinc on plants, soil microorganisms and soil enzymes. J Elem 18(4):769–796. https://doi.org/10.5601/jelem.2013.18.4.455

Wyszkowska J, Boros-Lajszner E, Borowik A, Baćmaga M, Kucharski J, Tomkiel M (2016) Implication of zinc excess on soil health. J Environ Sci Health Part B Pestic Food Contam Agric Wastes 51(5):261–270. https://doi.org/10.1080/10934529.2015.1128726

Xuan Y, Scheuermann EB, Meda AR, Hayen H, von Wirén N, Weber G (2006) Separation and identification of phytosiderophores and their metal complexes in plants by zwitterionic hydrophilic interaction liquid chromatography coupled to electrospray ionization mass spectrometry. J Chromatogr A 1136(1):73–81. https://doi.org/10.1016/j.chroma.2006.09.060

Xue Y, Yan W, Gao Y, Zhang H, Jiang L, Qian X, Cui Z, Zhang C, Liu S, Wang H, Li Z, Liu K (2021) Interaction effects of nitrogen rates and forms combined with and without zinc supply on plant growth and nutrient uptake in maize seedlings. Front Plant Sci 12(December):1–12. https://doi.org/10.3389/fpls.2021.722752

Yang Q, Shohag MJI, Feng Y, He Z, Yang X (2017) Transcriptome comparison reveals the adaptive evolution of two contrasting ecotypes of Zn/cd hyperaccumulator sedum alfredii Hance. Front Plant Sci 8(April):1–12. https://doi.org/10.3389/fpls.2017.00425

Yang Q, Ma X, Luo S, Gao J, Yang X, Feng Y (2018) SaZIP4, an uptake transporter of Zn/Cd hyperaccumulator sedum alfredii Hance. Environ Exp Bot 155(June):107–117. https://doi.org/10.1016/j.envexpbot.2018.06.021

Yang Y, Li Y, Chen W, Wang M, Wang T, Dai Y (2020) Dynamic interactions between soil cadmium and zinc affect cadmium phytoavailability to rice and wheat: regional investigation and risk modeling. Environ Pollut 267:115613. https://doi.org/10.1016/j.envpol.2020.115613

Yoneyama T, Ishikawa S, Fujimaki S (2015) Route and regulation of zinc, cadmium, and iron transport in rice plants (Oryza sativa L.) during vegetative growth and grain filling: metal transporters, metal speciation, grain Cd reduction and Zn and Fe biofortification. Int J Mol Sci 16(8):19111–19129. https://doi.org/10.3390/ijms160819111

Yoshida S, Tanaka A (1969) Zinc deficiency of the rice plant in calcareous soils. Soil Sci Plant Nutr 15(2):75–80. https://doi.org/10.1080/00380768.1969.10432783

Zalewska M (2012) Response of perennial ryegrass (lolium perenne L.) to soil contamination with zinc. J Elem 17(2):329–343. https://doi.org/10.5601/jelem.2012.17.2.14

Zhao FJ, Jiang RF, Dunham SJ, McGrath SP (2006) Cadmium uptake, translocation and tolerance in the hyperaccumulator Arabidopsis halleri. New Phytol 172(4):646–654. https://doi.org/10.1111/j.1469-8137.2006.01867.x

Zhao K, Liu X, Xu J, Selim HM (2010) Heavy metal contaminations in a soil-rice system: identification of spatial dependence in relation to soil properties of paddy fields. J Hazard Mater 181(1–3):778–787. https://doi.org/10.1016/j.jhazmat.2010.05.081

Zhong X, Chen Z, Li Y, Ding K, Liu W, Liu Y, Yuan Y, Zhang M, Baker AJM, Yang W, Fei Y, Wang Y, Chao Y, Qiu R (2020) Factors influencing heavy metal availability and risk assessment of soils at typical metal mines in eastern China. J Hazard Mater 400(June 2020):123289. https://doi.org/10.1016/j.jhazmat.2020.123289

Zhou G, An Q, Liu Z, Wan Y, Bao W (2023) Systematic analysis of NRAMP family genes in Areca catechu and its response to Zn/Fe deficiency stress. Int J Mol Sci 24(8). https://doi.org/10.3390/ijms24087383

Physiological, Biochemical, and Molecular Plant Responses to Zinc Deficiency and Excess

Harmanjit Kaur, Km Sandhya Devi, Bhawna Sunkaria, and Harpreet Kaur

Abstract

Zinc (Zn) is the next most widespread transition heavy metal following iron (Fe) and a necessary micronutrient implicated in diverse physiological and biochemical processes in plants. However, Zn accessibility to plants is hindered by its immobile behavior and hostile soil conditions, leading to its deficit despite being high quantities present in soil. Nowadays, Zn insufficiency is the most prevalent micronutrient scarcity in plants, humans, and animals, with 25–30% of global population being affected by it. Shortage of Zn leads to loss in yield and in severe cases even death of plant. Nevertheless, simultaneously, problem of Zn toxicity exists in few areas largely because of environmental contamination derived from farming and industrial activities. Excess Zn causes initiation of oxidative burst and obstructs several plant metabolic processes that cause cellular alterations. Physiological, biochemical, and molecular mechanisms that permit plants to flourish in Zn deficit and polluted soils are inadequately explored and comprehend. Thus, this chapter intends to review and discuss the various features of different morpho-physiological, biochemical, and molecular responses in plants raised under Zn deficit as well as toxic conditions to understand how plants respond differentially in two extreme Zn conditions.

H. Kaur (✉) · K. S. Devi
Department of Botany, University of Allahabad, Prayagraj, Uttar Pradesh, India
e-mail: harmanjitkaur@allduniv.ac.in

B. Sunkaria
Yadavindra Department of Science, Punjabi University Guru Kashi Campus, Talwandi Sabo, Bathinda, Punjab, India

H. Kaur
Department of Agriculture, Jain College of Pharmacy, Abohar, Punjab, India

S. Srivastava, H. Kaur (eds.), *Zinc in Soil-Plant Continuum*,
https://doi.org/10.1007/978-981-96-4253-3_4

Keywords

Photosynthesis · Zinc toxicity · Nitrogen fixation · Phytohormones · Protein synthesis · Zinc deficiency

1 Introduction

Zinc (Zn), belonging to the set of vital elements, is essential in trace amounts to living beings for natural metabolic activities (Sturikova et al. 2018). It is an important micronutrient for plants whose essentiality was initially reported in maize (Mazé 1915) and then in dwarf sunflower and barley (Sommer and Lipman 1926). Conversely, Zn can turn out to be lethal to plants beyond supra-optimum levels. Therefore, supply of an optimal amount of Zn to plants is imperative in order to sustain regular metabolic processes exclusive of Zn scarcity or excess (Sturikova et al. 2018). Zinc is largely found as divalent Zn^{2+} ions in plants and soils. For optimal growth, crops require a moderate or adequate concentration of Zn that ranges between 20 and 60 mg/kg dry mass (de Almeida et al. 2020). According to Kabata-Pendias (2011), plant homeostatic levels of Zn range from 30 to 200 mg/kg dry weight (DW), with 25–150 mg/kg being adequate, whereas Zn more than 400 mg/kg DW decreases yield.

In general, Zn participates in several important functions in plants like enzymatic activation, diverse metabolic and physiological processes, and ionic balance (Noman et al. 2019; Yang et al. 2020; Zaheer et al. 2022). Bio-physicochemical functions of Zn in plants comprise gene induction and regulation, protein synthesis, carbohydrate and phosphate metabolisms, an important factor for maintaining structural integrity of ribosomes, and configurational and functional participation in membranes (Noman et al. 2019; Sturikova et al. 2018). Additionally, Zn is a component of metalloenzymes or acts as a cofactor for over 300 enzymes, for example, hydrolases, oxidoreductases, lyases, transferases, ligases, isomerases, anhydrases, dehydrogenases, oxidases, and peroxidases (Gautam et al. 2017). Among these metalloenzymes some are involved in DNA, RNA, and protein synthesis, while few are involved in energy metabolism (Castillo-González et al. 2018). Zinc is also necessary for chlorophyll biosynthesis, functioning of pollen grains, germination, and fertilization (Panda 2017). It plays a vital role in regulation of nitrogen metabolism, photosynthesis, cell proliferation, and synthesis of auxins in plants (Kaur and Garg 2021).

Paucity of Zn is the most abundant trace nutrient deficiency in many agro-climatic zones worldwide (Humayan Kabir et al. 2014). Zinc insufficiency is a grave issue in farming soils across the globe since the total Zn present in soils is found in minerals coupled with absorption by other soil fractions, resulting in limited amount of Zn available for plant uptake (Ali et al. 2022a, b). Zinc has been reported to be deficit in about 75% of cultivated soils of sub-Sahara Africa; 50% of farming soils in Turkey and India, respectively; 45% of agricultural soils in Western Australia; and 33% of soils in China (Alloway 2009; Hacisalihoglu 2020). Indian soils are deficient in Zn (<0.06 mg/kg), and among states, Zn deficiency in soils varies, spanning from 2% in Delhi to 67% in Bihar (SHC 2023). Four states exhibited above 50% Zn shortage in soils: Karnataka having 62%, Maharashtra with 53%, Bihar containing 51%, and Rajasthan 48% (SHC 2023). Shukla et al. (2021) examined over 240,000 soils across India from 2012 to 2018 under AICRP-MSPE program and reported 39% Zn-deficit soils in Uttar Pradesh. Since Zn plays an important role in several cellular processes, its deficiency can hinder key plant metabolic processes implicated in regular functioning and even detoxification mechanisms. Therefore, plants growing in Zn-deficit environment display growth retardation and yield losses (Xie et al. 2019; Zhao and Wu 2017).

On the other hand, Zn toxicity is exhibited by crop plants growing in regions contaminated with Zn-affluent leftover of agricultural practices and industries, thereby affecting several plant bio-physicochemical processes (Bankaji et al. 2019; Sidhu et al. 2020). Surplus Zn can impair plants by negatively affecting growth and well-being, such as seedling emergence, root and stem length, disturbed ionic homeostasis, alteration in enzymatic activities, and cytostructural disorganization (Garg and Singh 2018; Bankaji et al. 2019; Blasco et al. 2019; Fatemi et al.,2020; Goodarzi et al. 2020; Šimon et al. 2021). Zn imbalances can also cause oxidative burst resulting in ROS accretion (Sagardoy et al. 2009: Tsonev and Cebola Lidon 2012). With this backdrop, this chapter aims to highlight various physiological (including morphological), biochemical, and molecular responses in planta exposed to Zn deficiency and excess in order to understand how plants behave differently under this two extreme nutrient conditions.

2 Physiological Plant Responses

2.1 Zn-Deficit Conditions

In general, lack of Zn disturbs the vital processes of plants like transpiration, photosynthesis, and many other critical physiological processes, resulting in growth impairment and chlorosis of leaves (Hamzah Saleem et al. 2022) (Fig. 1). Zinc deficiency also disturbs the nutrient uptake since it prevents nutrients (P, N, K, Mg, Fe, Ca, etc.) from being absorbed and being distributed to various plant parts (Sagardoy et al. 2009; Garg and Singh 2018), eventually contributing to Zn deficiency in human diet. Additionally, Zn deficiency disrupts the metabolic equilibrium of ROS and causes oxidative burst, thereby decreasing crop productivity (Zhang et al. 2023). Khanif and Saleem (2013) reported Zn deficit symptoms in rice, for example, loss of chlorophyll in mid vein at the bottom of the youngest leaf in 24 weeks following sowing of seeds or transplanting, manifestation of brown-colored spots on mature leaves which enlarged, coalesced, and gave the leaves a brown color. The authors further suggested that under mild Zn deficiency, rice plants may recuperate after 4–6 weeks; however ripeness is slowed down, and produce particularly of sensitive cultivars is diminished. They further reported turgidity loss, where plants fell and floated on the water surface. The basal foliage became pale

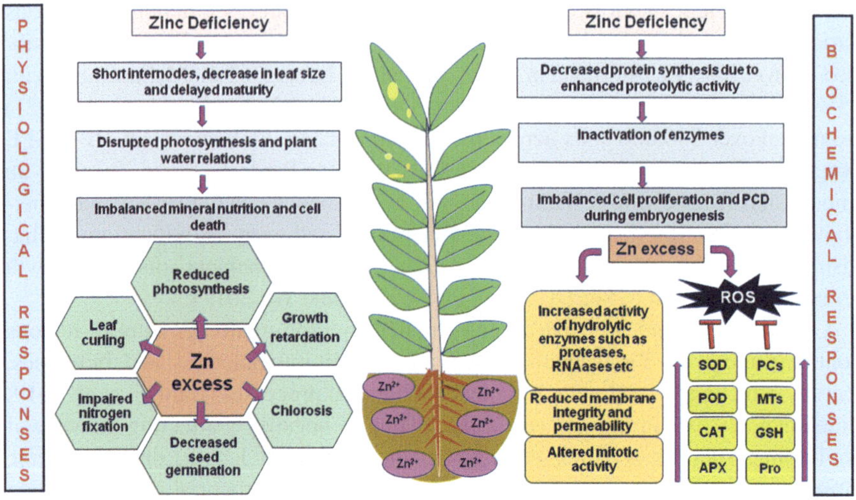

Fig. 1 Physiological and biochemical responses of plants under Zn deficiency and excess. *APX* ascorbate peroxidase, *CAT* catalase, *GSH* glutathione, *MT* metallothionien, *POD* peroxidase, *PC* phytochelatins, *PCD* programmed cell death, *Pro* proline, *ROS* reactive oxygen species, *SOD* superoxide dismutase

green, and subsequent to 3–7 days, the leaves became chlorotic (Khanif and Saleem 2013). In India 50% of arable soils in which groundnut is raised display Zn-deficit conditions, causing substantial yield losses (Singh et al. 2004). Typical symptoms of Zn deficiency reported in maize plants included chlorotic streaks and purple tinges on the periphery and leaf sheath (Mattiello et al. 2015). Further, there were considerable reductions in stomatal conductance, net photosynthesis, maximum efficiency of PS-I (estimated by Fv/Fm), biomass (dry mass), and lessened vascular bundle ratio along with greater stomatal density under Zn deficiency conditions (Mattiello et al. 2015). In mung bean, the visible symptoms of Zn shortage reported were small internodes, reduction in leaf dimension, and slowed maturity (Samreen et al. 2017). Acute Zn-deficit signs have been reported in several crop species. For instance, acute Zn paucity causes necrosis of root tip ("dieback"), whereas sublethal Zn shortage produces interveinal chlorosis (known as leaf mottling), formation of bronze or reddish-brown traces (bronzing), and varied auxin deficiency-mediated responses like internode reduction (known as rosetting), downward and inward twisting of leaf lamina, and decrease in leaf size (Broadley et al. 2007). Hajiboland and Amirazad (2010) studied the impact of Zn insufficiency in *Brassica oleracea* L. var. capitata f. rubra plants raised in nutrient solution under regulated conditions and observed that Zn deficiency negatively affected the number, dry mass, and surface area of leaves more than root dry weight. Further, chlorophyll luminance analysis revealed photoinhibition followed by decreased stomatal conductance and diminished CO_2 availability at carbonation sites although chloroplast was not critically harmed under Zn-starved conditions. Chlorophyll a and a/b ratio declined, but anthocyanins were build up under Zn paucity (Hajiboland and Amirazad 2010). Similarly, due to Zn shortage in the soil, tomato genotypes (e.g., Marglobe and Ratan) exhibited significant yield losses (Akther et al. 2020). Zn starvation negatively affected plant dry mass and photosynthetic attributes such as quantum efficiency of photosystems, chlorophyll levels, and photosynthetic performance index in Zn-deficit susceptible genotype (Ratan). In addition, Zn deficiency resulted in enhanced electrolyte outflow, cell lethality percentage, and hydrogen peroxide levels in Ratan genotype, while all these parameters were not noticeable in marglobe (Zn-deficit resistant genotype) (Akther et al. 2020). Zinc shortage has been reported to suppress the induction of FLOWERING LOCUS T (FT), a key gene regulating flowering in *Arabidopsis*. This suppression or loss of FT led to an increase in rosette width by retarding the transformation to flowering (Chen and Ludewig 2018). Further, Zn deficiency disrupts the development of male and female sex organs and process of pollination, leading to a significant decline in plant productivity. This reduction could be linked to decreased synthesis of indole acetic acid (IAA). Additionally, the diminished seed formation in response to Zn shortage might result from increased abscisic acid levels, which can cause early floral bud and leaf abscission, or from impairments in anther and pollen growth and function (Alloway 2004). Zinc deficiency leads to a reduction in protein synthesis (Marschner 1995) as a result of RNA disintegration (Cakmak et al. 1989), diminished RNA polymerase activity, distorted ribosomal structure, and a reduction in ribosome number (Sadeghzadeh 2013). Low carbonic anhydrase (CA) activity under Zn deficiency in plants leads to a minimal conversion of bicarbonate (HCO_3^-) into carbon dioxide (CO_2) and water (H_2O) (Hopkinson et al. 2011: Tavallali et al. 2009). Thus, Zn deficiency reduces the plant's ability to utilize HCO_3^- and decreases photosystem II (PS II) activity because of accumulation of surplus HCO_3^-. Additionally, shortage of Zn inhibits other Zn-dependent enzymes, like glutamate dehydrogenase and alcohol dehydrogenase, further hindering growth of plants (Hafeez et al. 2013). Sekimoto et al. (1997) in their investigation observed reduction in GA levels in Zn-deficit maize plants and confirmed its direct impact on biomass loss. In terms of plant growth under Zn deficiency, phytohormones such as gibberellins (GAs), IAA, and cytokinins (CKs) affected the growth of *Lactuca sativa* in a negative manner, while in *Brassica oleracea* these growth regulators sustained plant development as they increased or maintained their levels, suggesting the hormonal response under Zn-deficit conditions to be plant specific (Navarro-León et al. 2016). Zinc-deficit foliage of mandarin orange plants exhibited deformity in vacuolar ultrastructure and minute stomatal size along with decrease in number of stomata (Subba et al. 2014). Similarly, nitrogen and carbohydrate metabolites reduced in foliage of tea in response to Zn deficiency (Zhang et al. 2017). Senescence related to maturity is indicated by enhanced lipid peroxidation and ROS linked to decreased antioxidative potential (Vanacker 2006). Premature senescence has been reported in plants under Zn-deficit conditions (Talukdar and Aarts 2007). Similarly, barley plants raised under Zn shortage revealed elevated AsA/DHA than controls (Paradisone et al. 2022).

2.2 Zn Excess Conditions

Though low levels of Zn are essential for physiology of crops, its high amounts similar to other heavy metals cause alterations in physio-biochemical processes responsible for healthy well-being of plants, leading to toxicity (Antoniadis et al. 2017; Bankaji et al. 2019; Wolf and Paulino 2020; Aghajanzadeh et al. 2020). Surplus Zn impacts the functioning of amylases and the movement of sugars toward embryo axis, leading to a reduced percentage of germinated seeds, as has been reported in *Cicer arietinum* (Sharma et al. 2009). Zinc excess incites paucity of other essential nutrients because of related ionic radii, thereby hampering their uptake and distribution within the plant system (Bankaji et al. 2019). For example, excess Zn has been reported to cause distinct Fe-deficit symptoms in plants grown under low Fe levels. Further, Zn and Fe interaction has been found to affect Cu and Mn uptake in soybean plants (de Oliveira et al. 2023). On account of their chemical resemblance, Zn and Fe exhibit antagonistic relationship, contending for uptake and translocation in several plant species (Fan et al. 2021; Rai et al. 2021; de Oliveira et al. 2023). Surplus Zn also results in the accretion of harmful heavy metals such as lead (Pb) and cadmium (Cd) in plants, which is detrimental for plant well-being and in turn affects health of humans (Stanton et al. 2022). Excess level of Zn has been reported to cause various alterations like reduced growth, decreased structural integrity, and photosynthetic and respiratory rate and provoke oxidative injury through increased production of ROS in several plants (De Oliveira and Tibbett 2018; Mateos-Naranjo et al. 2018; Goodarzi et al. 2020; Kaur and Garg 2021; Li et al. 2022; Natasha et al. 2022) (Fig. 1). At morphological level, surplus Zn results in slow growth, curling and chlorosis of leaves, and leaf tip necrosis (AbdElgawad et al. 2020). Surplus Zn has an effect on foliage with regard to area and number of leaves which can be due to either reduced cell division or elongation or both (Glinska et al. 2016). Cell elongation may be impaired due to reduced cell wall elasticity or decreased turgor pressure (Liščáková et al. 2022). In sweet potato, excess Zn inhibited the plant growth through imbalanced mineral nutrition, damaging the cellular organization and decreasing photosynthesis (Meng et al. 2023). In rice plants, Zn toxicity has been reported to cause drooping of leaves because of loss of turgidity, chlorophyll loss at the base of leaves, impeded plant development, "bronzing" of leaves, and sometimes collapsing of rice seedlings (Khanif and Saleem 2013). In terms of nitrogen metabolism, Zn toxicity has been documented to decrease nitrate (NO_3) absorption and transport along with total nitrogen levels, thus leading to disturbance in nitrogen metabolism, for instance, in *Brassica* species (Blasco et al. 2019). Furthermore, too much Zn either causes disruption in the chloroplasts or a restricted amount of NO_3 at the location where nitrate reductase (NR) is synthesized or even both, which in turn hinders the enzyme's function (Sidhu 2016). Surplus Zn also impacts the process of nodulation, i.e., number and weight of nodules along with nitrogen fixing potential (leghemoglobin levels, nitrogenase activity) in many legume species (Souza et al. 2020). Perhaps Zn contamination causes damage to symbiotic plasmids and thus affects the ability of bacteria to nodulate legume plants, as the genes responsible for nodule formation and nitrogen fixation are localized on these plasmids (Wang et al. 2018). In case of plant-water relations, Zn toxicity alters water relations or reduces the relative water content (RWC) in tissues (Garg and Singh 2018). A reduction in water level indicates loss of turgidity, which sequentially slows down the process of cell extension and eventually prevents the growth and generation of biomass. Enhanced osmotic potential in the neighboring medium and reduced water absorption are the main causes of the drop in water potential in leaves after elevated Zn levels (Mir et al. 2015). At higher Zn levels, decreased stomatal conductance was related with decline in root hydraulic activity and photosynthetic rate in *Brassica rapa* (Fatemi et al. 2020). Membranes of thylakoid inflate up as well as complexes are spaced out as a result of the charge shielding impact of Zn ions, which lowers the efficiency of energy passing on. Furthermore, under high Zn concentration, the stomatal size and density were found to be smaller in *Beta vulgaris* (Sagardoy et al. 2010). In an alternative study, Sagardoy et al. (2009) reported smaller-sized stomata with more rounded shape in highly Zn-stressed plants than those present in controls. As per Paunov et al. (2018), negative impacts of Zn on photosynthetic process in *Triticum durum* could be because of decline in electron transfer flux from QA to QB at acceptor region of PSII which in turn lowers the flow of electrons from PSII to PSI. Correspondingly, Szopinski et al. (2019) witnessed a decrease in electron transport rate and proportion of active reaction centers in *A. halleri* along with *A. arenosa* in response to Cd and Zn toxicity. Distinct effects on PSII photochemistry related with antagonist behavior of Zn with Mn during photolysis of water, hindering the electron flux and release of O_2, have been reported in *Polypogon monspeliensis* (Ouni et al. 2016). An excess of electrons from oversupplied electron transport chain to O_2 may follow the displacement of Mg^{2+} from Rubisco, which could lead to an underutilization of NADPH as well as ATP in CO_2 fixation. Additionally, plants such as *Populus deltoides* and *Sedum alfredii* experience the disintegration of their chloroplast membrane, diminution in grana and thylakoids, increase in starch grains and plastoglobuli, and disarray of their chloroplasts due to excess concentrations of Zn (Stoláriková-Vaculíková et al. 2015). Surplus Zn affects the uptake of Mg^{2+} ions, which participates in synthesis of chlorophyll, and hence decreased chlorophyll content is a key indicator of Zn phytotoxicity (Zhang et al. 2020). Reduced chlorophyll under excess Zn results in lesser seed production, ultimately causing chlorosis and even plant death (Adele et al. 2021). Under Zn toxicity, it has also been reported that ratio of chlorophyll a/ b decreases, suggesting that chlorophyll a is more sensitive to Zn toxicity compared to chlorophyll b (Garg and Singh 2018). This might be owing to enhancement in PSII light harvesting complexes (LHC II) than reaction centers, given that LHC II contains more chlorophyll b compared to chlorophyll-associated proteins of PSII. The negative relationship between excess Zn and chlorophyll could be attributed to decline in chlorophyll biosynthetic enzymes such as protochlorophyllide reductase and δ-aminolevulinic acid dehydratase (ALA-D) through Zn obstructing the functional sulfhydryl region, resulting in lesser synthesis of 5-aminolevulinic acid (ALA) or by enhanced activity of chlorophyll disintegrating enzymes like chlorophyllase or both, as was demonstrated in *P. monspeliensis* by Ouni et al. (2016). Surplus concentration of Zn has been reported to reduce the activity of enzymes involved in dark reactions of photosynthesis (Kaznina and Titov 2017). Surplus Zn causes injury to the mitochondria (e.g., tea plant) and results in reduced levels of nicotinamide, hence hastening NAD^+ degradation and lessening the energy metabolism (Zhang et al. 2017). It has been demonstrated that too much Zn leads to negative geotropism in tomato roots and inhibits auxin activity (Kosesakal et al. 2009). Excess concentrations of Zn can lead to a slowdown in cell division by prolonging the length of mitotic stages and whole mitotic process itself (Mahmoudi et al. 2021).

3 Biochemical Plant Responses

3.1 Zinc-Deficit Conditions

In general, Zn deficiency decreases protein synthesis in plant cells and affects enzymes, cell membrane, membrane proteins, and chlorophyll leading to chlorosis in leaves, ultimately inhibiting photosynthesis and plant growth (Cakmak 2000) (Fig. 1). Additionally, Zn paucity could result in enhanced proteolytic activity, as has been reported in *Picea* embryos (Helmersson et al. 2008). Disruption of Zn homeostasis disturbs the balance between programmed cell death (PCD) and cell proliferation which is essential for embryogenesis in plants (Fig. 1). Thus Zn has been identified as a key factor regulating plant cell fate decisions (Helmersson et al. 2008). It has been observed that Zn deficiency negatively affects plant metabolism via alteration in cellular redox equilibrium and production of oxidative stress. For example, in *Mulberry* plants the CAT activity increased as a result of Zn toxicity (Tewari et al. 2008). Contrarily, Kaznina et al. (2021) observed enhancement in mRNA levels of *HvPRX07* gene, but decrease in peroxidase (POX) and superoxide dismutase (SOD) activities under Zn-deficit conditions than those at favorable Zn content in leaves of *Hordeum vulgare* L. cv. Nur. Besides antioxidant enzymes, plants synthesize compatible solutes like proline, which acts as a metal chelator and osmolyte, neutralize ROS, and impart protection against ROS-triggered cellular damage in plants. In this context, Zn deficiency resulted in enhanced synthesis of proline to counteract the increased generation of superoxide radical ($_1O^2$), MDA, and H_2O_2 levels in *Citrus reticulata Blanco* seedlings (Subba et al. 2014). Similarly, under Zn deficiency the proline pool was found to increase in *Phaseolus vulgaris* (Michael and Krishnasway 2011).

3.2 Zn Excess Conditions

Zn stress induces catabolic activities by promoting activity of hydrolytic enzymes like protease and RNase which reduce the content of proteins and RNA in plants (Sidhu 2016) (Fig. 1). Further, Zn toxicity induces senescence in plants by increasing protein degradation, which is believed to be linked to the attachment of Zn to sulfhydryl groups present in proteins, leading to malformed structure of proteins

(Singh et al. 2018). Few studies have revealed that elevated ROS levels impact protein functions, either through the oxidation of side chain containing amino acids (lysine, histidine, and cysteine) or by reactions with aldehydic products from glycosylation or lipid peroxidation leading to the formation of carbonyl clusters in protein molecules (Madian and Regnier 2010). Such alterations in proteins can result in chemical fragmentation or protein aggregation owing to covalent cross- linking reactions as well as increasing surface hydrophobicity, thereby causing loss in protein functionality (Madian and Regnier 2010). Additionally, these proteins are targeted in a selective manner by proteases or other catabolic enzymes activated by excess Zn. In lettuce plants both cysteine protease and cystatin accumulate at high Zn level exposure (Lucini and Bernardo 2015). In contrast Siddiqui and Bano (2019) observed a reduction in the inhibitory action of phytocystatin upon Zn attachment attributing this to the failure of protein function due to conformational alterations, which negatively impacted the growth of garlic plants. Similarly, growth of *Camellia sinensis* was found to reduce in response to surplus Zn (Mukhopadhyay et al. 2013). Excess Zn has also been demonstrated to modify the mitotic cycle (Rout and Das 2003), affects membrane structure and permeability (Stoyanova and Doncheva 2002), and could even be lethal for cells (Wang et al. 2009a). Like Zn deficiency, Zn excess also can trigger oxidative stress in plants through enhanced production of reactive oxygen species (ROS) such as hydroxyl radicals (·OH), superoxide anion ($O^{\cdot-}_2$), and hydrogen peroxide (H_2O_2) (Sagardoy et al. 2009; Tsonev and Lidon 2012) and reactive nitrogen species (RNS) (Feigl et al. 2015). For instance, elevated ROS levels and oxidative burst hindered the growth of *Jatropha curcas* through interference with normal activities of cell (Luo et al. 2010). Coffee plants when exposed to varied Zn doses exhibited change in cellular homeostasis and elevated ROS generation (Li et al. 2013). Nitric oxide (NO) production was found to increase in underground parts of Zn-stressed *Brassica juncea* in dose-dependent fashion (Feigl et al. 2015). In the root tips of *Solanum nigrum*, Zn-induced Fe (iron) deficiency was linked to enhanced NO production (Xu et al. 2010). Nitrate reductase is the primary mode of NO production (enzymatic) in roots, yet its enzymatic activity remained unaffected by excess Zn levels in *Brassica roots (*Bartha et al. 2005). Oxidative burst triggered by high doses of Zn promotes lipid peroxidation, a process in which free radicals attack membrane lipids, resulting in generation of malondialdehyde (MDA) (Goodarzi et al. 2020). High concentration of ROS induced by excess Zn damages cells, destabilizes membranes, and also damages DNA (Apel and Hirt 2004). Zinc causes oxidative damage by reacting with crucial functional groups in biomolecules, as it strongly binds with oxygen, nitrogen, or sulfur atoms. As a result, the sulfur in the cysteine residues of enzymes is replaced by Zn, leading to enzyme deactivation (Nieboer and Richardson 1980). It also replaces other ions at their catalytic sites (Schutzendubel and Polle 2002). Surplus Zn can considerably change the nature of many genes involved in several metabolic processes, among which various genes encoding proteins are related to defense against oxidative burst (Van de Mortel et al. 2006). Many investigations have demonstrated excess Zn-mediated increase in antioxidant enzymatic activities (SOD, APX, CAT, POD, and GST) along with antioxidant metabolites [glutathione GSH) and ascorbate] in plants (Madhava Rao and Sresty 2000; Cuypers et al. 2002; Tewari et al. 2008). The higher activity of antioxidant enzymes is because of their role in the neutralization of excess ROS produced by Zn toxicity, aimed at regulating cellular homeostasis. In *Hydrilla verticillata* the high Zn concentration increased GST activity (Wang et al. 2009b). Barrameda-Medina et al. (2014) demonstrated that in *Brassica oleracea* and *Lactuca sativa*, the non-enzymatic and enzymatic defense systems alone were insufficient to confer strong tolerance to excess Zn-mediated oxidative stress. In these species, the accretion of phytochelatins (PCs) played a critical role in survival rather than enhancing tolerance against Zn toxicity. *B. oleracea* and *L. sativa* relied more on the detoxifying action of PCs than on conventional antioxidant pathways to manage Zn excess and improve plant growth. Wei et al. (2022) reported that the ascorbate-glutathione (AsA-GSH) cycle, along with non-protein thiols (NPTs) and phytochelatins (PCs), played pivotal roles in mitigating Zn toxicity in wheat seedlings. In general, GSH, PCs, and metallothioneins (MTs) bind with Zn and form chelates that reduce the harmful impacts of Zn in plants (Grennan 2011; Pochodylo, and Aristilde 2017). *Dunaliella tertiolecta* exposed to Zn excess demonstrated elevated GSH production (Tsuji et al. 2003). Contrarily, Zlobin et al. (2017) reported that GSH did not play a significant role in canola root cells in terms of Zn chelation. In tobacco plants, Zn excess caused alterations in the concentration of free amino acids (glutamic acid, aspartate, glutamine, cysteine) (Daş et al. 2016), which are vital for intracellular osmoregulation, helping plants adapt to changes in their environment by balancing osmotic pressure. In oregano (*Origanum vulgare* L.), proline levels were considerably elevated in plants exposed to high concentrations of Zn (1500 and 3000 ppm) (Kulbat-Warycha et al. 2020). The accretion of proline in response to Zn toxicity may result from a reduction in the activity of enzymes responsible for proline breakdown, leading to increased proline level, or from heightened functionality of enzymes involved in proline biosynthesis (Kholodova et al. 2011). The enhancement in glycine betaine and proline in *Lactuca sativa* did not indicate a tolerance mechanism instead; it was likely an indication of Zn toxicity (Paradisone et al. 2015). Another defense mechanism through which plants protect themselves from Zn toxicity is through the cell wall which acts as a main obstacle against stress. Lignin is the chief constituent of the secondary cell wall which forms an interlocking network to toughen the cell wall and is made up of several extremely branched polymers of phenylpropanoid groups. In this context, genes linked to synthesis of xyloglucan endotransglucosylase (responsible for disintegrating pectin in the cell wall) were reported to be downregulated in Zn-stressed *B. juncea* plants to protect the cell wall (Feigl et al. 2019). The researchers explained that during the course of Zn toxicity, pectin in the cell wall binds to extra Zn, callose fixes it in place and makes sure that surplus Zn does not enter into the cytoplasm, and in this way plants are protected from excess Zn, but at the same time, it may hinder plant growth (Feigl et al. 2019).

4 Molecular Plant Responses

In rhizosphere, Zn is found as crystalline structure in Fe-Mg minerals, for example, sulfide (ZnS), and is mostly adsorbed in inter-convertible forms such as $ZnOH^+$, Zn^{2+}, and $ZnCl^+$ and onto clay particles and organic substances (Moreira et al. 2018). In general, root region absorbs Zn in the form of Zn^{2+} ions; however at elevated pH it is absorbed as ZnOH (Li 2019). Certain transporter families, for instance, P-type ATPase family, zinc-regulated transporter/iron-regulated transporter-like proteins [ZRT/IRT-like protein (ZIP) family], cation diffusion facilitator (CDF) family, and natural resistance-associated macrophage protein (NRAMP) family, participate in Zn^{2+} absorption, translocation, and regulating homeostasis (Krämer et al. 2007).

4.1 Zn-Deficit Conditions

Transcription factors, namely, bZIP19 and bZIP23, are considered to be the main regulators of Zn-deficit responses in *Arabidopsis thaliana* (Fig. 2a). Both these are part of basic region leucine-zipper (bZIP) family of transcription factors in eukaryotes. These are distinguished by an extremely conserved bZIP domain constituting a basic area which targets to the nucleus, implicated in precise DNA binding, along with a leucine-zipper for protein-protein interface and dimer formation that are crucial for bZIP transcription factor activity (Vinson et al. 2006; Amoutzias et al. 2006).

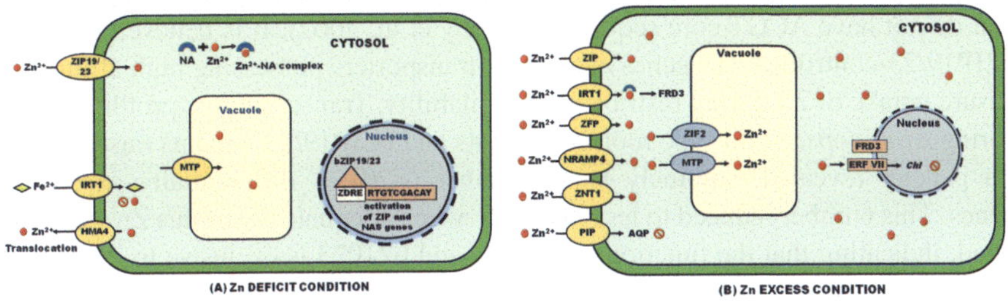

Fig. 2 (**a**) Under Zn deficiency condition, transcription factors like bZIP19/23 localized in the nucleus bind to hexamer DNA ZDRE (Zn deficiency response element) region in the promoter of the target genes (such as ZIP and NAS) and activate the transcription process. The NAS enzyme produces Zn ligand called NA which contributes to long distance Zn distribution, whereas Zn-NA complexes contribute to cytosolic Zn buffer due to activation of ATPase (HMA4). IRT-1 also plays a major role in prevention of excess uptake of non-Fe metal by triggering ubiquitination and degradation. (**b**) Under Zn excess condition, transporters like ZIP and ZNT present on plasma membrane mediate Zn influx from outside. PIP located on the plasma membrane suppresses the expression of aquaporin (AQP) genes which results in reduction of hydraulic conductance in roots. ZIF2 and TgMTP1 located on tonoplast mediate Zn uptake in vacuoles. Sometimes Zn competes with other metals like Fe for its Fe site of PCO (plant cysteine oxidase) in IRT-1 transporter to oxidize cysteine and diminishes the activity of ERF VII transcriptional factor which stimulates anaerobic gene, and these genes repress the chlorophyll biosynthesis. Along with this during competition, other genes like FRD3, NRAMP4, and ZFP4 also get activated and alter the functioning of plant cell

The bZIP family of *Arabidopsis* was sub-divided based on similarities in sequence and functional attributes, and an amended classification categorizes 78 members divided into 13 groups (designated A–M) (Jakoby et al. 2002; Dröge-Laser et al. 2018). bZIP19, bZIP23, and bZIP24 constitute F group of bZIP transcription factors (termed as F-bZIP), which are distinguished by a motif localized at N terminal region. This F-bZIP motif constitutes two small areas rich in amino acids, histidine (His) and cysteine (Cys) (Assunção et al. 2013). These three F-bZIP transcription factors are located in the nucleus and function as homodimers (Yang et al. 2009; Inaba et al. 2015; Lilay et al. 2019, 2021). bZIP19 and bZIP23 have 69% mutual similarity in amino acid sequence, with bZIP19 having 28% and bZIP23 having 32% similarity with bZIP24, but bZIP24 gets activated during salinity stress (Yang et al. 2009) and is not involved in Zn-deficit responses (Lilay et al. 2019). On the contrary, bZIP19 and bZIP23 are necessary for Zn-deficit responses in plants. In contrast to wild-type plants of *Arabidopsis*, bZIP19 bZIP23 dual mutant does not succeed in coping with Zn-deficit stress, resulting in Zn-deficit hypersensitive phenotype (Assunção et al. 2010). Under Zn-deficit conditions, bZIP19 and bZIP23 combine with Zn Deficiency Response Elements (ZDRE) present in the promoter region of their target genes (several ZIP and nicotianamine synthase (NAS) genes) and trigger transcription (Assunção et al. 2010). The ZDRE comprises a 10-bp imperfect palindromic arrangement, with consensus (RTGTCGACAY) and its detection corresponded to a unique bZIP-DNA combining sequence in plants. In general, bZIP transcription factors in plants have a preference to bind to hexamer DNA sequences comprising ACGT core, as located in A-, G-, or C-box (T/G/CACGTG/C/A), although there are few cases of non-palindromic binding regions that do not have ACGT-core sequence (Jakoby et al. 2002). It is believed that few bZIP19/23-controlled ZIP genes encode Zn transporters possessing high affinity to ensure uptake of Zn during restricted Zn availability. Transcriptomic profiling comparing *Arabidopsis* wild-type roots and shoots and bZIP19/23 mutants raised under Zn paucity revealed extremely small number (<20) of differentially expressed genes. This number reduced to less than 50% when adequate or surplus Zn was supplied, indicating that the function of bZIP19 and bZIP23 is exclusive to Zn-deficit response. These differentially expressed genes consisted of genes associated with Zn homeostasis, i.e., members of ZIP family and NAS enzymes that synthesize Zn ligand NA (Castro et al. 2017). NA participates in inter-cellular and long-distance Zn allocation (Clemens 2019). It is suggested that molecules of NA and Zn-NA complexes play a role in cytosolic Zn buffering under Zn-deficit conditions. ZIP proteins were categorized during proteomic profiling comparing *Arabidopsis* bZIP19 and wild-type roots in response to Zn-deficit condition as well (Inaba et al. 2015). This set of bZIP19/23-mediated genes provided important insights on plant's response toward Zn deficiency. In *Arabidopsis*, among the 15 members of ZIP family, 8 (AtZIP1/3/4/5/9/10/12 and AtIRT3) are managed by bZIP19 and bZIP23. They contain ZDREs in their promoter region, and their expression is upregulated by Zn-deficit levels and is subdued in bZIP19/23 mutant, where it could be reinstated upon efficient harmonization with bZIP19 or bZIP23 (Assunção et al. 2010; Lilay et al. 2019). Excluding ZIP5, all other ZIP members were found to be involved in uptake of Zn by heterologous yeast complementation (Lin et al. 2009; Assunção et al. 2010; Milner et al. 2013; Lee et al. 2021). The ZIP4 promoter, comprising two ZDREs, was effectively employed as bait in yeast-one-hybrid selection experiment, resulting in the recognition of bZIP19 and bZIP23 (Assunção et al. 2010). Moreover, ZIP4 promoter region combined with GUS in bZIP19/23 mutant setting suggested that ZIP4 is controlled by bZIP19/23 only (Castro et al. 2017). This permitted applying pZIP4::GUS as a Zn-deficit marker in plants to acquire deep insights into the controlling mechanism of bZIP19 and bZIP23. The physiological role of ZIP transporters singly in maintaining micronutrient homeostasis in plants is usually not well clear, which may be due to their potential functional redundancy (Sinclair and Krämer 2012; Lee et al. 2021). One more instance of post-translational modulation of a member of ZIP family is from *Arabidopsis* IRT1 transporter, which is involved in Fe uptake and is transcriptionally activated in roots in response to Fe deficiency (Vert et al. 2002). This transporter is post-translationally controlled by non-Fe metals such as Mn, Zn, and Co (which are transported by IRT1as well), and due to their direct attachment to cytosolic His-rich region of IRT1, ubiquitination, endocytosis, and disintegration of IRT1 are triggered. This mechanism of metal sensing present in IRT1 checks the extra absorption of non-Fe metals, especially when Fe is needed by plant (Dubeaux et al. 2018). Additionally significant may be the combination of bZIP19/23-mediated Zn-deficit response with other micro- and macronutrient complex networks. Lysophosphatidylcholine acyltransferase (LPCAT1) was recognized as a chief determinant of phosphate (Pi) accretion in *Arabidopsis* shoots under Zn paucity conditions (Kisko et al. 2018). In the categorized pathway, it was bZIP23 and not bZIP19 which associates with a new motif sequence found in LPCAT1 5'UTR, and it is suggested that this blocks the transcription of LPCAT1 during limited Zn supply, which consecutively results in enhanced Pi transporter gene expression PHT1;1 and elevated buildup of Pi in shoots. Expression of LPCAT1 is stimulated under Zn adequacy, as compared to Zn paucity, and it is stimulated under Zn-deficit conditions in bZIP23 mutant (Kisko et al. 2018). Elevated levels of ZIP19 and ZIP23 expression were found in *A. thaliana* grown in Zn-restricted culture medium (Humayan Kabir et al. 2014), which indicates that the stimulation of ZIP19 and ZIP23 in *A. halleri* along with *N. caerulescens* could be linked with low Zn level in rhizosphere because of increased activity of HMA4 (implicated in Zn transport from xylem to aerial parts) (Merlot et al. 2018; Lin et al. 2016), whereas soil deficient in Zn expressed ZIP1 and ZIP3 proteins in roots showing the evidence of Zn transport in plants from soil, while expression of ZIP4 in both root and shoot suggested intracellular transport of Zn in plant tissue.

4.2 Zn Excess Conditions

Cation diffusion facilitator (CDF) acts as metal tolerance/transport protein (MTP) in excess Zn situations and functions in transporting metal out of the cytosol (Ricachenevsky et al. 2013). In excess of Zn levels, MTP3 present on vacuolar membrane (Zhao et al. 2018) harmonizes the yeast mutant zrc1cot1 (Lu et al. 2009), whereas MTP2 is present on the membrane of endoplasmic reticulum (Sinclair et al. 2018), and MTP5 and MTP12 are localized at Golgi body in cucumber and *Arabidopsis thaliana* (Migocka et al. 2018). Upregulation of MTP1, MTP8, and MTP11 by high Zn was also confirmed in the shoots of the Zn hyperaccumulators *N. caerulescens* and *A. halleri* (Talke et al. 2006; Hanikenne and Nouet 2011; Moreira et al. 2018). Several transporters of ZRT-IRT-like protein or ZIP family participate in radial translocation of Zn to the stellar region in roots, for example, in *N. caerulescens*; NcZNT1 acts as a major means to transport Zn across the endodermal cells (Van de Mortel et al. 2006). Constitutive over-activation of NcZNT1, a plasma membrane-localized transporter, increased the resistance to surplus Zn exposure and enhanced the accretion of Zn in *A. thaliana* in comparison to wild-type plants (Lin et al. 2016). Expression of ZIP19 and ZIP23 was considerably stimulated in roots of both *N. caerulescens* and *A. halleri* under optimum Zn (Lin et al. 2016). Remy et al. (2014) demonstrated that zinc-induced facilitator 2 (ZIF2) is mainly present at the vacuolar membrane of root cortical cells and is involved in Zn uptake. Yang et al. (2018) recognized SaZIP4 gene as the main Zn transporter in hyperaccumulation mechanism of Zn in *Sedum alfredii.* A subcellular investigation revealed that ZIP4 is localized in cytoplasmic membrane and contributes to Zn uptake in aerial parts as well as roots, while ZIP6, present in the cell membrane of shoots, supports Zn uptake since its expression was mainly observed in aerial parts of *A. halleri* (Becher et al. 2004; Gupta et al. 2016). In *N. goesingense*, TgMTP1 protein enhances Zn tolerance by accumulating high quantities in tonoplast of shoots (Gustin et al. 2009). In addition to MTP1, ZNT5 is also a significant participant involved in Zn sequestration into the storage cells of epidermis (Küpper and Kochian 2010; Kozhevnikova et al. 2017).

Kholodova et al. (2011) observed acute deactivation of aquaporin (AQP) genes whose transcripts were found in the cell membrane (McPIP1;1, McPIP2;1, and McPIP2;3) or vacuolar membrane (McTIP1;2 and McTIP2;2) in *Mesembryanthemum crystallinum* under Cu/Zn stress. In the same way, PIP2 (plasma membrane intrinsic proteins) AQP was observed to decline in lettuce in response to Zn toxicity (Lucini and Bernardo 2015). The decline in root built-in water transfer feature in barley under surplus Zn was attributed to decrease in root and aerial organs' hydraulic conductivity due to inactivation of AQPs (HvPIP1;2, HvPIP2;4, and HvPIP2;5) (Fig. 2b) and enhanced synthesis of suberin in endodermis (Gitto and Fricke 2018). Fatemi et al. (2020) observed decline in root hydraulic activity and participation of all PIP1 isoforms, PIP2;6, and PIP2;7 in perceiving prolonged response to Zn supplementation in *Brassica rapa*. Surplus Zn activated the transcriptional behavior of Fe-mediated genes, i.e., FRO2 (ferric reduction oxidase 2) and BTS, indicating that Zn interferes with Fe homeostasis at Fe sensing stage. Study conducted by Erenoglu (2019) revealed for the first time the existence of IRT-like proteins in wheat. In spite of this, IRT-1-mediated Fe deficit caused by surplus Zn has not been confirmed and therefore cannot be generalized for all plant species. Activation of genes, like FERRIC REDUCTASE DEFECTIVE3 (FRD3, AT3G08040) related to MATE (multidrug and toxic compound extrusion) efflux family, NRAMP4 (AT3G13100), metal-linked domain-comprising protein, C2H2-type Zn finger proteins (ZFPs), and various Zn finger proteins (e.g., At3g19580, At5g54040, At1g76410, At3g52800, At5g18550, At2g19810, At5g04340, and At5g59820), may indicate the molecular responses to surplus Zn-mediated P deficiency in *A. thaliana* (Pineau et al. 2012; Bouain et al. 2014). Also, PHO1 (phosphate transporter-exporter), PHO1;H3, and PHR1 genes connect Zn and P homeostasis and nutrition in *A. thaliana* (Khan et al. 2014). Carbonare et al. (2019) demonstrated competition of Zn^{2+} with Fe (II) for binding to cysteine oxidase (PCO) and diminishing its activity, which lead to ERF VII transcription factor stabilization followed by activation of anaerobic genes [Hypoxia Responsive Attenuator1 (Pop_HRA1), Alcohol Dehydrogenase (Pop_ADH), and PCO (Pop_PCO)] in poplar clone roots as a broad response to Zn toxicity. At molecular level, ground-breaking work of Di Baccio et al. (2011) related to photosynthesis revealed inactivation of genes encoding transcripts involved in light reaction, like LHC II proteins along with polypeptides associated with ATP synthases, PSI, PSII, and cytochrome b6/f complex connected with redox chain, and electron carriers related with ferredoxin as well as inactivation of big subunit of Rubisco and stimulation of small subunit in Zn-stressed *Populus×euramericana* clone I-214. Through comparison of transcriptional responses of poplar roots under Zn toxicity (Ariani et al. 2015) and flooding conditions (Kreuzwieser et al. 2009), Romeo et al. (2014) reported a considerable coincide between Zn-modulated genes and oxygen deprivation genes in poplar. The oxygen starvation-like response stimulated under Zn toxicity was considered to be a component of plant adaptive response toward surplus Zn, before being activated by decrease in root oxygen availability. The connection between these responses was notwithstanding the Zn tolerance of poplar variety, as revealed by its presence in two poplar cultivars differentiated by varied sensitivity to Zn stress (Romeo et al. 2014). The transcriptional response to oxygen deprivation in *Arabidopsis* is controlled largely by ERF-VII transcription factors, namely, RAP2.2 and RAP2.12, whose aerobic unreliability is guaranteed by N-degron pathway (van Dongen and Licausi 2015). It has been reported that PCOs need O_2 and Fe^{2+} to oxidize Cys (White et al. 2017, 2018). Zn^{2+} may outperform Fe^{2+} in terms of PCO binding in vivo as well. Definitely, Fe starvation on long-term basis also resulted in stabilization of an ERF-VII and stimulated oxygen starvation-responsive genes in *Arabidopsis* seedlings. Romeo et al. (2014) gave two justifications regarding toxic impacts of Zn observed in their study. Firstly, Zn can compete with Fe for uptake by the roots with the help of low-affinity transporters, like IRT1 (Sinclair and Krämer 2012). Secondly, Zn-mediated Fe deprivation can inhibit chlorophyll biosynthesis, as Fe is needed for synthesis of heme and it is expected that Zn replaces Mg^{2+} in porphyrin ring of chlorophyll molecule (Tripathy and Pattanayak 2012). If not, the stable ERF-VII proteins, stimulated by surplus Zn or flooding, could contribute to active suppression of chlorophyll biosynthetic genes, as was reported by Abbas et al. (2015).

5 Conclusions

Zinc plays a major role in the basic operations of plants because of its essential nature. At the same time, it is harmful for plants at un-optimum levels, i.e., inadequate or excess above the favorable range. Both Zn deficiency and toxicity affect various morpho-physiological, biochemical, and molecular attributes, ultimately decreasing crop produce which endangers food safety and health of mankind universally. The root cell walls and vacuoles are considered to be the key regions of Zn subcellular allocation. Additionally, Zn and other nutrients share common transporters responsible for making way into the roots. Future investigations should be directed on developing the most effective method for Zn supplementation that would assist in maximizing Zn availability for absorption and accumulation by plants. Alternatively, environment-friendly and cost-effective mitigation strategies for decreasing Zn toxicity stress in plants should be developed in order to enhance crop yield and meet food requirements for ever-increasing human population.

References

Abbas M, Berckhan S, Rooney DJ, Gibbs DJ, Vicente Conde J, Sousa Correia C, Bassel GW, Marín-de la Rosa N, León J, Alabadí D et al (2015) Oxygen sensing coordinates photomorphogenesis to facilitate seedling survival. Curr Biol 25:1483–1488

AbdElgawad H, Zinta G, Hamed BA, Selim S, Beemster G, Hozzein WN et al (2020) Maize roots and shoots show distinct profiles of oxidative stress and antioxidant defense under heavy metal toxicity. Environ Pollut 258:113705

Adele NC, Ngwenya BT, Heal KV, Mosselmans JFW (2021) Role of plant growth promoting bacteria in driving speciation gradients across soil-rhizosphere-plant interfaces in zinc-contaminated soils. Environ Pollut 279:116909

Aghajanzadeh TA, Prajapati DH, Burow M (2020) Differential partitioning of thiols and glucosinolates between shoot and root in Chinese cabbage upon excess zinc exposure. J Plant Physiol 244:153088

Akther MS, Das U, Tahura S, Prity SA, Islam M, Kabir AH (2020) Regulation of Zn uptake and redox status confers Zn deficiency tolerance in tomato. Sci Hortic 273:109624

Ali B, Saleem MH, Ali S, Shahid M, Sagir M, Tahir MB et al (2022a) Mitigation of salinity stress in barley genotypes with variable salt tolerance by application of zinc oxide nanoparticles. Front Plant Sci 13:973782

Ali M, Wang X, Haroon U, Chaudhary HJ, Kamal A, Ali Q et al (2022b) Antifungal activity of Zinc nitrate derived nano Zno fungicide synthesized from Trachyspermum ammi to control fruit rot disease of grapefruit. Ecotoxicol Environ Saf 233:113311

Alloway B (2004) Zinc in soil and crop nutrition. In: Zinc in plant nutrition. International Zinc Association, Brussels

Alloway BJ (2009) Soil factors associated with zinc deficiency in crops and humans. Environ Geochem Health 31:537–548

Amoutzias G, Veron A, Weiner J et al (2006) One billion years of bZIP transcription factor evolution: conservation and change in dimerization and DNA-binding site specificity. Mol Biol Evol 24:827–835

Antoniadis V, Levizou E, Shaheen SM, Ok YS, Sebastian A, Baum C et al (2017) Trace elements in the soil-plant interface: phytoavailability, translocation, and phytoremediation–a review. Earth-Sci Rev 171:621–645

Apel K, Hirt H (2004) Reactive oxygen species: metabolism, oxidative stress, and signal transduction. Annu Rev Plant Biol 55(1):373–399

Ariani A, Di Baccio D, Romeo S, Lombardi L, Andreucci A, Lux A, Horner DS, Sebastiani L (2015) RNA sequencing of Populus 3 canadensis roots identifies key molecular mechanisms underlying physiological adaption to excess zinc. PLoS One 10:e0117571

Assunção AGL, Herrero E, Lin Y-F et al (2010) Arabidopsis thaliana transcription factors bZIP19 and bZIP23 regulate the adaptation to zinc deficiency. Proc Natl Acad Sci USA 107:10296–10301

Assunção AGL, Persson DP, Husted S et al (2013) Model of how plants sense zinc deficiency. Metallomics 5:1110–1116

Bankaji I, Pérez-Clemente RM, Caçador I, Sleimi N (2019) Accumulation potential of Atriplex halimus to zinc and lead combined with NaCl: effects on physiological parameters and antioxidant enzymes activities. S Afr J Bot 123:51–61

Barrameda-Medina Y, Montesinos-Pereira D, Romero L, Blasco B, Ruiz JM (2014) Role of GSH homeostasis under Zn toxicity in plants with different Zn tolerance. Plant Sci 227:110–121

Bartha B, Kolbert Z, Erdei L (2005) Nitric oxide production induced by heavy metals in Brassica juncea L. Czern. and Pisum sativum L. Acta Biol Szegediensis 49(1–2):9–12

Becher M, Talke IN, Krall L, Krämer U (2004) Cross-species microarray transcript profiling reveals high constitutive expression of metal homeostasis genes in shoots of the zinc hyperaccumulator Arabidopsis halleri. Plant J 37:251–268

Blasco B, Navarro-León E, Ruiz JM (2019) Study of Zn accumulation and tolerance of HMA4 TILLING mutants of Brassica rapa grown under Zn deficiency and Zn toxicity. Plant Sci 287(3):110201

Bouain N, Shahzad Z, Rouached A (2014) Phosphate and zinc transport and signalling in plants: toward a better understanding of their homeostasis interaction. J Exp Bot 65:5725–5741

Broadley MR, White PJ, Hammond JP, Zelko I, Lux A (2007) Zinc in plants. New Phytol 173(4):677–702

Cakmak I (2000) Possible roles of zinc in protecting plant cells from damage by reactive oxygen species. New Phytol 146:185–205. Go to original source... Go to PubMed

Cakmak I, Marschner H, Bangerth F (1989) Effect of zinc nutritional status on growth, protein metabolism and levels of indole-3-acetic acid and other phytohormones in bean (*Phaseolus vulgaris* L.). J Exp Bot 40(3):405–412

Carbonare LD, White MD, Shukla V, Francini A, Perata P, Flashman E, Sebastiani L, Licausi F (2019) Zinc excess induces a hypoxia-like response by inhibiting cysteine oxidases in Poplar roots. Plant Physiol 180:1614–1628

Castillo-González J, Ojeda-Barrios D, Hernández-Rodríguez A, González-Franco AC, Robles-Hernández L, López-Ochoa GR (2018) Zinc metalloenzymes in plants. Interciencia 43:242–248

Castro PH, Lilay GH, Munoz-Merida A et al (2017) Phylogenetic analysis of F-bZIP transcription factors indicates conservation of the zinc deficiency response across land plants. Sci Rep 7:3806

Chen X, Ludewig U (2018) Biomass increase under zinc deficiency caused by delay of early flowering in *Arabidopsis*. J Exp Bot 69(5):1269–1279

Clemens S (2019) Metal ligands in micronutrient acquisition and homeostasis. Plant Cell Environ 42:2902–2912

Cuypers A, Vangronsveld J, Clijsters H (2002) Peroxidases in roots and primary leaves of Phaseolus vulgaris L. copper and zinc Phytotoxicity: a comparison. J Plant Physiol 159:869–876

Daş ZA, Dimlioğlu G, Bor M, Özdemir F (2016) Zinc induced activation of GABA-shunt in tobacco (Nicotiana tabacum L.). Environ Exp Bot 122:78–84

de Almeida HJ, Carmona VMV, Inocêncio MF, Neto AEF, Filho ABC, Mauad M (2020) Soil type and zinc doses in agronomic biofortification of lettuce genotypes. Agronomy 2020(10):124

De Oliveira VH, Tibbett M (2018) Tolerance, toxicity and transport of Cd and Zn in Populus trichocarpa. Environ Exp Bot 155:281–292

de Oliveira NT, Namorato FA, Rao S, de Souza Cardoso AA, de Rezende PM, Guilherme LRG et al (2023) Iron counteracts zinc-induced toxicity in soybeans. Plant Physiol Biochem 194:335–344

Di Baccio D, Galla G, Bracci T, Andreucci A, Barcaccia G, Tognetti R, Sebastiani L (2011) Transcriptome analyses of Populus × euramericana clone I-214 leaves exposed to excess zinc. Tree Physiol 31:1293–1308

Dröge-Laser W, Snoek BL, Snel B, Weiste C (2018) The Arabidopsis bZIP transcription factor family – an update. Curr Opin Plant Biol 45:36–49

Dubeaux G, Neveu J, Zelazny E, Vert G (2018) Metal sensing by the IRT1 transporter-receptor orchestrates its own degradation and plant metal nutrition. Mol Cell 69:953–964.e5

Erenoglu EB (2019) Iron deficiency-induced zinc uptake by bread wheat. J Plant Nutr Soil Sci 000:1–6

Fan X, Zhou X, Chen H, Tang M, Xie X (2021) Cross-talks between macro-and micronutrient uptake and signaling in plants. Front Plant Sci 12:663477

Fatemi H, Zaghdoud C, Nortes PA, Carvajal M, Martínez-Ballesta MDC (2020) Differential aquaporin response to distinct effects of two Zn concentrations after foliar application in Pak Choi (Brassica rapa L.) plants. Agronomy 10(3):450

Feigl G, Lehotai N, Molnár Á, Ördög A, Rodríguez-Ruiz M, Palma JM, Corpas FJ, Erdei L, Kolbert Z (2015) Zinc induces distinct changes in the metabolism of reactive oxygen and nitrogen species (ROS and RNS) in the roots of two Brassica species with different sensitivity to zinc stress. Ann Bot 116(4):613–625. https://doi.org/10.1093/aob/mcu246

Feigl G et al (2019) Zinc-induced root architectural changes of rhizotron-grown B. napus correlate with a differential nitro-oxidative response. Nitric Oxide 90:55–65. https://doi.org/10.1016/j.niox.2019.06.003

Garg N, Singh S (2018) Arbuscular mycorrhiza *Rhizophagus irregularis* and silicon modulate growth, proline biosynthesis and yield in *Cajanus cajan* L. Millsp. (pigeonpea) genotypes under cadmium and zinc stress. J Plant Growth Regul 37:46–63

Gautam S, Rathoure AK, Chhabra A, Pandey SN (2017) Effects of nickel and zinc on biochemical parameters in plants-a review. Octa J Environ Res 5(1)

Gitto A, Fricke W (2018) Zinc treatment of hydroponically grown barley plants causes a reduction in root and cell hydraulic conductivity and isoform-dependent decrease in aquaporin gene expression. Physiol Plant 164:176–190

Glinska S, Gapinska M, Michlewska S, Skiba E, Kubicki J (2016) Analysis of Triticum aestivum seedling response to the excess of zinc. Protoplasma 253:367–377

Goodarzi A, Namdjoyan S, Soorki AA (2020) Effects of exogenous melatonin and glutathione on zinc toxicity in safflower (Carthamus tinctorius L.) seedlings. Ecotoxicol Environ Saf 201:110853

Grennan AK (2011) Metallothioneins, a diverse protein family. Plant Physiol 155:1750–1751

Gupta N, Ram H, Kumar B (2016) Mechanism of Zinc absorption in plants: uptake, transport, translocation and accumulation. Rev Environ Sci Bio Technol 15:89–109

Gustin JL, Loureiro ME, Kim D, Na G, Tikhonova M, Salt DE (2009) MTP1-dependent Zn sequestration into shoot vacuoles suggests dual roles in Zn tolerance and accumulation in Zn-hyperaccumulating plants. Plant J 57:1116–1127

Hacisalihoglu G (2020) Zinc (Zn): The last nutrient in the alphabet and shedding light on Zn efficiency for the future of crop production under suboptimal Zn. Plan Theory 9(11):1471

Hafeez BMKY, Khanif YM, Saleem M (2013) Role of zinc in plant nutrition-a review. Am J Exp Agric 3(2):374–391

Hajiboland R, Amirazad F (2010) Growth, photosynthesis and antioxidant defense system in Zn-deficient red cabbage plants. Plant Soil Environ 56(5):209–217

Hamzah Saleem M, Usman K, Rizwan M, Al Jabri H, Alsafran M (2022) Functions and strategies for enhancing zinc availability in plants for sustainable agriculture. Front Plant Sci 13:1033092

Hanikenne M, Nouet C (2011) Metal hyperaccumulation and hypertolerance: a model for plant evolutionary genomics. Curr Opin Plant Biol 14:252–259

Helmersson A, von Arnold S, Bozhkov PV (2008) The level of free intracellular zinc mediates programmed cell death/cell survival decisions in plant embryos. Plant Physiol 147(3):1158–1167

Hopkinson BM, Dupont CL, Allen AE, Morel FM (2011) Efficiency of the CO2-concentrating mechanism of diatoms. Proc Natl Acad Sci 108(10):3830–3837

Humayan Kabir A, Swaraz AM, Stangoulis J (2014) Zinc-deficiency resistance and biofortification in plants. J Plant Nutr Soil Sci 177(3):311–319
Inaba S, Kurata R, Kobayashi M et al (2015) Identification of putative target genes of bZIP19, a transcription factor essential for Arabidopsis adaptation to Zn deficiency in roots. Plant J 84:323–334
Jakoby M, Weisshaar B, Dröge-Laser W et al (2002) bZIP transcription factors in Arabidopsis. Trends Plant Sci 7:106–111
Kabata-Pendias A (2011) Trace elements in soils and plants, 4th edn. Taylor and Francis, Boca Raton
Kaur H, Garg N (2021) Zinc toxicity in plants: a review. Planta 253(6):129
Kaznina NM, Titov AF (2017) Effect of zinc deficiency and excess on the growth and photosynthesis of winter wheat. J Stress Physiol Biochem 13(4):88–94
Kaznina NM, Batova YV, Repkina NS (2021) Effect of zinc deficiency and excess on the antioxidant enzymes activity in barley seedling leaves
Khan GA, Bouraine S, Wege S, Li Y, de Carbonnel M, Berthomieu P, Poirier Y, Rouached H (2014) Coordination between zinc and phosphate homeostasis involves the transcription factor PHR1, the phosphate exporter PHO1, and its homologue PHO1;H3 in Arabidopsis. J Exp Bot 65:871–884
Khanif BMKY, Saleem M (2013) Role of zinc in plant nutrition-a review. Am J Exp Agric 3(2):374–391
Kholodova V, Volkov K, Abdeyeva A, Kuznetsov V (2011) Water status in Mesembryanthemum crystallinum under heavy metal stress. Environ Exp Bot 71:382–389
Kisko M, Bouain N, Saf A et al (2018) LPCAT1 controls phosphate homeostasis in a zinc-dependent manner. elife 7:32077
Kosesakal T, Unal M, Oz GC (2009) Influence of zinc toxicity on gravitropic response of tomato (Lycopersicon esculentum Mill.) roots. Fresenius Environ Bull 18(12):2402–2407
Kozhevnikova AD, Seregin I, Gosti F, Schat H (2017) Zinc accumulation and distribution over tissues in Noccaea caerulescens in nature and in hydroponics: A comparison. Plant Soil 411:5–16
Krämer U, Talke IN, Hanikenne M (2007) Transition metal transport. FEBS Lett 581:2263–2272
Kreuzwieser J, Hauberg J, Howell KA, Carroll A, Rennenberg H, Millar AH, Whelan J (2009) Differential response of gray poplar leaves and roots underpins stress adaptation during hypoxia. Plant Physiol 149:461–473
Kulbat-Warycha K, Georgiadou EC, Mańkowska D, Smolińska B, Fotopoulos V, Leszczyńska J (2020) Response to stress and allergen production caused by metal ions (Ni, Cu and Zn) in oregano (*Origanum vulgare* L.) plants. J Biotechnol 324:171–182
Küpper H, Kochian LV (2010) Transcriptional regulation of metal transport genes and mineral nutrition during acclimatization to cadmium and zinc in the Cd/Zn hyperaccumulator, Thlaspi caerulescens (Ganges population). New Phytol 185:114–129
Lee S, Lee J, Ricachenevsky FK et al (2021) Redundant roles of four ZIP family members in zinc homeostasis and seed development in Arabidopsis thaliana. Plant J 108:1162–1173
Li C (2019) Absorption and translocation of Zn foliar fertilisers. Master's thesis, The University of Queensland, Brisbane
Li X, Yang Y, Jia L, Chen H, Wei X (2013) Zinc-induced oxidative damage, antioxidant enzyme response and proline metabolism in roots and leaves of wheat plants. Ecotoxicol Environ Saf 89:150–157
Li SP, Zeng LS, Su ZL (2022) Wheat growth, photosynthesis and physiological characteristics under different soil Zn levels. J Integr Agric 21(7):1927–1940
Lilay GH, Castro PH, Campilho A, Assunção AGL (2019) The Arabidopsis bZIP19 and bZIP23 activity requires zinc deficiency – insight on regulation from complementation lines. Front Plant Sci 9:1955
Lilay GH, Persson DP, Castro PH et al (2021) Arabidopsis bZIP19 and bZIP23 act as zinc sensors to control plant zinc status. Nat Plants 7:137–143
Lin Y-F, Liang H-M, Yang S-Y et al (2009) Arabidopsis IRT3 is a zinc regulated and plasma membrane localized zinc/iron transporter. New Phytol 182:392–404

Lin Y-F, Hassan Z, Talukdar S, Schat H, Aarts MG (2016) Expression of the ZNT1 zinc transporter from the metal hyperaccumulator Noccaea caerulescens confers enhanced zinc and cadmium tolerance and accumulation to Arabidopsis thaliana. PLoS One 11:e0149750

Liščáková P, Nawaz A, Molnárová M (2022) Reciprocal effects of copper and zinc in plants. Int J Environ Sci Technol 19(9):9297–9312

Lu M, Chai J, Fu D (2009) Structural basis for autoregulation of the zinc transporter YiiP. Nat Struct Mol Biol 16:1063–1067

Lucini L, Bernardo L (2015) Comparison of proteome response to saline and zinc stress in lettuce. Front Plant Sci 6:240

Luo ZB, He XJ, Chen L, Tang L, Gao S, Chen F (2010) Effects of zinc on growth and antioxidant responses in *Jatropha curcas* seedlings. Int J Agric Biol 12:119–124

Madhava Rao KV, Sresty TV (2000) Antioxidative parameters in the seedlings of pigeonpea (Cajanus cajan (L.) Millspaugh) in response to Zn and Ni stresses. Plant Sci 157(1):113–128

Madian AG, Regnier FE (2010) Proteomic identification of carbonylated proteins and their oxidation sites. J Proteome Res 9:3766–3780

Mahmoudi H, Salah IB, Zaouali W, Zorrig W, Smaoui A, Ali T et al (2021) Impact of zinc excess on germination, growth parameters and oxidative stress of sweet basil (*Ocimum basilicum* L.). Bull Environ Contam Toxicol 106:899–907

Marschner H (1995) Mineral nutrition of higher plants. Academic, London

Mateos-Naranjo E, Pérez-Romero JA, Redondo-Gómez S, Mesa-Marín J, Castellanos EM, Davy AJ (2018) Salinity alleviates zinc toxicity in the saltmarsh zinc-accumulator Juncus acutus. Ecotoxicol Environ Saf 163:478–485

Mattiello EM, Ruiz HA, Neves JC, Ventrella MC, Araújo WL (2015) Zinc deficiency affects physiological and anatomical characteristics in maize leaves. J Plant Physiol 183:138–143

Mazé P (1915) Détermination des éléments minéraux rares nécessaires au développement du maïs. C R Hebd Seances Acad Sci 160:211–214

Meng Y, Xiang C, Huo J, Shen S, Tang Y, Wu L (2023) Toxicity effects of zinc supply on growth revealed by physiological and transcriptomic evidences in sweet potato (Ipomoea batatas (L.) Lam). Sci Rep 13(1):19203

Merlot S, de la Torre VSG, Hanikenne M (2018) Physiology and molecular biology of trace element hyperaccumulation. In: Agromining: farming for metals. Springer, Berlin, pp 93–116

Michael PI, Krishnaswamy M (2011) The effect of zinc stress combined with high irradiance stress on membrane damage and antioxidative response in bean seedlings. Environ Exp Bot 74:171–177

Migocka M, Małas K, Maciaszczyk-Dziubinska E, Posyniak E, Migdal I, Szczech P (2018) Cucumber Golgi protein CsMTP5 forms a Zn-transporting heterodimer with high molecular mass protein CsMTP12. Plant Sci 277:196–206

Milner MJ, Seamon J, Craft E, Kochian LV (2013) Transport properties of members of the ZIP family in plants and their role in Zn and Mn homeostasis. J Exp Bot 64:369–381

Mir BA, Khan TA, Fariduddin Q (2015) 24-epibrassinolide and spermidine modulate photosynthesis and antioxidant systems in Vigna radiata under salt and zinc stress. Int J Adv Res 3(5):592–608

Moreira A, Moraes LA, dos Reis AR (2018) The molecular genetics of zinc uptake and utilization efficiency in crop plants. In: Hossain MA, Kamiya T, Burritt DJ, Tran LSP, Fujiwara T (eds) Plant micronutrient use efficiency. Elsevier, Amsterdam, pp 87–108

Mukhopadhyay M, Das A, Subba P, Bantawa P, Sarkar B, Ghosh PD, Mondal TK (2013) Structural, physiological and biochemical profiling of tea plants (*Camellia sinensis* (L.) O. Kuntze) under zinc stress. Biol Plant 57:474–480

Natasha N, Shahid M, Bibi I, Iqbal J, Khalid S, Murtaza B et al (2022) Zinc in soil-plant-human system: a data-analysis review. Sci Total Environ 808:152024

Navarro-León E, Albacete A, de la Torre-González A, Ruiz JM, Blasco B (2016) Phytohormone profile in Lactuca sativa and Brassica oleracea plants grown under Zn deficiency. Phytochemistry 130:85–89

Nieboer E, Richardson DH (1980) The replacement of the nondescript term 'heavy metals' by a biologically and chemically significant classification of metal ions. Environ Pollut B 1(1):3–26

Noman A, Aqeel M, Khalid N, Islam W, Sanaullah T, Anwar M et al (2019) Zinc finger protein transcription factors: integrated line of action for plant antimicrobial activity. Microb Pathog 132:141–149

Ouni Y, Mateos-Naranjo E, Abdelly C, Lakhdar R (2016) Interactive effect of salinity and zinc stress on growth and photosynthetic responses of the perennial grass, Polypogon monspeliensis. Ecol Eng 95:171–179

Panda S (2017) Physiological impact of Zinc nanoparticle on germination of rice (Oryza sativa L) seed. J Plant Sci Phytopathol 1:062–070

Paradisone V, Barrameda-Medina Y, Montesinos-Pereira D, Romer L, Esposito S, Ruiz JM (2015) Roles of some nitrogenous compounds protectors in the resistance to zinc toxicity in Lactuca sativa cv. Phillipus and Brassica oleracea cv. Bronco. Acta Physiol Plant 37:137

Paradisone V, Navarro-León E, Albacete A, Ruiz JM, Esposito S, Blasco B (2022) Improvement of the physiological response of barley plants to both Zinc deficiency and toxicity by the application of calcium silicate. Plant Sci 319:111259

Paunov M, Koleva L, Vassilev A, Vangronsveld J, Goltsev V (2018) Effects of different metals on photosynthesis: cadmium and zinc affect chlorophyll fluorescence in durum wheat. Int J Mol Sci 19:787

Pineau C, Loubet S, Lefoulon C, Chalies C, Fizames C, Lacombe B, Ferrand M, Loudet O, Berthomieu P, Richard O (2012) Natural variation at the FRD3 MATE transporter locus reveals crosstalk between Fe homeostasis and Zn tolerance in Arabidopsis thaliana. PLoS Genet 8:e1003120

Pochodylo AL, Aristilde L (2017) Molecular dynamics of stability and structures in phytochelatin complexes with Zn, Cu, Fe, Mg, and Ca: implications for metal detoxification. Environ Chem Lett 15:495–500

Rai S, Singh PK, Mankotia S, Swain J, Satbhai SB (2021) Iron homeostasis in plants and its crosstalk with copper, zinc, and manganese. Plant Stress 1:100008

Remy E, Cabrito TR, Batista RA, Hussein MA, Teixeira MC, Athanasiadis A, Sá-Correia I, Duque P (2014) Intron retention in the 50 UTR of the novel ZIF2 transporter enhances translation to promote zinc tolerance in Arabidopsis. PLoS Genet 10:e1004375

Ricachenevsky FK, Menguer PK, Sperotto RA, Williams LE, Fett JP (2013) Roles of plant metal tolerance proteins (MTP) in metal storage and potential use in biofortification strategies. Front Plant Sci 4:1144

Romeo S, Francini A, Ariani A, Sebastiani L (2014) Phytoremediation of Zn: Identify the diverging resistance, uptake and biomass production behaviours of poplar clones under high zinc stress. Water Air Soil Pollut 225:1–12

Rout GR, Das P (2003) Effect of metal toxicity on plant growth and metabolism: I. Zinc. Agronomie 23:3–11

Sadeghzadeh B (2013) A review of zinc nutrition and plant breeding. J Soil Sci Plant Nutr 13(4):905–927

Sagardoy RUTH, Morales FERM, López-Millán AF, Abadía ANUNCIACIÓN, Abadía JAVIER (2009) Effects of zinc toxicity on sugar beet (Beta vulgaris L.) plants grown in hydroponics. Plant Biol 11(3):339–350

Sagardoy R, Vazquez S, Florez-Sarasa ID, Albacete A, Ribas-Carbo M, Flexas J, Abadia J, Morales F (2010) Stomatal and mesophyll conductances to CO_2 are the main limitations to photosynthesis in sugar beet (*Beta vulgaris*) plants grown with excess zinc. New Phytol 187:145–158. https://doi.org/10.1111/j.1469-8137.2010.03241.x

Samreen T, Shah HU, Ullah S, Javid M (2017) Zinc effect on growth rate, chlorophyll, protein and mineral contents of hydroponically grown mungbeans plant (Vigna radiata). Arab J Chem 10:S1802–S1807

Schutzendubel A, Polle A (2002) Plant responses to abiotic stresses: heavy metal-induced oxidative stress and protection by mycorrhization. J Exp Bot 53(372):1351–1365

Sekimoto H, Hoshi M, Nomura T, Yokota T (1997) Zinc Deficiency Affects the Levels of Endogenous Gibberellins in *Zea mays* L. Plant Cell Physiol 38:1087–1090

Sharma S, Sharma P, Datta SP, Gupta V (2009) Morphological and biochemical response of Cicer arietinum var. Pusa-256 towards an excess of zinc concentration. Life Sci J 7(1):95–98

SHC Soil health card Ministry of Agriculture and Farmers Welfare, Government of India (2023). https://www.soilhealth.dac.gov.in/legacy

Shukla AK, Behera SK, Prakash C, Tripathi A, Patra AK, Dwivedi BS et al (2021) Deficiency of phyto-available sulphur, zinc, boron, iron, copper and manganese in soils of India. Sci Rep 11:19760. https://doi.org/10.1038/s41598-021-99040-2

Siddiqui MF, Bano B (2019) Probing the binding interaction of zinc and cadmium with garlic phytocystatin: implication of the abiotic stress on garlic phytocystatin. Int J Biol Macromol 133:945–956

Sidhu GPS (2016) Physiological, biochemical and molecular mechanisms of zinc uptake, toxicity and tolerance in plants. J Glob Biosci 5:4603–4633

Sidhu GPS, Bali AS, Singh HP, Batish DR, Kohli RK (2020) Insights into the tolerance and phytoremediation potential of Coronopus didymus L. (Sm) grown under zinc stress. Chemosphere 244:125350

Šimon M et al (2021) Proteomic investigation of Zn-challenged rice roots reveals adverse effects and root physiological adaptation. Plant Soil 460:69–88. https://doi.org/10.1007/s11104-020-04772-1

Sinclair SA, Krämer U (2012) The zinc homeostasis network of land plants. Biochim Biophys Acta 1823:1553–1567

Sinclair SA, Senger T, Talke IN, Cobbett CS, Haydon MJ, Krämer U (2018) Systemic upregulation of MTP2-and HMA2-mediated Zn partitioning to the shoot supplements local Zn deficiency responses. Plant Cell 30:2463–2479

Singh AL, Basu MS, Singh NB (2004) Mineral disorders of groundnut. Director, National Research Centre for Groundnut

Singh RP, Mishra S, Jha P, Raghuvanshi S, Jha PN (2018) Effect of inoculation of zinc-resistant bacterium Enterobacter ludwigii CDP-14 on growth, biochemical parameters and zinc uptake in wheat (Triticum aestivum L.) plant. Ecol Eng 116:163–167

Sommer AL, Lipman CB (1926) Evidence on the indispensable nature of zinc and boron for higher green plants. Plant Physiol 1:231–249

Souza SCR, Souza LA, Schiavinato MA, de Oliveira Silva FM, de Andrade SAL (2020) Zinc toxicity in seedlings of three trees from the Fabaceae associated with arbuscular mycorrhizal fungi. Ecotoxicol Environ Saf 195:110450

Stanton C, Sanders D, Krämer U, Podar D (2022) Zinc in plants: integrating homeostasis and biofortification. Mol Plant 15(1):65–85

Stoláriková-Vaculíkováa M, Romeo S, Minnocci A, Luxová M, Vaculík M, Lux A, Sebastiani L (2015) Anatomical, biochemical and morphological responses of poplar Populus deltoides clone Lux to Zn excess. Environ Exp Bot 109:235–243

Stoyanova Z, Doncheva S (2002) The effect of zinc supply and succinate treatment on plant growth and mineral uptake in pea plant. Braz J Plant Physiol 14:111–116

Sturikova H, Krystofova O, Huska D, Adam V (2018) Zinc, zinc nanoparticles and plants. J Hazard Mater 349:101–110

Subba P, Mukhopadhyay M, Mahato SK, Bhutia KD, Mondal TK, Ghosh SK (2014) Zinc stress induces physiological, ultra-structural and biochemical changes in mandarin orange (Citrus reticulata Blanco) seedlings. Physiol Mol Biol Plants 20:461–473

Szopinski M, Sitko K, Gieron Z, Rusinowski S, Corso M, Hermans C, Verbruggen N, Małkowski E (2019) Toxic effects of Cd and Zn on the photosynthetic apparatus of the Arabidopsis halleri and Arabidopsis arenosa pseudo-metallophytes. Front Plant Sci 10:748

Talke IN, Hanikenne M, Kr€amer U. (2006) Zinc-dependent global transcriptional control, transcriptional deregulation, and higher gene copy number for genes in metal homeostasis of the hyperaccumulator Arabidopsis halleri. Plant Physiol 142:148–167

Talukdar S, Aarts MGM (2007) Arabidopsis thaliana and Thlaspi caerulescens respond com- parably to low zinc supply. Plant Soil 306(2007):85–94
Tavallali V, Rahemi M, Maftoun M, Panahi B, Karimi S, Ramezanian A, Vaezpour M (2009) Zinc influence and salt stress on photosynthesis, water relations, and carbonic anhydrase activity in pistachio. Sci Hortic 123(2):272–279
Tewari RK, Kumar P, Sharma PN (2008) Morphology and physiology of zinc-stressed mulberry plants. J Plant Nutr Soil Sci 171:286–294
Tripathy BC, Pattanayak GK (2012) Chlorophyll biosynthesis in higher plants. In: Eaton-Rye JJ, Tripathy BC, Sharkey TD (eds) Photosynthesis: advances in photosynthesis and respiration. Springer, Dordrecht, pp 63–94
Tsonev T, Cebola Lidon FJ (2012) Zinc in plants-an overview. Emir J Food Agric 24(4)
Tsuji N, Hirayanagi N, Iwabe O, Namba T, Tagawa M, Miyamoto S, Miyasaka H, Takagi M, Hirata K, Miyamoto K (2003) Regulation of phytochelatin synthesis by zinc and cadmium in marine green alga, Dunaliella tertiolecta. Phytochemistry 62(3):453–459
van de Mortel JE, Almar Villanueva L, Schat H, Kwekkeboom J, Coughlan S, Moerland PD, Loren V, van Themaat E, Koornneef M, Aarts MG (2006 Nov) Large expression differences in genes for iron and zinc homeostasis, stress response, and lignin biosynthesis distinguish roots of Arabidopsis thaliana and the related metal hyperaccumulator Thlaspi caerulescens. Plant Physiol 142(3):1127–1147
van Dongen JT, Licausi F (2015) Oxygen sensing and signaling. Annu Rev Plant Biol 66:345–367
Vanacker H (2006) Roles for redox regulation in leaf senescence of pea plants grown on different sources of nitrogen nutrition. J Exp Bot 57:1735–1745
Vert G, Grotz N, Dédaldéchamp F et al (2002) IRT1, an Arabidopsis transporter essential for iron uptake from the soil and for plant growth. Plant Cell 14:1223–1233
Vinson C, Acharya A, Taparowsky EJ (2006) Deciphering B-ZIP transcription factor interactions in vitro and in vivo. Biochim Biophys Acta 1759:4–12
Wang C, Zhang SH, Wang PF, Hou J, Zhang WJ, Li W, Lin ZP (2009a Jun) The effect of excess Zn on mineral nutrition and antioxidative response in rapeseed seedlings. Chemosphere 75(11):1468–1476
Wang C, Zhang SH, Wang PF, Qian J, Hou J, Zhang WJ, Lu J (2009b) Excess Zn alters the nutrient uptake and induces the antioxidative responses in submerged plant Hydrilla verticillata (L.f.) Royle. Chemosphere 76:938–945
Wang X, Zhao L, Zhang L, Wu Y, Chou M, Wei G (2018) Comparative symbiotic plasmid analysis indicates that symbiosis gene ancestor type affects plasmid genetic evolution. Lett Appl Microbiol 67(1):22–31
Wei C, Jiao Q, Agathokleous E, Liu H, Li G, Zhang J et al (2022) Hormetic effects of zinc on growth and antioxidant defense system of wheat plants. Sci Total Environ 807:150992
White MD, Klecker M, Hopkinson RJ, Weits DA, Mueller C, Naumann C, O'Neill R, Wickens J, Yang J, Brooks-Bartlett JC et al (2017) Plant cysteine oxidases are dioxygenases that directly enable arginyl transferase-catalysed arginylation of N-end rule targets. Nat Commun 8:14690
White MD, Kamps JJAG, East S, Taylor Kearney LJ, Flashman E (2018) The plant cysteine oxidases from Arabidopsis thaliana are kinetically tailored to act as oxygen sensors. J Biol Chem 293:11786–11795
Wolf M, Paulino AT (2020) Phytotoxicity increase induced by zinc accumulation in Cichorium intybus. Bull Environ Contam Toxicol 105(3):405–410
Xie R, Zhao J, Lu L, Ge J, Brown PH, Wei S, Wang R, Qiao Y, Webb SM, Tian S (2019) Efficient phloem remobilization of Zn protects apple trees during the early stages of Zn deficiency. Plant Cell Environ 42(12):3167–3181
Xu J, Yin H, Li Y, Liu X (2010) Nitric oxide is associated with long-term zinc tolerance in Solanum nigrum. Plant Physiol 154(3):1319–1334
Yang O, Popova OV, Süthof U et al (2009) The Arabidopsis basic leucine zipper transcription factor AtbZIP24 regulates complex transcriptional networks involved in abiotic stress resistance. Gene 436:45–55

Yang Q, Ma X, Luo S, Gao J, Yang X, Feng Y (2018) SaZIP4, an uptake transporter of Zn/Cd hyperaccumulator Sedum alfredii Hance. Environ Exp Bot 155:107–117

Yang M, Li Y, Liu Z, Tian J, Liang L, Qiu Y et al (2020) A high activity zinc transporter OsZIP9 mediates zinc uptake in rice. Plant J 103(5):1695–1709

Zaheer IE, Ali S, Saleem MH, Yousaf HS, Malik A, Abbas Z et al (2022) Combined application of zinc and iron-lysine and its effects on morpho-physiological traits, antioxidant capacity and chromium uptake in rapeseed (Brassica napus L.). PLoS One 17(1):e0262140

Zhang Y, Wang Y, Ding Z, Wang H, Song L, Jia S, Ma D (2017) Zinc stress affects ionome and metabolome in tea plants. Plant Physiol Biochem 111:318–328

Zhang H, Xu Z, Guo K, Huo Y, He G, Sun H et al (2020) Toxic effects of heavy metal Cd and Zn on chlorophyll, carotenoid metabolism and photosynthetic function in tobacco leaves revealed by physiological and proteomics analysis. Ecotoxicol Environ Saf 202:110856

Zhang J, Yang T, Zhang C, Zhang T, Pu L, Zhao W (2023) Effects of exogenous zinc on the physiological characteristics and enzyme activities of Passiflora edulis Sims f. edulis seedlings. PeerJ 11:e16280

Zhao K, Wu Y (2017) Effects of Zn deficiency and bicarbonate on the growth and photosynthetic characteristics of four plant species. PLoS One 12(1):e0169812

Zhao SB, Li YG, Zhao B, Lin H (2018) Identification and expression analyses of Mt MTP3, a zinc transporter of CDF family in Medicago truncatula. J Huazhong Agric Univ 37:52–60

Zlobin IE, Kartashov AV, Shpakovski GV (2017) Different roles of glutathione in copper and zinc chelation in Brassica napus roots. Plant Physiol Biochem 118:333–341

Molecular Aspects of Zinc Homeostasis in Plants: Zinc Deficit and Surplus Conditions

Komal Dubey and Sudhakar Srivastava

Abstract

Zinc (Zn), classified as a micronutrient, plays a critical role in plant health, soil quality, and plant-microbe interactions. Despite high total zinc content in soils, many regions globally experience low plant-available zinc, and this affects crop quality and subsequently compromises human health. The optimum zinc nutrition of plants and zinc content of grains can be achieved if the detailed mechanisms of zinc uptake and transport to different organs and cellular organelles are known. A number of zinc transporters along with some zinc regulatory elements have been identified and characterized in plants over the years. It is understood that plants sense free zinc ion concentrations in cellular environment and accordingly activate or deactivate various zinc-responsive genes for maintaining optimum zinc levels. However, there are several unknown facts. For example, there are some zinc-responsive genes whose expression is linked to zinc levels, but the signal regulating them is not known. Additionally, there are other genes whose expression is not linked to zinc status and they may be subjected to other regulatory elements. This chapter aims to summarize up-to-date information with regard to zinc homeostasis in both zinc-deficient and zinc-sufficient conditions in plants.

Keywords

bZIP19/bZIP23 · HMA2/4 · Hyperaccumulators · mRNA degradation · Unfolded protein response

K. Dubey · S. Srivastava (✉)
Plant Stress Biology Laboratory, Institute of Environment and Sustainable Development, Banaras Hindu University, Varanasi, Uttar Pradesh, India
e-mail: sudhakar.iesd@bhu.ac.in

S. Srivastava, H. Kaur (eds.), *Zinc in Soil-Plant Continuum*,
https://doi.org/10.1007/978-981-96-4253-3_5

1 Introduction

Zinc (Zn), classified as a micronutrient, plays a critical role in plant health, soil quality, and plant-microbe interactions. In its oxidized form, Zn(II), zinc functions as a vital cofactor in numerous enzymes including carbonic anhydrase and alcohol dehydrogenase and regulatory proteins. It is crucial for the structural Zn-finger domains involved in DNA binding and protein interactions (Sriram and Lonchyna 2009). Unlike transition metals like iron and copper, which can directly accept or donate electrons, zinc is relatively safe around sensitive macromolecules like DNA (Lane and Morel 2000). Evolution has made zinc increasingly integral, with the Zn metalloproteome comprising about 5–6% of prokaryotic proteins and around 9% of eukaryotic proteins (Andreini and Bertini 2009). In protein binding sites, a number of possible coordination geometries may exist in which Zn may be present; however, the most common conformation is distorted tetrahedral (Stanton et al. 2022). Zinc-dependent processes in plants include auxin regulation, photosystem II repair, CO_2 concentration maintenance in the mesophyll, and various other physiological functions (Gupta et al. 2016).

Despite high total zinc content in soils, many regions globally experience low plant-available zinc, affecting up to 50% of cereal-growing areas (Cakmak 2008). Certain soil types, such as sandy, peat, and calcareous soils, are naturally low in zinc because of high phosphorus and silicon concentrations (Alloway 2008). A shortage of zinc can severely impact plant growth causing stunted development due to chlorophyll deficiency. Moreover, zinc uptake into phloem-fed fruits, seeds, and tubers is often limited, leading to low zinc levels in these edible parts even when zinc is adequately supplied (White and Broadley 2011). Consequently, zinc deficiency affects approximately 31% of the global population (Zlobin 2021). The dietary zinc deficiency remains a significant health issue, particularly in Africa and Asia (Kumssa et al. 2015). The disease symptoms for zinc deficiency develop rapidly. Earlier it was thought that these symptoms cannot be due to nutritional deficiency due to rapid progression of the symptoms and early deterioration of the plant. Evidence to support the need of zinc by plants was also hindered by the fact that it was difficult to free nutrient solutions from trace amounts of zinc. Owing to the requirement of trace amounts of zinc by plants, deficit conditions could not be achieved. Earliest evidence to prove the importance of zinc for normal growth of certain fungi was given by Raulin in 1869 (Clemens 2022).

Zinc is relatively immobile within plants, with deficiency symptoms first appearing in newly emerging leaves. Stunted growth is one of the most common indicators of zinc deficiency, but it is not a specific symptom. In addition to deficiency, zinc excess can also cause toxicity to plants, and a number of studies have demonstrated negative effects of zinc on plants (Kaur et al. 2023). Zinc toxicity, although less common, can interfere with the absorption of essential elements like iron and manganese and stimulate the production of reactive oxygen species (ROS). The concentration of total zinc in leaves varies between 30 and 100 mg Zn kg^{-1} DW (dry weight). Generally, beyond 100 mg kg^{-1}, zinc toxicity begins to appear in plants, while deficiency symptoms are observed below 15 mg Zn kg^{-1} DW (Marschner

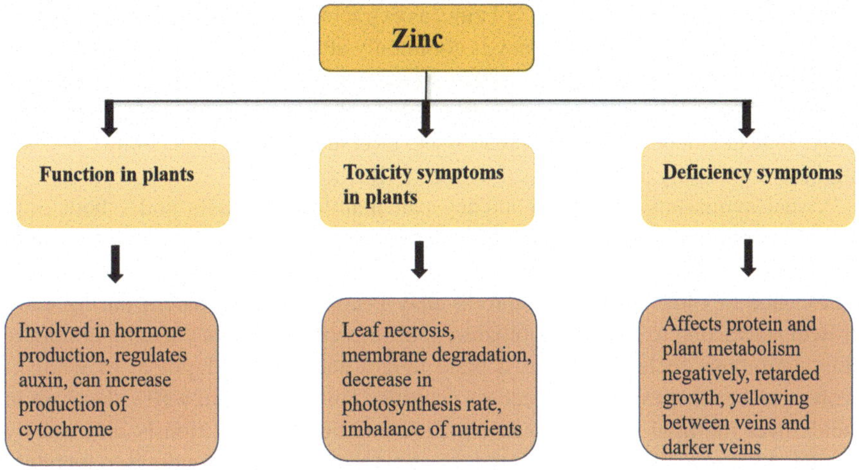

Fig. 1 The overview of Zn functions and its toxicity and deficiency symptoms

1995; White and Broadley 2011). However, total zinc concentrations can be very high 0.3–3 mM (Blindauer and Schmid 2010) but the total Zn concentration is not of much relevance in understanding the physiology of zinc deficit and surplus condition. It is the free zinc ion concentration in pM range, that is not bound by zinc-binding proteins and ligands (glutathione, nicotianamine, phosphate ions etc.), which is crucial to maintain availability of Zn (Zlobin 2021). In the presence of excess zinc, free Zn ion concentrations increase to nM levels, whereas in deficiency, free zinc ion can be almost completely depleted (Zlobin et al. 2019). Since both the deficiency and excess of zinc can be detrimental to plant growth and development (Fig. 1), proper regulation of intracellular zinc concentrations and its chemical forms is crucial for all organisms. The essentiality of zinc and the toxic response at excess levels have prompted the evolution of a sophisticated network of zinc transporters, low-molecular-weight ligands, and regulatory factors. This network ensures accurate zinc delivery to specific proteins, cellular compartments, and tissues (Eide 2009). Therefore, enhancing our understanding of zinc homeostasis in plants is essential for improving soil zinc acquisition and increasing zinc levels in edible plant parts (Assunção et al. 2010).

2 Zinc-Deficit Condition in Plants

Several factors can adversely affect the availability of zinc in soils for plant uptake, including low total zinc content, high pH levels exceeding 7, low redox potential, extended periods of flooding, microbial activity in the rhizosphere, high organic matter, elevated calcium carbonate and bicarbonate levels, high iron and manganese oxide content, a high magnesium-to-calcium exchangeable ratio, etc. (Zhang et al. 2012; Moreno-Lora and Delgado 2020). Zinc deficiency in soils is characterized by concentrations below 0.5 mg kg^{-1} of zinc, as extracted by diethylene triamine pentaacetic acid (DTPA). This deficiency is commonly attributed to high soil pH and low total zinc levels globally (Qadar 2002). More than half of the zinc-deficient soils worldwide are used for cereal cultivation, adversely impacting agricultural productivity. Zinc deficiency is a widespread issue, affecting nearly every country, including both developing and developed nations (Alloway 2009).

Visual symptoms of zinc deficiency can manifest in plants under both mild drought conditions and adequate soil moisture, including reduced biomass production, poor floral fertility, leaf chlorosis, increased shoot branching, and premature senescence of older leaves. Symptoms may become more apparent during early growth stages due to zinc immobilization. Severe zinc deficiency results in noticeable symptoms, including reduced plant height and the development of whitish-brown patches that eventually become necrotic. This can lead to a 50% decrease in yields over multiple growing seasons. Zinc and auxin production is intrinsically related. Although zinc is not essential for auxin synthesis, it is required to maintain the effectiveness of the active site. Zinc deficiency disrupts the formation of indole-3-acetic acid, impairing leaf and branch development. Protein synthesis is also affected negatively due to more production of ROS that happens due to decrease in activity of Cu/Zn superoxide dismutase (Cu/Zn-SOD). This results in membrane damage, chlorophyll degradation, and impaired photosynthesis and growth as ROS causes damage to the structural integrity of the cell. Premature bud development occurs during the deficiency due to increase in abscisic acid synthesis (Alloway 2008).

Chlorotic patches appear on older plant leaves, which progress from light green to yellow and ultimately to bleached areas (Alloway 2008). Yellow streaks and chlorotic bands on older plants also signal deficiency. In severe cases, immature leaves may fold. Wheat plants, for example, may exhibit dusty brown spots on upper leaves of stunted plants, with inhibited shoot growth compared to root growth, decreased tillering, increased spikelet sterility, chlorotic midribs near the base of younger leaves, and loss of turgor with brown blotches and streaks on lower leaves (Marschner 1995). While some plants can grow in highly zinc-deficient soils, the genetic basis for this ability is not well understood. In some wheat cultivars, zinc micronutrient use efficiency is associated not with increased root zinc uptake or translocation to the shoot, but with the ability to maintain the activity of zinc-dependent enzymes such as Cu/Zn-SOD and carbonic anhydrase under zinc-deficient conditions (Rashid et al. 1997).

3 Zinc Surplus Condition in Plants

The threshold for zinc toxicity differs among plant species and is influenced by the duration of exposure to zinc stress and the composition of the nutrient medium. Generally, plants accumulate zinc primarily in their roots as a survival strategy to maintain lower zinc levels in above-ground parts. Excessive zinc levels can severely impact photosynthesis and lead to phytotoxicity in plants. High zinc concentrations can inhibit auxin activity and cause negative geotropism in tomato roots (Kosesakal et al. 2009). Conversely, high zinc levels have also been found to upregulate genes related to auxin biosynthesis, leading to increased auxin accumulation in the root tips of *Arabidopsis thaliana* (Zhang et al. 2018). Zinc toxicity is generally marked by a reduction in plant yield when leaf Zn concentrations reach approximately 300–1000 mg kg^{-1}, with a typical critical concentration for phytotoxicity being around 500 mg kg^{-1}.

High levels of Zn can cause both structural and functional abnormalities in plants, impairing their overall performance. These effects vary depending on the plant species and their developmental stages. Plants exhibit various morphological, physiological, biochemical, and molecular responses to excess Zn. The most noticeable physiological effects of zinc toxicity are reduced germination vigor and biomass, which ultimately result in lower yield and poor quality of plant. The zinc is accumulated in the seed for germination and early seedling growth of the plant (Wang et al. 2020). In *Cicer arietinum*, excess zinc was found to inhibit amylase activity and sugar transport to the embryo axis, leading to fewer germinated seeds (Sharma et al. 2010).

Excess zinc can lead to chlorosis which may be followed by necrosis; this has been linked to a decrease in V-ATPase levels in *Arabidopsis thaliana* (Fukao et al. 2011). High zinc concentrations affect leaf area and number, likely due to inhibited cell division, elongation, or both (Glinska et al. 2016). Elevated zinc levels can damage mitochondrial organization, accelerate NAD+ breakdown, and diminish energy metabolism, thereby stunting overall plant growth (Zhang et al. 2017). High zinc concentrations also cause significant callose deposition in the root apical meristems of *Brassica juncea* and *B. napus*, potentially contributing to growth inhibition by reducing cell wall loosening and hindering symplastic transport (Feigl et al. 2015). In *A. thaliana*, zinc stress has been shown to disrupt root hair morphology, likely due to reduced levels of proteins like root hair defective 3 (RHD3) and morphogenesis of root hair 5 (MRH5), which are essential for root hair development (Fukao et al. 2011). Interactions between iron (Fe) and zinc are significant because both metals share common transporter proteins for their absorption. High levels of zinc can cause physiological iron deficiency, leading to stunted growth and chlorosis in plants (Bouain et al. 2014). There are well-documented cross-talks between zinc and nitrogen (N) and zinc and phosphorus (P) nutrition also (Cakmak et al. 2010; Bouain et al. 2014).

4 Zinc Uptake from the Soil

Plants absorb zinc from the soil through their roots via a process known as facilitated diffusion, which is driven by membrane potential and specific transporters. Zinc is primarily taken up as the divalent ion Zn^{2+}, although some absorption of organic ligand-Zn complexes may also occur. The uptake of divalent cations such as Zn^{2+} involves two main physiological strategies based on the type of ligand secreted by the plant roots. In Strategy I reductants, hydrogen ions, and organic acids are released which increase the solubility of zinc complexes (such as zinc phosphates and hydroxides) and release Zn^{2+} ions for absorption by root epidermal cells. Strategy II, which is specific to cereal roots, involves the excretion of phytosiderophores (phytometallophores) that form stable complexes with zinc and are subsequently taken up by root epidermal cells (Erenoglu et al. 2000). These phytosiderophores are non-protein amino acids; one of the most well-known examples is nicotianamine (Erenoglu et al. 2000). Some zinc-efficient cultivars are also known to exist. These are comparably more tolerant to zinc deficiency. It is commonly accepted that the zinc efficiency is due to better ability of the plant to obtain zinc and not due to increased tolerance to zinc deficiency (Ruel and Bouis 1998). The exact mechanism of how this is possible is not known, but the current knowledge suggests that this ability may be due to higher secretion of phytosiderophores by these plants (Zhang et al. 1989). Plants can also utilize symbiotic association with arbuscular mycorrhizal fungi (AMF) to uptake more zinc from soil (Coccina et al. 2019).

5 Zinc Homeostasis Mechanisms

5.1 The Role of Zinc Transporters

The transporters tightly regulate the uptake and the place of storage of zinc. Hence, the regulation of zinc homeostasis in plants is done with a tight control of zinc absorption and accumulation with the help of specific zinc transporters (Table 1). The transporters mediate zinc uptake from roots, radial transport of zinc from epidermal cells to vascular tissues, root-to-shoot translocation, and intracellular zinc movement from cytoplasm to endoplasmic reticulum, vacuoles, chloroplast, etc. (Stanton et al. 2022). A role of several transporters is suggested including cation diffusion facilitator (CDF), heavy metal ATPase (HMA), and zinc-regulated, iron-regulated transporter-like proteins (ZIP). The members of zinc transporters have been characterized in various plants like *A. thaliana* and *Oryza sativa* (rice) revealing their importance in regulating zinc levels in cells and tissues.

CDFs transport metals in the organism from the cytoplasm to the intercellular organelles (Kambe et al. 2004; Haney et al. 2005). Twelve CDF members have been reported in *A. thaliana* (Mäser et al. 2001). A member of CDF, Metal Transport Protein (MTP1), also called as Zinc transporter of *Arabidopsis thaliana* (ZAT1), has been characterized (Desbrosses-Fonrouge et al. 2005). It has been shown that upon over-expression of MTP1, the plant becomes increasingly resistant to otherwise toxic levels of zinc. The resistance is conferred to the plant due to increased sequestration of zinc and not due to efflux mechanism. This can be said as the roots of such plants have accumulation of more zinc in comparison to the wild variety (Desbrosses-Fonrouge et al. 2005; Kobae et al. 2004). MTP1 is responsible for zinc transport into the vacuole, and the transporter is highly sensitive to high zinc levels (Kambe et al. 2004).

Heavy metal ATPase 2 (HMA2) and Heavy metal ATPase 4 (HMA4) play a very important role in zinc homeostasis. These two metal transporters are part of the

Table 1 Various transporters involved in Zn homeostasis

S. no.	Zn transporter	Experimental organism	Function	References
1.	ZNT1	*Thlaspi caerulescens*	Metal accumulation	Lasat et al. (2000)
2.	ZIP6	*Arabidopsis halleri*	More expressed in shoots of *Arabidopsis helleri* in comparison to shoots of *Arabidopsis thaliana*	Becher et al. (2004)
3.	ZIP9	*Arabidopsis halleri*	Zn detoxification	Weber et al. (2004)
4.	HMA3	*Arabidopsis halleri*	Increases Zn tolerance	Becher et al. (2004)
5.	MTP1	*Thlaspi goesingense*	Zn efflux from cell	Persans et al. (2001), Krämer et al., (1997)
6.	HMA1	*Hordeum vulgare*	The remobilization of zinc and copper from chloroplasts and their transport during grain filling takes place, with gene expression being controlled by an unknown mechanism that responds to changes in zinc nutrition levels	Mikkelsen et al. (2012)
7.	HMA2	*Arabidopsis thaliana*	Helps in transport of Zn from pericycle to xylem	Arsova et al. (2019)
8.	HMA3	*Noccaea caerulescens* and *Arabidopsis halleri*	Cd/Zn hyperaccumulators exhibit a high basal level of expression, facilitating the transport of zinc into the vacuole lumen	Mishra et al. (2017)
9.	bZIP19 bZIP23	*Arabidopsis thaliana*	A double mutant of both bZIP19 and bZIP23 shows heightened sensitivity to zinc deficiency; both have been shown to bind with Zn^{2+} Mutations in just one of the bZIP genes have minimal impact on the plant	Lilay et al. (2021)
10.	MTP1	This transporter protein is found in both plants and animals	*MTP1* is also referred to as zinc transporter of *Arabidopsis thaliana (ZAT1)*; when MTP1 is overexpressed, plants exhibit resistance to toxic concentrations of zinc	Desbrosses-Fonrouge et al. (2005)
11.	MTP2	Mainly found in plants	Responsible for transporting zinc to the lumen of ER. An unknown factor regulates transcription in response to changes in cellular zinc levels	Arsonova et al. (2019), Zlobin et al. (2020)

(continued)

Table 1 (continued)

S. no.	Zn transporter	Experimental organism	Function	References
12.	MTP3	*Arabidopsis thaliana*	Responsible for transporting zinc to the lumen of vacuole. Transcription is activated by excess zinc through an unidentified mechanism	Arrivvault et al. (2006)
13.	MTP5	*Cucumis sativus*	Responsible for transporting zinc to the lumen of Golgi apparatus. Both *MTP5* and *MTP5* form complex with each other to function as Zn transporters	Migocka et al. (2018)
14.	MTP12	*Cucumis sativus* and *Arabidopsis thaliana*	Responsible for transporting zinc to the lumen of Golgi apparatus by forming complex with *MTP5*	Migocka et al. (2018), Zlobin et al. (2020)
15.	MTP1	*Poplar*	Over expression provides plants with zinc tolerance	Blaudez et al. (2003)

P1B-type ATPase family, which consists of eight members in *Arabidopsis*. It was an interesting observation that single mutants for hma2, hma3 (heavy metal ATPase 3), and hma4, as well as the double mutant hma3 hma4, did not show any noteworthy growth issues in the soil. However, the hma2 hma4 double mutants exhibited chlorosis, and plants were unable to produce seeds (Hussain et al. 2004). Despite lower zinc levels in the aerial parts, the roots of these mutants contained about twice as much zinc as the wild type, indicating a possible defect in zinc translocation from roots to shoots (Hussain et al. 2004; Verret et al. 2004). hma2 and hma4 mediate movement of zinc from pericycle into the xylem and subsequent long-distance mass flow-driven transport to shoots. Under conditions of severe Zn deficiency, AtMTP2 plays a role in root-to-shoot zinc transport by mediating the movement of zinc from cytosol to endoplasmic reticulum from where it is moved out by another transporter ZTP29 into pericycle. This mechanism aids in symplastic radial transport of zinc within roots and avoids zinc consumption in the route by metalloproteins (Sinclair et al. 2018).

The ZIP and iron-regulated transporter (IRT) families are crucial for zinc transport and consist of over 100 members across various organisms, with 16 found in *Arabidopsis*. In *Arabidopsis*, ZIP1, ZIP2, ZIP3, and ZIP4 proteins can each functionally compensate for a yeast strain deficient in zinc uptake (Grotz et al. 1998). The soybean homolog of ZIP1, known as GmZIP1, plays a role in the symbiotic relationship between soybean and *Bradyrhizobium japonicum* (Moreau et al. 2002). Besides complementing the zinc uptake defect in yeast, GmZIP1 is localized to the peribacteroid membrane (also known as symbiosome membrane), and its mRNA is specifically found in nodules. Notably, GmZIP1 appears to transport zinc in a direction opposite to other ZIP family members since it facilitates zinc uptake into the symbiosome, which is analogous to transport out of the cell. In plants, the mRNAs of ZIP1, ZIP3, and ZIP4 increase in response to zinc deficiency, with ZIP4 also being regulated by copper (Wintz et al. 2003).

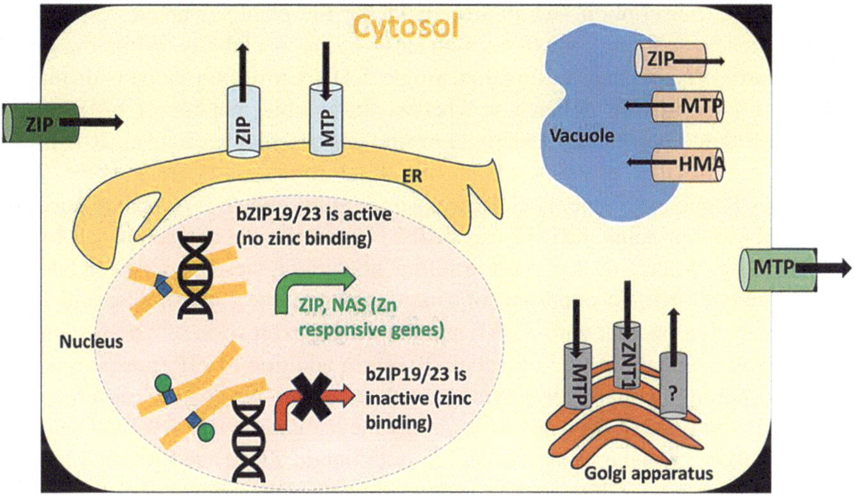

Fig. 2 The diagram depicts the transport of zinc (Zn) through various transporters, with black arrows indicating the direction of Zn movement within the cell. The primary emphasis is on the Zn-sensors, bZIP19/23, located in the nucleus, which gets activated in the absence of Zn. These sensors also detect Zn by binding to a Zn-sensing motif in their N-terminal region. When zinc levels are adequate, free or loosely bound Zn ions from the "free" Zn pool attach to the Cys/His-rich Zinc Sensor Motif (ZSM) found in the bZIP19 and bZIP23 transcription factors. Therefore, when Zn is abundant, this binding inactivates bZIP19/23, disrupting their control over the expression of genes that respond to Zn availability. In the absence of zinc, bZIP19/23 interacts with zinc transporters and other proteins via zinc deficiency response elements (ZDRE) and activate their expression

OsZIP4 in rice accumulates in both zinc-deficient shoots and roots (Ishimaru et al. 2005). When expressed in yeast mutants lacking plasma membrane zinc uptake, OsZIP4 restores the yeast's growth under zinc-limited conditions, suggesting its role in aiding zinc accumulation under deficiency. OsZIP4 likely transports zinc across the plasma membrane into the cytoplasm. Additionally, a protein similar to ZIP2 has been studied in *Medicago truncatula*. MtZIP2 can complement a yeast strain defective in zinc uptake and is located at the plasma membrane. Figure 2 illustrates the anticipated direction and location of zinc transport within the plant cell.

6 Zinc Deficiency Adaptations in Plant

To adapt to zinc deficiency conditions, plants need to sense free zinc ion levels in cells and to regulate the expression of transporters accordingly. Through a yeast-one-hybrid screening utilizing promoter fragments of the *Arabidopsis* ZIP4 zinc transporter gene, which is strongly induced under zinc-deficient conditions, two closely related members of the basic-region leucine-zipper (bZIP) transcription factor family in *Arabidopsis*, bZIP19 and bZIP23, were identified (Assunção et al.

2013). These transcription factors are crucial for the plant's adaptive response to zinc deficiency, enabling growth even under low zinc availability. While the genes exhibit partial redundancy in function, single T-DNA insertion mutants displayed mild sensitivity to zinc deficiency, whereas the double mutants of bZIP19 and bZIP23 exhibited heightened sensitivity to zinc scarcity (Assunção et al. 2010). The bZIP19 and bZIP23 proteins specifically bind to a palindromic 10-bp Zinc Deficiency Response Element (ZDRE) sequence motif, which is present in genes of zinc responsive proteins. bZIP19 and bZIP23 can sense the cytosolic free Zn ion concentration. When zinc is present in sufficient amount, these bind zinc and do not interact with ZDRE. In condition of zinc deficiency, these loose zinc and form homodimers that interact with ZDRE and induce the expression of zinc-responsive gene (Assunção et al. 2013; Lilay et al. 2019). In addition to ZIP transporters, the genes encoding nicotianamine synthases, NAS2 and NAS4, are differentially expressed in response to zinc deficiency under the regulation of bZIP19 and bZIP23. However, there are other genes which do not contain ZDRE and respond to zinc deficiency. In rice, OsZIP gene does not contain any ZDREs, but it does respond to zinc concentrations (Castro et al. 2017). Hence, there shall be other zinc regulators also that need to be delineated.

The homeostasis of zinc and the expression of zinc responsive genes are linked to the free zinc ion concentration in cells. This fact becomes important considering the root-to-shoot zinc translocation and organeller movement that needs to continue in conditions of both zinc excess and deficiency. Hence, there are transporters that need active energy input to drive zinc against the concentration gradient in cytoplasm both for xylem loading and transport to subcellular organelles. HMA2, HMA4, and MTP2 constitute such energy requiring transporters (Fujiwara et al. 2015). The expression of HMA2 and MTP2 can be induced even when there is an excess of free zinc ion in the cytoplasm by unidentified signal which is different from bZIP19/bZIP23 (Sinclair et al. 2018). However, local zinc level can also affect the expression of MTP2 (Zlobin et al. 2020).

7 Surplus Zinc Adaptions: Hyperaccumulators

Hyperaccumulators are plants capable of accumulating metal levels that would be toxic to non-hyperaccumulating plants. For example, *Thlaspi caerulescens* and *Arabidopsis halleri* are known to hyperaccumulate zinc and cadmium (Brown et al. 1995). Zinc transporters are more active in hyperaccumulators; for example, ZNT1 in *T. caerulescens* is closely related to ZIP4 in *A. thaliana* (Pence et al. 2000). It has been observed that ZNT1 mRNA is present in high levels in *T. caerulescens* regardless of zinc availability, suggesting significant role of ZNT1 in metal accumulation in this plant (Lasat et al. 2000). In contrast, in the non-hyperaccumulating species *T. arvense*, ZNT1 mRNA is only present under zinc-deficient conditions (Pence et al. 2000).

In *A. halleri*, the expression levels of ZIP6 and ZIP9 are significantly higher compared to *A. thaliana* (Becher et al. 2004). RT-PCR study was able to confirm that ZIP6 is expressed 23–24 times more in the shoots and 4–9 times higher in the roots of *A. halleri* when compared to *A. thaliana*. In the same study, AhHMA3 was identified as being highly expressed in the shoots of *A. halleri*. In the roots of *A. halleri*, ZIP9 expression was found to be 43 times higher than in *A. thaliana* (Weber et al. 2004). The CDF family member AhCDF1–3 (also called AhMTP1), similar to MTP1/ZAT1 in *A. thaliana*, is expressed about 10- to 20-fold higher in the shoots of *A. halleri* than in *A. thaliana* (Becher et al. 2004). AhCDF1-3 is localized to the vacuolar membrane in both yeast and *Arabidopsis* protoplasts. While the activity of this protein might be higher in *A. halleri* compared to non-hyperaccumulating *A. thaliana*, the increased expression seems partly due to its presence in multiple copies in *A. halleri*, whereas *A. thaliana* has a single copy. Similarly, high basal expression of HMA4 in *A. halleri* and *Noccaea caerulescens* is due to increase in copy number of the gene (Hanikenne et al. 2008; Cracium et al. 2012). The high tolerance of hyperaccumulator plants to cytosolic zinc concentrations is attributable to their ability to transport zinc out of the cytoplasm and store it in the vacuole (Tanaka et al. 2013). A vacuolar zinc transporter MTP1 is regulated as per zinc availability for the purpose. MTP1 has a His-rich loop on the cytosolic side that can bind four zinc ions, and this binding changes its conformation to allow zinc transport into vacuole (Tanaka et al. 2013).

8 The Potential Mechanisms for Regulating Zinc Homeostasis

Plants must adjust their capacity for zinc uptake in roots and its transport to aerial parts to cope with fluctuations in zinc availability in the soil. This is achieved by upregulation and downregulation of zinc responsive genes in short timescales (Claus et al. 2013) to prevent drastic changes in zinc levels. This mechanism keeps plants safe in both Zn toxic and deficient conditions. Rapid changes in gene expressions mean that the turnover of their mRNAs occurs in quick time. Recent research by Zlobin et al. (2020) suggested that an increase in free Zn ion levels caused by zinc supplementation in zinc-deficient *Arabidopsis* cells resulted in rapid degradation of mRNAs for several zinc homeostasis genes (ZIP1, ZIP4, and MTP2) and protein levels of several zinc transporters declined rapidly (Arsova et al. 2019), providing indirect evidence of mechanisms actively downregulating zinc acquisition machinery in plants.

In condition of zinc deficiency, zinc recycling from organelles and storing it can be helpful. Nutrient recycling is a unique process adopted by plants to manage deficient conditions, and autophagy has an important role in it (Chen et al. 2019). Though not known with certainty, a few experiments demonstrate that autophagy might have a role in zinc recycling. When autophagy is disrupted, the sensitivity of *Arabidopsis* mutants to zinc deficiency has been found to increase (Shinozaki et al. 2020). The induction of autophagy has been proposed to be linked to local zinc ion concentrations in zinc-deficient conditions (Shinozaki et al. 2020). A method by which plants respond to zinc deficiency is by increasing their uptake, which is mediated by transporters. However, another important factor in this mechanism is the induced secretion of root exudates that help make zinc bioavailable to plants (Hussain et al. 2018).

Protein activity, alongside ion transporters and mRNA degradation, may play a role in detecting zinc status within plant cells. In conditions of zinc deficiency, zinc-requiring proteins may not get sufficient zinc and may become improperly folded. This can lead to presence of improperly structured apoproteins as has been observed in yeast cells (MacDiarmid et al. 2013; Wang et al. 2018). The build-up of misfolded proteins triggers specialized stress responses known as unfolded protein response (UPR) in the endoplasmic reticulum and the cytosolic protein response (CPR) in the cytosol (Angelos et al. 2017; Lin et al. 2018). A direct link between UPR and zinc transport into endoplasmic reticulum has been established (Ellis et al. 2004; Wang et al. 2010; Nguyen et al. 2013). Thus, zinc status of the cells may induce both UPR and CPR, though their specific roles are not yet fully understood.

9 Conclusions

This chapter provides a comprehensive overview of zinc and its complex interactions with plants. The chapter highlights the importance of zinc in plants, followed by an examination of the mechanisms of zinc homeostasis and the effects of both zinc deficiency and excess in different plants. Despite considerable advances in recent years in understanding how plants absorb zinc from the soil and distribute it internally, many gaps still remain. To effectively engineer plants that can enhance human nutrition, a comprehensive understanding of all metal transport mechanisms is essential. It is important to note that within a multicellular organism like a plant, the zinc requirements of different cells may vary. For instance, the needs of photosynthetically active cells are likely different from those of non-photosynthetically active cells. Nevertheless, all cells must achieve proper metal homeostasis to function cohesively. Additionally, a deeper understanding is needed of the quantitative regulation of zinc homeostasis genes, including their transcription and mRNA degradation and how these relate to changes in physiological zinc status.

References

Alloway BJ (2008) Zinc in soils and crop nutrition. International Zinc Association and International Fertilizer Association, p 16

Alloway BJ (2009) Soil factors associated with zinc deficiency in crops and humans. Environ Geochem Health 31:537–548

Andreini C, Bertini I (2009) Metalloproteomes: a bioinformatic approach. Acc Chem Res 42:1471–1479

Angelos E, Ruberti C, Kim SJ, Brandizzi F (2017) Maintaining the factory: the roles of the unfolded protein response in cellular homeostasis in plants. Plant J 90:671–682

Arrivault S, Senger T, Kramer U (2006) The Arabidopsis metal tolerance protein AtMTP3 maintains metal homeostasis by mediating Zn exclusion from the shoot under Fe deficiency and Zn oversupply. Plant J 46(5):861–879

Arsova B, Amini S, Scheepers M, Baiwir D, Mazzucchelli G, Carnol M, Bosman B, Motte P, de Pauw E, Watt M, Hanikenne M (2019) Resolution of the proteome, transcript and ionome dynamics upon Zn re-supply in Zn-deficient arabidopsis. bioRxiv 600569

Assunção AG, Herrero E, Lin YF, Huettel B, Talukdar S, Smaczniak C, Immink RGH, van Eldik M, Fiers M, Schat H, Aarts MG (2010) Arabidopsis thaliana transcription factors bZIP19 and bZIP23 regulate the adaptation to zinc deficiency. Proc Natl Acad Sci USA 107:10296–10301

Assunçao AG, Persson DP, Husted S, Schjørring JK, Alexander RD, Aarts MG (2013) Model of how plants sense zinc deficiency. Metallomics 5(9):1110–1116

Becher M, Talke IN, Krall L, Krämer U (2004) Cross-species microarray transcript profiling reveals high constitutive expression of metal homeostasis genes in shoots of the zinc hyperaccumulator *Arabidopsis halleri*. Plant J 37:251–268

Blaudez D, Kohler A, Martin F, Sanders D, Chalot M (2003) Poplar metal tolerance protein 1 confers zinc tolerance and is an oligomeric vacuolar zinc transporter with an essential leucine zipper motif. Plant Cell 15:2911–2928

Blindauer CA, Schmid R (2010) Cytosolic metal handling in plants: determinants for zinc specificity in metal transporters and metallothioneins. Metallomics 2:510–529

Bouain N, Shahzad Z, Rouached A, Khan GA, Berthomieu P, Abdelly C, Poirier Y, Rouached H (2014) Phosphate and zinc transport and signalling in plants: toward a better understanding of their homeostasis interaction. J Exp Bot 65:5725–5741

Brown SL, Chaney RL, Angle JS, Baker AJM (1995) Zinc and cadmium uptake by hyperaccumulator Thlaspi caerulescens grown in nutrient solution. Soil Sci Soc Am J 59(1):125–133

Cakmak I (2008) Enrichment of cereal grains with zinc: agronomic or genetic biofortification? Plant Soil 302(1–2):1–17

Cakmak I, Kalayci M, Kaya Y, Torun AA, Aydin N, Wang Y, Arisoy Z, Erdem H, Yazici A, Gokmen O et al (2010) Biofortification and localization of zinc in wheat grain. J Agric Food Chem 58:9092–9102

Castro PH, Lilay GH, Munoz-Merida A, Schjoerring JK, Azevedo H, Assunçao AG (2017) Phylogenetic analysis of F-bZIP transcription factors indicates conservation of the zinc deficiency response across land plants. Sci Rep 7(1):1–14

Chen Q, Shinozaki D, Luo J, Pottier M, Have M, Marmagne A, Reisdorf-Cren M, Chardon F, Thomine S, Yoshimoto K, Masclaux-Daubresse C (2019) Autophagy and nutrients management in plants. Cells 8(11):1426

Claus J, Bohmann A, Chavarría-Krauser A (2013) Zinc uptake and radial transport in roots of Arabidopsis thaliana: a modelling approach to understand accumulation. Ann Bot 112(2):369–380

Clemens S (2022) The cell biology of zinc. J Exp Bot 73:1688–1698

Coccina A, Cavagnaro TR, Pellegrino E, Ercoli L, McLaughlin MJ, Watts-Williams SJ (2019) The mycorrhizal pathway of zinc uptake contributes to zinc accumulation in barley and wheat grain. BMC Plant Biol 19:133

Craciun AR, Meyer CL, Chen J, Roosens N, De Groodt R, Hilson P, Verbruggen N (2012) Variation in HMA4 gene copy number and expression among Noccaea caerulescens populations presenting different levels of Cd tolerance and accumulation. J Exp Bot 63(11):4179–4189

Desbrosses-Fonrouge AG, Voigt K, Schroder A, Arrivault S, Thomine S, Kramer U (2005) Arabidopsis thaliana MTP1 is a Zn transporter in the vacuolar membrane which mediates Zn detoxification and drives leaf Zn accumulation. FEBS Lett 579:4165–4174

Eide DJ (2009) Homeostatic and adaptive responses to zinc deficiency in Saccharomyces cerevisiae. J Biol Chem 284:18565–18569

Ellis CD, Wang F, MacDiarmid CW, Clark S, Lyons T, Eide DJ (2004) Zinc and the Msc2 zinc transporter protein are required for endoplasmic reticulum function. J Cell Biol 166(3):325–335

Erenoglu B, Eker S, Cakmak I, Derici R, Romheld V (2000) Effect of iron and zinc deficiency on release of phytosiderophores in barley cultivars differing in zinc efficiency. J Plant Nutr 23:1645–1656

Feigl G, Lehotai N, Molnár A, Ördög A, Rodríguez-Ruiz M, Palma JM et al (2015) Zinc induces distinct changes in the metabolism of reactive oxygen and nitrogen species (ROS and RNS) in the roots of two Brassica species with different sensitivity to zinc stress. Ann Bot 116(4):613–625

Fujiwara T, Kawachi M, Sato Y, Mori H, Kutsuna N, Hasezawa S, Maeshima M (2015) A high molecular mass zinc transporter MTP12 forms a functional heteromeric complex with MTP5 in the Golgi in Arabidopsis thaliana. FEBS J 282(10):1965–1979

Fukao Y, Ferjani A, Tomioka R, Nagasaki N, Kurata R, Nishimori Y et al (2011) iTRAQ analysis reveals mechanisms of growth defects due to excess zinc in Arabidopsis. Plant Physiol 155(4):1893–1907

Glińska S, Gapińska M, Michlewska S, Skiba E, Kubicki J (2016) Analysis of Triticum aestivum seedling response to the excess of zinc. Protoplasma 253:367–377

Grotz N, Fox T, Connolly EL, Park W, Guerinot ML, Eide D (1998) Identification of a family of zinc transporter genes from Arabidopsis that respond to zinc deficiency. Proc Natl Acad Sci USA 95:7220–7224

Gupta N, Ram H, Kumar B (2016) Mechanism of zinc absorption in plants: uptake, transport, translocation and accumulation. Rev Environ Sci Biotechnol 15(1):89–109

Haney CJ, Grass G, Franke S, Rensing C (2005) New developments in the understanding of the cation diffusion facilitator family. J Ind Microbiol Biotechnol 32:215–226

Hanikenne M, Talke IN, Haydon MJ, Lanz C, Nolte A, Motte P, Kroymann J, Weigel D, Kramer U (2008) Evolution of metal hyperaccumulation required cisregulatory changes and triplication of HMA4. Nature 453(7193):391–395

Hussain D, Haydon MJ, Wang Y, Wong E, Sherson SM, Young J, Camakaris J, Harper JF, Cobbett CS (2004) P-type ATPase heavy metal transporters with roles in essential zinc homeostasis in Arabidopsis. Plant Cell 16:1327–1339

Hussain A, Zahir ZA, Asghar HN, Ahmad M, Jamil M, Naveed M, Akhtar MFUZ (2018) Zinc solubilizing bacteria for zinc biofortification in cereals: a step toward sustainable nutritional security. In: Role of Rhizospheric microbes in soil. Springer, Singapore, pp 203–227

Ishimaru Y, Suzuki M, Kobayashi T, Takahashi M, Nakanishi H, Mori S, Nishizawa NK (2005) OsZIP4, a novel zinc-regulated zinc transporter in rice. J Exp Bot 56:3207–3214

Kambe T, Yamaguchi-Iwai Y, Sasaki R, Nagao M (2004) Overview of mammalian zinc transporters. Cell Mol Life Sci 61:49–68

Kaur H, Kaur H, Kaur H, Srivastava S (2023) The beneficial roles of trace and ultratrace elements in plants. Plant Growth Regul 2023(100):219–236

Kobae Y, Uemura T, Sato MH, Ohnishi M, Mimura T, Nakagawa T, Maeshima M (2004) Zinc transporter of Arabidopsis thaliana AtMTP1 is localized to vacuolar membranes and implicated in zinc homeostasis. Plant Cell Physiol 45:1749–1758

Kosesakal T, Unal M, Oz GC (2009) Influence of zinc toxicity on gravitropic response of tomato (Lycopersicon esculentum Mill.) roots. Fresenius Environ Bull 18(12):2402–2407

Krämer U, Smith RD, Wenzel WW, Raskin I, Salt DE (1997) The role of metal transport and tolerance in nickel hyperaccumulation by Thlaspi goesingense Hálácsy. Plant Physiol 115:1641–1650

Kumssa DB, Joy EJ, Ander EL, Watts MJ, Young SD, Walker S, Broadley MR (2015) Dietary calcium and zinc deficiency risks are decreasing but remain prevalent. Sci Rep 5(1):1–11

Lane TW, Morel FM (2000) A biological function for cadmium in marine diatoms. Proc Natl Acad Sci 97(9):4627–4631

Lasat MM, Pence NS, Garvin DF, Ebbs SD, Kochian LV (2000) Molecular physiology of zinc transport in the Zn hyperaccumulator Thlaspi caerulescens. J Exp Bot 51(342):71–79

Lilay GH, Castro PH, Campilho A, Assunçao AG (2019) The Arabidopsis bZIP19 and bZIP23 activity requires zinc deficiency – insight on regulation from complementation lines. Front Plant Sci 9:1955

Lilay GH, Persson DP, Castro PH, Liao F, Alexander RD, Aarts MG, Assunção AG (2021) Arabidopsis bZIP19 and bZIP23 act as zinc sensors to control plant zinc status. Nat Plants 7(2):137–143

Lin KF, Tsai MY, Lu CA, Wu SJ, Yeh CH (2018) The roles of Arabidopsis HSFA2, HSFA4a, and HSFA7a in the heat shock response and cytosolic protein response. Bot Stud 59:1–9

MacDiarmid CW, Taggart J, Kerdsomboon K, Kubisiak M, Panascharoen S, Schelble K, Eide DJ (2013) Peroxiredoxin chaperone activity is critical for protein homeostasis in zinc-deficient yeast*◆. J Biol Chem 288(43):31313–31327

Marschner H (1995) Mineral nutrition of higher plants, 2nd edn. Academic Press, London

Mäser P., S. Thomine, J.I. Schroeder, J.M. Ward, K. Hirschi, H. Sze, I.N. Talke, A. Amtmann, F.J.M. Matthuis, D. Sanders, J.F. Harper, J. Tchieu, M. Gribskov, M.W. Persans, D.E. Salt, S.A. Kim, M.L. Guerinot, Phylogenetic relationships within cation transporter families of Arabidopsis, Plant Physiol 126 (2001) 1646–1667

Migocka M, Małas K, Maciaszczyk-Dziubinska E, Posyniak E, Migdal I, Szczech P (2018) Cucumber Golgi protein CsMTP5 forms a Zn-transporting heterodimer with high molecular mass protein CsMTP12. Plant Sci 277:196–206

Mikkelsen MD, Pedas P, Schiller M, Vincze E, Mills RF, Borg S et al (2012) Barley HvHMA1 is a heavy metal pump involved in mobilizing organellar Zn and Cu and plays a role in metal loading into grains. PLoS One 7(11):e49027

Mishra S, Mishra A, Küpper H (2017) Protein biochemistry and expression regulation of cadmium/zinc pumping ATPases in the hyperaccumulator plants Arabidopsis halleri and Noccaea caerulescens. Front Plant Sci 8:835

Moreau S, Tohomson RW, Kaiser BN, Trevaskis B, Guerinot ML, Udvardi MK, Puppo A, Day DA (2002) GmZIP1 encodes a symbiosis specific transporter in soybean. J Biol Chem 277:4738–4746

Moreno-Lora A, Delgado A (2020) Factors determining Zn availability and uptake by plants in soils developed under Mediterranean climate. Geoderma 376:114509

Nguyen TSL, Kohno K, Kimata Y (2013) Zinc depletion activates the endoplasmic reticulum-stress sensor Ire1 via pleiotropic mechanisms. Biosci Biotechnol Biochem 77(6):1337–1339

Pence NS, Larsen PB, Ebbs SD, Letham DL, Lasat MM, Garvin DF et al (2000) The molecular physiology of heavy metal transport in the Zn/Cd hyperaccumulator Thlaspi caerulescens. Proc Natl Acad Sci 97(9):4956–4960

Persans MW, Nieman K, Salt DE (2001) Functional activity and role of cation-efflux family members in Ni hyperaccumulation in Thlaspi goesingense. Proc Natl Acad Sci USA 98:9995–10000

Qadar A (2002) Selecting rice genotypes tolerant to zinc deficiency and sodicity stresses. I. Differences in zinc, iron, manganese, copper, phosphorus concentrations, and phosphorus/zinc ratio in their leaves. J Plant Nutr 25(3):457–473

Rashid A, Rafique E, Bughio N, Yasin M (1997) Micronutrient deficiencies in rainfed calcareous soils of Pakistan. IV. Zinc nutrition of sorghum. Commun Soil Sci Plant Anal 28(6–8):455–467

Ruel MT, Bouis HE (1998) Plant breeding: a long term strategy for the control of zinc deficiency in vulnerable populations. Am J Clin Nutr 68(Suppl. 2):488S–494S

Sharma S, Sharma P, Datta SP, Gupta V (2010) Morphological and biochemical response of Cicer arietinum L. var. pusa-256 towards an excess of zinc concentration. Life Sci J 7(1):95–98

Shinozaki D, Merkulova EA, Naya L, Horie T, Kanno Y, Seo M, Ohsumi Y, Masclaux-Daubresse C, Yoshimoto K (2020) Autophagy increases zinc bioavailability to avoid light-mediated reactive oxygen species production under zinc deficiency. Plant Physiol 182(3):1284–1296

Sinclair SA, Senger T, Talke IN, Cobbett CS, Haydon MJ, Kraemer U (2018) Systemic upregulation of MTP2-and HMA2-mediated Zn partitioning to the shoot supplements local Zn deficiency responses. Plant Cell 30(10):2463–2479

Sriram K, Lonchyna VA (2009) Micronutrient supplementation in adult nutrition therapy: practical considerations. JPEN J Parenter Enteral Nutr 33(5):548–562

Stanton C, Sanders D, Krämer U, Podar D (2022) Zinc in plants: integrating homeostasis and biofortification. Mol Plant 15(1):65–85

Tanaka N, Kawachi M, Fujiwara T, Maeshima M (2013) Zinc-binding and structural properties of the histidine-rich loop of Arabidopsis thaliana vacuolar membrane zinc transporter MTP1. FEBS Open Bio 3:218–224
Verret F, Gravot A, Auroy P, Leonhardt N, David P, Nussaume L, Vavasseur A, Richaud P (2004) Overexpression of AtHMA4 enhances root-to shoot translocation of zinc and cadmium and plant metal tolerance. FEBS Lett 576:306–312
Wang M, Xu Q, Yu J, Yuan M (2010) The putative Arabidopsis zinc transporter ZTP29 is involved in the response to salt stress. Plant Mol Biol 73:467–479
Wang Y, Weisenhorn E, MacDiarmid CW, Andreini C, Bucci M, Taggart J et al (2018) The cellular economy of the Saccharomyces cerevisiae zinc proteome. Metallomics 10(12):1755–1776
Wang M, Kong F, Liu R, Fan Q, Zhang X (2020) Zinc in wheat grain, processing, and food. Front Nutr 7:124
Weber M, Harada E, Vess C, Roepenack-Lahaye EV, Clemens S (2004) Comparative microarray analysis of Arabidopsis thaliana and Arabidopsis halleri roots identifies nicotianamine synthase, a ZIP transporter and other genes as potential metal hyperaccumulation factors. Plant J 37(2):269–281
White PJ, Broadley MR (2011) Physiological limits to zinc biofortification of edible crops. Front Plant Sci 2:80
Wintz H, Fox T, Wu YY, Feng V, Chen W, Chang HS, Zhu T, Vulpe C (2003) Expression profiles of Arabidopsis thaliana in mineral deficiencies reveal novel transporters involved in metal homeostasis. J Biol Chem 278:47644–47653
Zhang YinFei ZY, Wang Yu WY, Ding ZhaoTang DZ, Wang Hui WH, Song LuBin SL, Jia SiSi JS, Ma DeXin MD (2017) Zinc stress affects ionome and metabolome in tea plants
Zhang F, Römheld V, Marschner H (1989) Effect of zinc deficiency in wheat on the release of zinc and iron mobilizing root exudates. Z Pflanzenernhr Bodenkd 152:205–210
Zhang YQ, Deng Y, Chen RY, Cui ZL, Chen XP, Yost R et al (2012) The reduction in zinc concentration of wheat grain upon increased phosphorus-fertilization and its mitigation by foliar zinc application. Plant Soil 361:143–152
Zhang P, Sun L, Qin J, Wan J, Wang R, Li S, Xu J (2018) cGMP is involved in Zn tolerance through the modulation of auxin redistribution in root tips. Environ Exp Bot 147:22–30
Zlobin IE (2021) Current understanding of plant zinc homeostasis regulation mechanisms. Plant Physiol Biochem 162:327–335
Zlobin IE, Kartashov AV, Nosov AV, Fomenkov AA, Kuznetsov VV (2019) The labile zinc pool in plant cells. Funct Plant Biol 46(9):796–805
Zlobin IE, Pashkovskiy PP, Kartashov AV, Nosov AV, Fomenkov AA, Kuznetsov VV (2020) The relationship between cellular Zn status and regulation of Zn homeostasis genes in plant cells. Environ Exp Bot 176:104104

Part II

Strategies to Regulate Zinc Levels in Plants

Strategies for Increasing Zinc Uptake Efficiency in Plants

Vishnu Mishra, Pragya Shukla, Kavita Shukla, Vishal Varshney, Jawahar Singh, and Sudhakar Srivastava

Abstract

Zinc (Zn) is a pivotal micronutrient for both plants and animals and plays an important role in diverse biochemical and physiological processes. In plants, Zn acts as a crucial co-factor for various enzymes and is essential for cellular functions, growth, development, and overall yield. Zn deficiency, which is common worldwide, leads to reduced crop yields, poor produce quality, and malnutrition in humans and animals. Thus, keeping these problems in mind, we compiled recent advancements and strategies to enhance Zn uptake efficiency in plants and underscores the significance of Zn as a micronutrient in enzymatic and physio-

V. Mishra (✉)
National Institute of Plant Genome Research (NIPGR), New Delhi, India

Department of Plant and Soil Sciences, Delaware Biotechnology Institute (DBI), University of Delaware (UD), Newark, DE, USA
e-mail: vmishra@udel.edu

P. Shukla
Molecular Biology and Biotechnology Division, CSIR-National Botanical Research Institute, Lucknow, Uttar Pradesh, India

Department of Botany, Institute of Science, Banaras Hindu University, Varanasi, Uttar Pradesh, India

K. Shukla · S. Srivastava
Plant Stress Biology Laboratory, Institute of Environment and Sustainable Development, Banaras Hindu University, Varanasi, Uttar Pradesh, India

V. Varshney
Department of Botany, Govt. Shaheed GendSingh College, Charama, Chhattisgarh, India

Institut de Biologie Moléculaire des Plantes du CNRS, Strasbourg, France

J. Singh
University of Cambridge, Sainsbury Laboratory (SLCU), Cambridge, UK

S. Srivastava, H. Kaur (eds.), *Zinc in Soil-Plant Continuum*,
https://doi.org/10.1007/978-981-96-4253-3_6

logical processes. Additionally, we have summarized the current understanding of Zn absorption and transport in plants and described various strategies used to improve Zn utilization efficiency (ZUE) in plant systems. It covers approaches such as soil and fertilizer management, the use of both conventional and enhanced Zn sources, and rhizosphere management to leverage beneficial microbial interactions. The focus is on creating Zn-efficient crop genotypes through traditional breeding and advanced genetic engineering methods. Moreover, molecular techniques, including transgenic tools and molecular markers, have been explored to enhance ZUE. Thus, this chapter integrates insights from agronomy, soil science, plant physiology, and biotechnology, providing a holistic framework for tackling Zn deficiency, ensuring sustainable agriculture, and enhancing food security and human health.

Keywords
CAX · Nanotechnological approaches · Nutrient homeostasis · P-type ATPases · Transgenics · ZIP

1 Introduction

Zinc serves as a crucial micronutrient for plants, implicated in various physiological and biochemical functions. Zn is a key component of many enzymes and proteins, influencing the synthesis of chlorophyll, protein metabolism, and membrane integrity (Mousavi et al. 2013; Krishna et al. 2017). Despite its importance, Zn deficiency is a widespread problem affecting crop production and nutritional quality, especially in Zn-deficient soils. This deficiency not only hampers plant growth and productivity but also leads to Zn malnutrition in animals including humans, as plants are primary dietary sources of this micronutrient (Srivastava et al. 2015; Natasha et al. 2022). Therefore, improving Zn uptake efficiency in plants is critical for both agricultural sustainability and human health. Zn is vital for numerous plant functions, including enzyme activation, protein synthesis, and hormonal regulation. It acts as a catalytic or structural component in over 300 enzymes and proteins, facilitating processes such as photosynthesis, DNA synthesis, and stress response (Singh et al. 2005; Nandal and Solanki 2021; Hamzah Saleem et al. 2022). Zn also plays a role in the stabilization of ribosomal fractions and the activation of enzymes involved in carbohydrate metabolism, such as alcohol dehydrogenase and carbonic anhydrase (Stanton et al. 2022; Saxena et al. 2023). Zn deficiency can lead to chlorosis/yellowing, retarded growth, and enhanced susceptibility to diseases, significantly impacting crop yield and production (Krishna et al. 2017; Stanton et al. 2022; Bastakoti 2023). Zn deficiency is a prevalent issue, affecting almost half of the world's cultivated soils. Factors contributing to this deficiency include high soil pH, high phosphorus (P) levels, low organic matter, and climatic conditions. Regions with Zn-deficient soils, such as parts of Latin America, South Asia, and Sub-Saharan Africa, often experience reduced agricultural productivity and increased prevalence of Zn deficiency in the human population (Alloway 2008; de Valença and Bake 2016; Peramaiyan et al. 2022). This dual effect on crop productivity and nutritional value underscores the need for effective strategies to improve Zn uptake and accumulation in plants. Several strategies have been explored to increase Zn uptake efficiency in plants, ranging from traditional agricultural practices to advanced biotechnological approaches (Hefferon 2019; Verma et al. 2021). These strategies can be broadly categorized into soil management techniques, plant breeding, and genetic engineering (Hefferon 2019; Verma et al. 2021). Improving soil Zn availability through management practices is one of the primary approaches to address Zn deficiency. These practices include the application of Zn fertilizers, organic amendments, and soil pH management. Zn fertilizers, such as Zn sulfate and Zn chelates, are commonly used to supplement Zn in deficient soils (Srivastava et al. 2015; Haroon et al. 2021; Verma et al. 2021). Organic amendments, like compost and manure, can enhance soil organic matter content and microbial activity, improving Zn solubility and availability to plants. Adjusting soil pH through liming or acidification can also influence Zn availability, as Zn solubility decreases in alkaline soils (Krishna et al. 2017; Stanton et al. 2022). Breeding crops with enhanced Zn uptake and accumulation capabilities is a sustainable and long-term solution to Zn deficiency (Sadeghzadeh 2013; Hefferon 2019).

Conventional breeding techniques involve selecting and crossing plant varieties with superior Zn uptake traits (Krishna et al. 2017). Modern approaches, like marker-assisted selection (MAS) and genomic selection, enable the identification and incorporation of specific genetic loci associated with Zn efficiency into breeding programs (Krishna et al. 2017). For instance, identifying and utilizing quantitative trait loci (QTLs) associated with Zn uptake and translocation can accelerate the development of Zn-efficient crop varieties (Krishna et al. 2017). Genetic engineering offers a precise and targeted approach to enhancing Zn uptake and accumulation in plants. This process integrates genes crucial for regulating Zn transport and maintaining its equilibrium. Transgenic plants overexpressing Zn transporter (TRP) genes, such as members of the ZIP family, have shown increased Zn uptake and accumulation (Ramesh et al. 2004; Li et al. 2015b, 2019; Krishna et al. 2017). Additionally, manipulating genes involved in the synthesis of metal chelators, such as nicotianamine (NA) and phytosiderophores (PS), can enhance Zn solubility and mobility in the rhizosphere, facilitating uptake by plant roots (Srivastava et al. 2015; Gupta et al. 2016). Furthermore, advanced genome editing tools, such as CRISPR/Cas9, offer further potential for precisely modifying genes to improve Zn efficiency without introducing foreign DNA (Yağız et al. 2022). Enhancing Zn uptake efficiency in plants is crucial for addressing Zn deficiency in agriculture and human nutrition. Integrating soil management techniques, plant breeding, and genetic engineering offers a comprehensive approach to developing Zn-efficient crop varieties. Continued research and innovation in these areas are essential for achieving sustainable agricultural practices and improving global food security and nutrition. By leveraging these strategies, we can mitigate the impact of Zn deficiency on crop production and contribute to better health outcomes for populations worldwide. In this chapter, we investigate plant responses to Zn deficiency, focusing on physiological and biochemical strategies. Additionally, we emphasized the role of TRPs in controlling Zn transport and balance, along with traditional and innovative approaches to enhance Zn uptake in plants. These include breeding, genetic modifications, transgenic techniques, root structure modifications, biostimulants, and nanotechnology applications. Furthermore, we discuss novel methods such as inoculating arbuscular mycorrhizal fungi, integrating organic ligands with nanofertilizers, and utilizing chelators to improve Zn uptake and distribution.

2 Significance of Zinc, Deficiency, Toxicity Symptoms, and Physiology

Globally, Zn deficiency presents a major agricultural challenge, leading to lower crop yields and diminished nutritional value. Soil conditions such as low Zn content, alkaline pH (>7), low redox potential, long-term submergence or extended waterlogging, nature of microbial activity in the rhizosphere, rich organic content, presence of CaCO3 and bicarbonates (HCO3-), Fe content and manganese oxides, nutrient competitions, and varying ratios of Mg, Ca, Na, and P hinder Zn availability and uptake for the plants (Nandal and Solanki 2021; Amini et al. 2022; Hamzah Saleem et al. 2022). In Zn-deficient plants, protein synthesis is notably hindered due to Zn′s role in essential enzymes like glutamate dehydrogenase, ATPase, and ribonuclease, crucial for RNA and protein synthesis (Hafeez et al. 2013; Hamzah Saleem et al. 2022). However, the impact of Zn deficiency on protein and RNA synthesis differs across plant species (Hafeez et al. 2013; Hamzah Saleem et al. 2022).

Symptoms of Zn deficiency manifest as yellowing of young leaves, reduced leaf size, and stunted stem growth (Hamzah Saleem et al. 2022). Severe deficiency can cause bronzing of older leaves, impaired growth, leaf rolling, and wilting. Plants respond to Zn deficiency through various adaptive mechanisms, such as altering root architecture, forming symbiotic relationships with arbuscular mycorrhizal fungi, producing PS and organic acid anions, and enhancing safeguarding mechanisms against oxidative stress (Hamzah Saleem et al. 2022). These adaptations facilitate increased Zn uptake and transport mediated by Zn TRPs and influence the transcription of Zn-dependent activities.

On the other hand, excessive Zn exposure leads to toxicity in plants, disrupting nutrient uptake and transport due to ionic radius similarity with other essential nutrients. This disruption inhibits photosynthesis, stomatal conductance, and other metabolic processes, ultimately impairing plant growth and structural integrity (Chakraborty and Mishra 2020; Hamzah Saleem et al. 2022). Zn toxicity induces chlorosis and, in severe cases, plant death through oxidative stress, characterized by lipid peroxidation induced by reactive oxygen species (Goodarzi et al. 2020; Shukla et al. 2023). Despite these challenges, plants have developed defense mechanisms, to mitigate oxidative damage and restore cellular homeostasis under stress conditions (Hamzah Saleem et al. 2022). While most plants maintain Zn concentrations between 30 and 100 μg Zn^{2+} g^{-1} dry weight, Zn hyperaccumulators like Brassicaceae can accumulate over 10,000 μg Zn^{2+} g^{-1} dry weight without exhibiting toxicity symptoms (Hamzah Saleem et al. 2022). Nevertheless, Zn levels exceeding 300 μg

Zn^{2+} g^{-1} dry weight are detrimental to most plants (Hamzah Saleem et al. 2022). Numerous studies have underscored the critical role of Zn in regulating plant metabolism, highlighting the profound impact of both deficiency and excess on growth, photosynthesis, and overall physiological functions (Hamzah Saleem et al. 2022). Thus, understanding and managing Zn dynamics in soil-plant systems are essential for sustainable agriculture and food security.

3 Mechanistic Insights into Zinc Absorption and Translocation

Zinc is absorbed in the form of Zn^{2+} from the soil by root plasma membrane TRPs and transported to the shoot via the xylem (Hamzah Saleem et al. 2022). The uptake rate depends on root efficiency, Zn concentration at the root surface, and cell membrane permeability (Gupta et al. 2016; Kumar et al. 2016). These TRPs manage Zn absorption, internal transport, xylem loading/unloading, vacuolar sequestration, and remobilization (Gupta et al. 2016; Hamzah Saleem et al. 2022). To date, a copious number of Zn TRPs have been reported and categorized in plants. Thus, understanding the mechanisms of Zn uptake and translocation in plants involves exploring various TRP families and their interactions with other cations. These TRP families encompass ZRT/IRT-related protein (ZIP), vacuolar iron transporter (VIT), P-type ATPases, yellow stripe like (YSL), cation diffusion facilitator (CDF)/metal tolerance proteins (MTPs), plant cadmium resistance (PCR), cation/H+ exchanger (CAX), and natural resistance-associated macrophage protein (NRAMP) (Srivastava et al. 2015; Gupta et al. 2016; Amini et al. 2022; Hamzah Saleem et al. 2022; Song et al. 2024). Some of these key TRPs and their localizations are illustrated in Fig. 1. Each of these TRP families contributes to the intricate process of Zn homeostasis within plants. Several key TRP families play crucial roles in Zn uptake, transport, and maintaining physiological functions, as discussed below:

4 ZIP Family

The ZIP family includes over a hundred TRPs distributed across various phylogenetic levels, each possessing eight transmembrane domains with exposed amino and carboxyl ends on the plasma membrane surface (Guerinot 2000; Grotz and Guerinot 2006). Apart from Zn, ZIP TRPs facilitate the transport of multiple metal ions like manganese (Mn), copper (Cu), iron (Fe), nickel (Ni), cobalt (Co), and cadmium (Cd) (Pedas and Husted 2009; Krishna et al. 2017). Their roles have been studied in diverse plant species such as *Medicago truncatula*, maize, rice, *Arabidopsis*, cassava, wheat, grape, and soybean (Krishna et al. 2017). In *Arabidopsis*, many ZIP genes are upregulated under Zn deficiency, with ZIP1, ZIP2, ZIP3, and ZIP4 complementing Zn uptake-deficient yeast mutants(Krishna et al. 2017). Additionally, the iron-regulated transporter 3 (IRT3) in *Arabidopsis halleri* and *A. thaliana* also transports Zn and Fe, localized to the plasma membrane as confirmed by

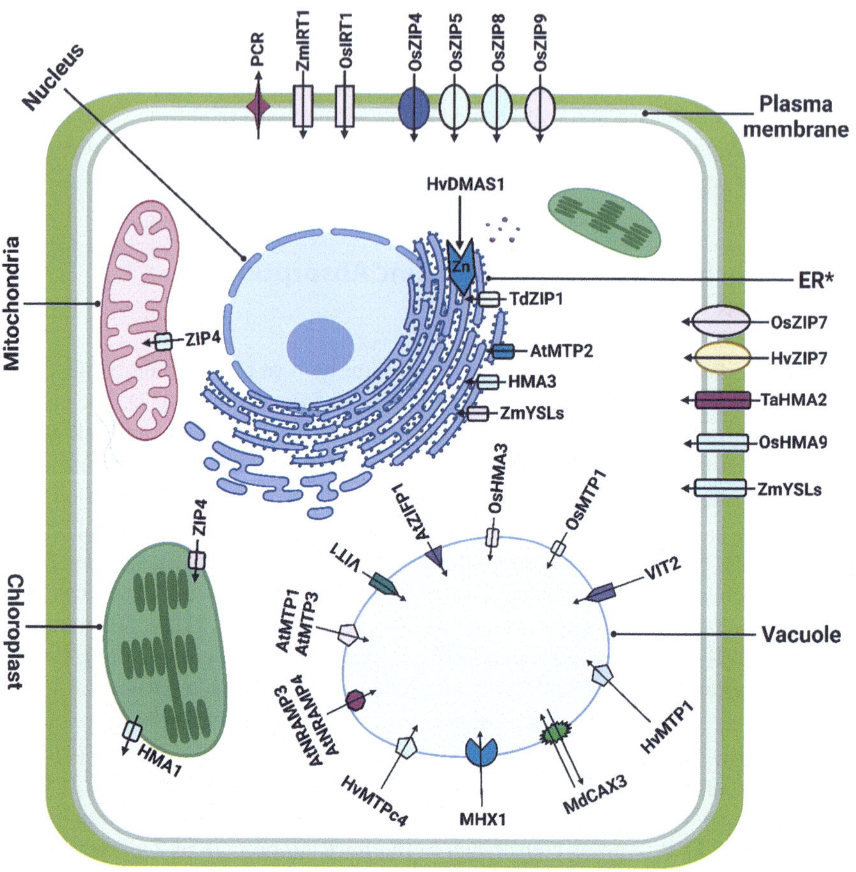

Fig. 1 The figure illustrates zinc (Zn) transporters and their localization within plant cells, highlighting their importance in Zn uptake efficiency. These transporter families, including ZIP, VIT, P-type ATPases, YSL, CDF, PCR, CAX, and NRAMP, are crucial for Zn uptake, transport, detoxification, and homeostasis. Their expression is regulated by soil Zn concentration to maintain proper Zn balance. These transporters ensure efficient Zn uptake and their management within plant cells

localization studies(Lin et al. 2009; Krishna et al. 2017). The ZIP TRPs also play roles in the uptake of other cations, such as Fe, Mn, and Cd (Hall and Guerinot 2006; Wang et al. 2012; Milner et al. 2013; Tan et al. 2020). This overlap can lead to competitive inhibition, where the presence of one cation affects the uptake of another, influencing the overall nutrient balance within the plant.

5 P-Type ATPase Family

P-type ATPases are enzymes that utilize ATP hydrolysis to transport cations across cellular membranes (Keeran et al. 2021). In Zn transport, these ATPases facilitate the sequestration of Zn into vacuoles and its movement from roots to shoots via the xylem (Krishna et al. 2017). It has been observed that heavy metal associated (HMA), HMA2, and HMA4 play crucial roles in Zn transport, significantly increasing Zn levels in roots, stems, and leaves of *Arabidopsis* (Hussain et al. 2004). Xylem loading is pivotal for Zn translocation within plants, and the HMA2 and HMA4 TRPs facilitate this process in xylem parenchymatous cells (Kraemer 2009). The detoxification function of *AtHMA1* has been studied in *Arabidopsis* using three different *hma1* knockout mutants (*hma1-1*, *hma1-2*, *hma1-3*) under Zn toxicity, confirming its role in Zn detoxification localized to the chloroplast envelope (Krishna et al. 2017). The functions of these P-type ATPase family members have been demonstrated in various plant species, including *Arabidopsis*, rice, and barley. Apart from Zn, P-type ATPases also transport other essential cations such as Cu and Co (Krishna et al. 2017). These ATPases are crucial for maintaining cellular ion homeostasis and detoxifying excess metals by compartmentalizing them into vacuoles.

6 CDF Family

The CDF family, also referred to as MTPs, mainly facilitates the movement of cations from the cytoplasm into vacuoles or outside the cell (Wang et al. 2021). This family includes TRPs that are key to Zn tolerance and homeostasis by preventing toxic accumulation of Zn in the cytoplasm (Gupta et al. 2016). In *A. thaliana*, AtMTP1 is a well-characterized member of the CDF family that localizes to the vacuolar membrane and facilitates the sequestration of Zn into vacuoles (Kobae et al. 2004). This mechanism not only helps in detoxifying excess Zn but also regulates its availability within the cell. CDF TRPs also play roles in the transporting Fe, Mn, Ni, Cd, and Co reflecting their broad substrate specificity and essential roles in metal homeostasis (Podar 2010).

7 PCR Family

The PCR TRP family, widely distributed across fungi, algae, plants, and animals, represents the largest gene family of metal TRPs (Song et al. 2010, 2011). These proteins primarily function as secondary TRPs, predominantly in epidermal cells and within the xylem of developing roots (Krishna et al. 2017). PCR proteins are involved in the detoxification and compartmentalization of heavy metals, including Zn and Cd (Lin and Aarts 2012; Hefferon 2019). These proteins help in the transport of metals into vacuoles, thereby reducing their cytoplasmic concentrations and mitigating potential toxicity. For instance, PCR2 plays a critical role in Zn redistribution and detoxification (Kawachi et al. 2011; Lin and Aarts 2012; Hefferon 2019; Liu et al. 2023). Loss-of-function mutants of *pcr2* exhibited compromised growth in both excessive and deficient Zn conditions (Song et al. 2010, 2011). It has been observed that root tissues of *pcr2* mutants accumulated higher Zn levels compared to control plants, whereas plants overexpressing *PCR2* showed reduced Zn accumulation in roots, suggesting PCR2 functions in Zn efflux from roots (Song et al. 2010, 2011). It has been shown that expression of PCR2 in yeast cells decreased intracellular Zn concentrations, supporting its role as a Zn-efflux TRP. PCR2 is primarily located in epidermal cells and xylem of young roots, shifting to epidermal cells in mature roots. Under Zn-limiting conditions, Zn accumulated in epidermal cells of *pcr2* roots, while in wild-type roots, it was concentrated in the stele (Song et al. 2010, 2011). It appears that the transport pathway of PCR2 is distinct from that of known Zn translocators like HMA2 and HMA4, as evidenced by more severe growth inhibition in *pcr2 hma4* and *pcr2 hma2hma4* mutants under both excess and deficient Zn conditions (Song et al. 2010, 2011). These findings underscore the pivotal role of PCR2 in Zn transport and maintaining optimal Zn levels in *Arabidopsis*.

8 CAX Family

CAX TRPs are typically involved in the exchange of cations across the vacuolar membrane, often using a proton gradient to drive the antiport of cations like Ca^{2+}, Zn^{2+}, and other metal ions (Cheng et al. 2005; Ghosh et al. 2021). These TRPs play a critical role in maintaining cellular ion balance and vacuolar pH (Cheng et al. 2005; Martins et al. 2017). CAX1 and CAX3 in *A. thaliana* are examples of TRPs that facilitate the vacuolar sequestration of Ca^{2+} and other cations, including Zn (Cheng et al. 2005). Additionally, these TRPs contribute to the broader metal homeostasis within the plant, influencing the uptake and distribution of Fe, Mn, Cd, and other cations (Socha and Guerinot 2014).

The TRPs discussed above do not function in isolation but rather in a coordinated manner to ensure optimal Zn uptake, translocation, and homeostasis. The interplay between different TRP families ensures that Zn is efficiently absorbed from the soil, translocated to various plant parts, and stored in appropriate cellular compartments. Therefore, by modulating these TRPs in plants, we can enhance Zn uptake efficiency in plants (Table 1).

9 Innovative Approaches for Optimizing Zinc Uptake and Utilization Efficiency (ZUUE) in Plants

As mentioned, Zn is vital for plants, essential for enzyme activation, protein synthesis, and hormone regulation. However, Zn deficiency is a common issue in different soil categories worldwide. It limits the crop productivity and quality. Therefore, enhancing Zn uptake efficiency in plants is a censorious goal for agricultural sustainability. Several strategies have been developed to address this issue, providing a

Table 1 Roles of various Zinc transporters and their role across different plant species

	Transporter name	Transporter family	Functional role	Plants	References
1	OsZIP1	ZIP	Zn regulation	Rice	Ishimaru et al. (2005), Huang et al. (2020)
2	OsZIP3	ZIP	Zn allocation	Rice	Ishimaru et al. (2005), Huang et al. (2020)
3	OsZIP4	ZIP	Zn allocation	Rice	Ishimaru et al. (2005), Huang et al. (2020)
4	OsZIP5	ZIP	Zn allocation	Rice	Ishimaru et al. (2005), Huang et al. (2020)
5	OsZIP6	ZIP	Zn regulation	Rice	Ishimaru et al. (2005), Huang et al. (2020)
6	OsZIP7	ZIP	Zn regulation	Rice	Ishimaru et al. (2005), Huang et al. (2020)
7	OsZIP8	ZIP	Zn absorption and distribution	Rice	Ishimaru et al. (2005), Huang et al. (2020)
8	OsZIP9	ZIP	Zn absorption and distribution	Rice	Ishimaru et al. (2005), Huang et al. (2020)
9	OsMTP1	MTP/CDF	Zn sequestration	Rice	Amini et al. (2022)
10	TaZIP3	ZIP	Zn transport	Wheat	Amini et al. (2022), Gong et al. (2022)
11	TaZIP6	ZIP	Zn transport	Wheat	Amini et al. (2022), Gong et al. (2022)
12	TaZIP7	ZIP	Zn transport	Wheat	Amini et al. (2022), Gong et al. (2022)
13	TaZIP9	ZIP	Zn transport	Wheat	Amini et al. (2022), Gong et al. (2022)
14	TaZIP13	ZIP	Zn transport	Wheat	Amini et al. (2022), Gong et al. (2022)
15	ZmZIP1	ZIP	Zn transportation	Maize	Li et al. (2013), Mondal et al. (2014), Ajeesh Krishna et al. (2020)
16	ZmZIP2	ZIP	Zn transportation	Maize	Li et al. (2013), Mondal et al. (2014), Ajeesh Krishna et al. (2020)
17	ZmZIP3	ZIP	Zn transportation	Maize	Li et al. (2013), Mondal et al. (2014), Ajeesh Krishna et al. (2020)
18	ZmZIP4	ZIP	Zn transportation	Maize	Li et al. (2013), Mondal et al. (2014), Ajeesh Krishna et al. (2020)
19	ZmZIP5	ZIP	Zn transportation	Maize	Li et al. (2013), Mondal et al. (2014), Ajeesh Krishna et al. (2020)

(continued)

Table 1 (continued)

	Transporter name	Transporter family	Functional role	Plants	References
20	ZmZIP6	ZIP	Zn transportation	Maize	Li et al. (2013), Mondal et al. (2014), Ajeesh Krishna et al. (2020)
21	ZmZIP7	ZIP	Zn transportation	Maize	Li et al. (2013), Mondal et al. (2014), Ajeesh Krishna et al. (2020)
22	ZmZIP8	ZIP	Zn transportation	Maize	Li et al. (2013), Mondal et al. (2014), Ajeesh Krishna et al. (2020)
23	ZmIRT1	ZIP	Zn transportation	Maize	Li et al. (2013)
24	ZmZLP1	ZIP	Transport of Zn from the ER to the cytoplasm	Maize	Li et al. (2013)
25	TdZIP1	ZIP	Zn transport	Wild emmer wheat	Gong et al. (2022)
26	HvZIP3	ZIP	Zn uptake and transport	Barley	Tiong et al. (2014), Ajeesh Krishna et al. (2020)
27	HvZIP5	ZIP	Zn transport	Barley	Tiong et al. (2014), Ajeesh Krishna et al. (2020)
28	HvZIP7	ZIP	Zn transport	Barley	Tiong et al. (2014), Ajeesh Krishna et al. (2020)
29	HvZIP8	ZIP	Zn transport	Barley	Tiong et al. (2014), Ajeesh Krishna et al. (2020)
30	HvZIP10	ZIP	Zn transport	Barley	Tiong et al. (2014), Ajeesh Krishna et al. (2020)
31	HvZIP13	ZIP	Zn transport	Barley	Tiong et al. (2014), Ajeesh Krishna et al. (2020)
32	MtZIP1	ZIP	Zn transport	Barrel Medic	López-Millán et al. (2004)
33	MtZIP5	ZIP	Zn transport	Barrel Medic	López-Millán et al. (2004)
34	MtZIP6	ZIP	Zn transport	Barrel Medic	López-Millán et al. (2004)
35	MeZIP	ZIP	Zn transport	Cassava	Bamrungsetthapong et al. (2010)
36	AtZIP1	ZIP	Zn transport	*Arabidopsis*	Palusińska et al. (2020)
37	AtZIP2	ZIP	Zn transport	*Arabidopsis*	Palusińska et al. (2020)
38	AtZIP3	ZIP	Zn transport	*Arabidopsis*	Krishna et al. (2017)
39	AtZIP4	ZIP	Zn transport	*Arabidopsis*	Mitra (2015)
40	AtZIP5	ZIP	Zn transport	*Arabidopsis*	Li et al. (2013)

(continued)

Table 1 (continued)

	Transporter name	Transporter family	Functional role	Plants	References
41	AtZIP6	ZIP	Zn transport	*Arabidopsis*	Krishna et al. (2017)
42	AtZIP7	ZIP	Zn transport	*Arabidopsis*	Krishna et al. (2017)
43	AtZIP8	ZIP	Zn transport	*Arabidopsis*	Krishna et al. (2017)
44	AtZIP9	ZIP	Zn transport	*Arabidopsis*	Li et al. (2013)
45	AtZIP10	ZIP	Zn transport	*Arabidopsis*	Krishna et al. (2017)
46	AtZIP11	ZIP	Zn transport	*Arabidopsis*	Krishna et al. (2017)
47	AtZIP12	ZIP	Zn transport	*Arabidopsis*	Li et al. (2013)
48	AtIRT1	ZIP	Zn and Fe transport	*Arabidopsis*	Krishna et al. (2017), Rodrigues et al. (2023)
49	AtIRT2	ZIP	Zn and Fe transport	*Arabidopsis*	Krishna et al. (2017), Jiang et al. (2021)
50	AtIRT3	ZIP	Zn and Fe transport	*Arabidopsis*	Krishna et al. (2017), Jiang et al. (2021)
51	GmZIP1	ZIP	Zn transport	*Soybean*	Moreau et al. (2002)
52	VvZIP3	ZIP	Zn transport	Grapes	Gainza-Cortés et al. (2012)
53	OsHMA1	P-type ATPase	Translocation of Zn	Rice	Tian et al. (2023)
54	OsHMA2	P-type ATPase	Translocation of Zn and Cd	Rice	Takahashi et al. (2012)
55	OsHMA3	P-type ATPase	Zn homeostasis	Rice	Cai et al. (2019)
56	OsHMA7	P-type ATPase	Zn regulation	Rice	Wairich et al. (2022)
57	OsHMA9	P-type ATPase	Zn regulation	Rice	Tian et al. (2023)
58	TaHMA2	P-type ATPase	Zn transport from roots to shoots	Wheat	Amini et al. (2022)
59	HvHMA2	P-type ATPase	Zn transport from roots to shoots	Barley	Mills et al. (2012)
60	TuMTP1	MTP/CDF	Zn transport and regulation	Red wild einkorn wheat	Wang et al. (2021)
61	HvMTP1	MTP/CDF	Zn transport	Barley	Amini et al. (2022)
62	PtdMTP1	MTP/CDF	Zn sequestration	*Populus* spp.	Blaudez et al. (2003)
63	AtMTP1	MTP/CDF	Zn transport	*Arabidopsis*	Kobae et al. (2004)
64	AtZAT1	MTP/CDF	Zn transport	*Arabidopsis*	Krishna et al. (2017)
65	AtMHX1	CAX	Zn transport	*Arabidopsis*	Hermans et al. (2004)

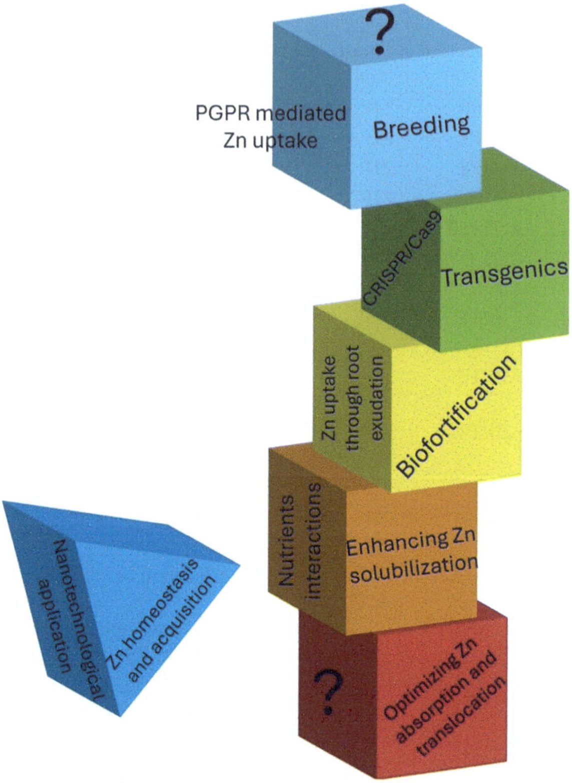

Fig. 2 Diagram illustrating various strategies that can be utilized to enhance Zn uptake efficiency in plants. A question mark (?) highlights strategies that are yet to be discovered and require further research to understand their role in Zn uptake

comprehensive approach to tackling Zn deficiency in plants and ensuring sustainable and productive agriculture (Fig. 2).

10 Breeding Approaches

Natural genetic variations lead to diverse genotypes and phenotypes in crops, helping them adapt to environmental stresses (Verma et al. 2021). These variations have been observed across crop species, including rice, barley, and wheat, affecting ZUUE. Thus, identifying significant natural variations or QTLs associated with high Zn efficiency can guide genetic modifications in crops (Dong et al. 2006; Verma et al. 2021). Thus, integrating basic research, molecular genetics, and QTL analysis facilitates the selection of traits for breeding to improve crop varieties with enriched nutrient content.

QTL identification reveals genetic differences modulating root system architecture (RSA) and nutrient transport, critical for nutrient acquisition in crops like maize and rice (Shimizu et al. 2004; Steele et al. 2006; Li et al. 2015a, 2016). Variability in Zn acquisition and utilization efficiency among plant species and genotypes of the same species suggests opportunities for screening or genetically modifying crops for improved Zn uptake (Verma et al. 2021). Research efforts, particularly in rice, aim to pinpoint QTLs for higher Zn concentration in grains, although specific genetic loci influencing Zn efficiency remain incompletely understood (Verma et al. 2021). Assessing QTLs associated with high Zn uptake and use efficiency promises to enhance Zn content and yield in crops (Verma et al. 2021). Genome-wide association studies (GWAS) employing sequencing and bioinformatics tools offer robust methods to identify QTLs and natural variations controlling agronomic traits (Yano et al. 2017; Gupta et al. 2022). These QTLs can be leveraged in crop breeding to enhance nutrient efficiency, including Zn uptake and utilization. GWAS presents a promising avenue for pinpointing genetic loci that govern Zn nutrition traits, paving the way for effective genetic improvements in crops (Gupta et al. 2022).

11 Transgenic Approaches

Genetic modification presents a promising approach for Zn concentration in crops. Transgenic strategies focus on improving ZUUE by manipulating TRP genes like ZIP, CDF, MTP, CAX, PCR, etc. For example, increasing the expression of *OsZIP1* has been shown to enhance Zn concentration in tobacco and finger millet (Yamunarani Ramegowda et al. 2013). Furthermore, recently it has been shown that increasing the activity of the Zn TRP *OsZIP9* facilitates Zn uptake in rice and silencing of *AtMTP3* enhances Zn accumulation and causes hypersensitivity in *Arabidopsis* (Arrivault et al. 2006; Huang et al. 2020). *HvZIP7* in barley increases Zn content in grains, whereas AtHMA4 overexpression in *A. thaliana* boosts shoot Zn levels. Modifying the NA synthesis pathway, such as through *HvNAS1*, *OsNAS2*, and *OsNAS3* overexpression, also enhances Zn accumulation in grains (Tiong et al. 2014; Lee et al. 2023). Previous research has demonstrated that Os*PCR1* affects grain weight and Zn accumulation in rice grains (Song et al. 2015). Additionally, the recently identified long-distance mobile mRNA *CAX3* regulates Zn compartmentalization in *Malus domestica* (Hao et al. 2022). These transgenic approaches offer powerful tools to enhance Zn uptake in plants. By targeting specific genes and pathways, scientists can develop crop varieties that are more efficient at absorbing, transporting, and utilizing Zn. These advancements not only improve plant health and yield but also contribute to addressing Zn deficiency in human diets, promoting better nutrition worldwide.

12 Potential Biotechnological Approaches

These methods include genetic engineering to introduce genes that enhance Zn uptake mechanisms in plants, such as TRPs and metal chelators (Stanton et al. 2022; Banerjee et al. 2023). Biotechnological tools also encompass genome editing techniques like CRISPR-Cas9, allowing precise modification of plant genomes to optimize Zn utilization pathways (Stanton et al. 2022; Banerjee et al. 2023). Additionally, biofortification programs aim to develop Zn-enriched crop varieties through selective breeding or genetic modification.

13 Boosting Zinc Uptake in Plants by Improving Absorption, Translocation, and Solubilization

Zinc uptake in plants is a complex process driven by active transport mechanisms from the rhizospheric soil, fueled by ATP generated during photosynthesis (Hamzah Saleem et al. 2022). This uptake occurs through both apoplastic and symplastic pathways across the epidermis, cortex, and endodermis, eventually reaching vascular tissues (Hamzah Saleem et al. 2022). Crucial TRP proteins like ZIP, HMA, MTP, YSL, and PCR families play integral roles in this journey (Hamzah Saleem et al. 2022; Krishna et al. 2023; Rani et al. 2023). ZIP proteins, characterized by eight transmembrane domains, facilitate Zn influx into the cytosol, whereas HMAs enable Zn efflux to the apoplast, and MTPs sequester Zn into intracellular compartments such as vacuoles and ER (Clemens 2022; Yin et al. 2023).

Regulation of ZIP family genes, governed by bZIP transcription factors (TFs) under Zn deficiency, illustrates their importance in maintaining Zn homeostasis. Interactions with mycorrhizal fungi, like *Glomus versiforme*, further modulate Zn TRP gene expression, impacting Zn concentrations within plant tissues (Burleigh et al. 2003). These interactions underscore the dynamic nature of Zn uptake regulation in response to environmental cues.

Functional diversity among ZIP TRPs is evident across plant species such as *Arabidopsis* and rice, influencing Zn distribution and accumulation in various plant parts (Gong et al. 2022; Huang and Yamaji 2024). Similarly, HMAs are pivotal for maintaining metal homeostasis, with mutations in AtHMA2 and AtHMA4 altering Zn distribution between roots and shoots, mirroring the role of OsHMAs in rice (Huang and Yamaji 2024). MTPs act as cation/proton antiporters, crucial for Zn partitioning in plants like *Arabidopsis* and barley, while YSL proteins facilitate Zn transport in complex with NA, vital for Zn loading into the xylem and phloem in rice (Verma et al. 2021; Yang et al. 2022).

Soil conditions significantly influence Zn availability, with factors like pH and mineral and organic matter content playing critical roles. Higher pH levels in soils often lead to decreased Zn availability due to the formation of insoluble Zn complexes (Haroon et al. 2021). Organic acids produced by plant roots and soil microorganisms can mitigate this by lowering soil pH, thereby enhancing Zn solubility and uptake (Hamzah Saleem et al. 2022). Growth-promoting bacteria like *Pseudomonas frederiksbergensis* and *Enterobacter hormaechei* further contribute by secreting organic acids that aid in Zn solubilization from various compounds, supporting higher Zn uptake in crops (Fahsi et al. 2021).

Additionally, plants deploy root exudation strategies to enhance Zn uptake, particularly under deficient conditions, secreting compounds such as PS (Northover et al. 2021, 2022). These specialized exudates exhibit a strong affinity for Zn ions, facilitating their mobilization to the root apoplast within the rhizosphere. In rice, the PA 2′-deoxymugineic acid (DMA) plays a critical role in Zn transport under deficient conditions, highlighting genetic variability in exudation kinetics across different rice genotypes (Northover et al. 2021, 2022). Thus, understanding these complex interactions and mechanisms is crucial for optimizing Zn bioavailability in agricultural soils and mitigating Zn deficiency in crops, underscoring the need for continued research into soil-plant-microbe interactions and their impact on nutrient uptake dynamics.

14 Nutrients Interactions: Impacting Zinc Absorption and Transport

The interaction between nutrients profoundly impacts the absorption and movement of Zn (Zn) within plants (Fariduddin et al. 2022; Lešková et al. 2022). Zn uptake primarily occurs as Zn^{2+} ions in the rhizospheric soil, alongside other divalent cations such as Fe, Cd, Ni, and Cu, which compete for binding sites on PS and TRP proteins (Fariduddin et al. 2022; Lešková et al. 2022). This competitive interaction can hinder Zn uptake efficiency (Ayyar and Appavoo 2020). Notably, TRPs like IRT1 and non-selective cation channels play roles in the transport of multiple divalent cations, influencing their uptake dynamics (Martín-Barranco et al. 2020). In the case of Fe-Zn interactions, both minerals utilize common TRPs, and deficiencies in one can affect the accumulation of the other (Gupta et al. 2016). For instance, under Fe deficiency, plants tend to accumulate more Zn, highlighting a regulatory link between these nutrients (Gupta et al. 2016). Furthermore, nitrogen (N) application has shown synergistic effects on Zn absorption and translocation rates. Elevated N levels enhance the activity of TRPs in the root zone, thereby promoting Zn translocation to shoots (Ji et al. 2022). This enhancement may also involve increased production of chelators like NA and PS, which facilitate Zn transport (Sinclair and Krämer 2012). Conversely, phosphate (Pi) levels in soil can influence Zn availability; high Pi content can lead to the formation of insoluble Zn phosphate complexes, reducing Zn uptake by roots (Lay-Pruitt et al. 2022). Conversely, Zn deficiency can lead to increased Pi accumulation in shoots, suggesting a complex interplay in Pi-Zn homeostasis (Lay-Pruitt et al. 2022). The transcriptional regulation underpinning these interactions involves key factors like PHR1 and PHO1, which are involved in Pi homeostasis and show interdependency with Zn deficiency responses (Lay-Pruitt et al. 2022). Additionally, PHO1;H3, a PHO1 homolog, acts like a negative modulator in Pi transport under Zn-deficient conditions (Khan et al. 2014). This intricate crosstalk highlights the importance of nutrient balance in optimizing plant growth and nutrient utilization efficiency, suggesting avenues for improving crop resilience and productivity through targeted nutrient management strategies.

15 Zinc Homeostasis and Acquisition

Zinc regulation in plants involves a complex interplay of membrane TRPs, Zn-binding proteins, and small ligands, crucial for maintaining optimal Zn concentrations in grains (Yoneyama et al. 2015; Aiqing et al. 2022). During vegetative growth, Zn initially accumulates in leaves and later translocates to grains during grain filling, reflecting its role in metabolically active tissues (Yoneyama et al. 2015). Zn distribution within rice plants occurs through intricate processes in both the phloem and xylem (Yoneyama et al. 2015). Efficient Zn transport from roots to shoots and among leaves significantly impacts grain Zn content, with a notable storage reservoir at the basal node affecting overall distribution (Yoneyama et al. 2015). In rice, NA plays a key role in Zn translocation via the phloem, influenced by genotypic variations and recently recognized interactions with cytokinin (Nishiyama et al. 2012; Gao et al. 2019; Verma et al. 2021). However, comprehensive understanding of genetic factors governing Zn uptake, transport, remobilization, and grain loading remains incomplete. Maintaining a critical Zn concentration, particularly around 20–25 mg kg^{-1} in dried rice leaves, is crucial for optimal growth, underscoring the need to manage internal Zn levels effectively (Verma et al. 2021). Thus, tailoring Zn management strategies for cultivars with varying Zn tolerance levels could enhance our understanding of Zn homeostasis mechanisms and improve agricultural practices related to Zn nutrition.

16 Nanotechnological Application for Efficient Zinc Uptake

Nanotechnology has emerged as a promising avenue to enhance Zn uptake in plants, addressing challenges related to nutrient deficiency and soil remediation (Fiol et al. 2021). Nanomaterials can facilitate controlled delivery and efficient utilization of Zn, thereby improving plant growth and productivity (Munir et al. 2018). One approach involves the use of Zn oxide nanoparticles (ZnO NPs), which can release Zn ions gradually in the rhizosphere, ensuring sustained availability for plant uptake (Hussain et al. 2018; Munir et al. 2018). These nanoparticles have been shown to enhance Zn bioavailability in crops like wheat and maize (Rehman et al. 2021; Srivastav et al. 2021).

Additionally, nanoscale Zn fertilizers encapsulated in polymers or nanocarriers offer targeted delivery to plant roots, minimizing nutrient loss and environmental impact (Shukla et al. 2024). For instance, Zn nanoparticles coated with biodegradable polymers have demonstrated improved uptake efficiency in various plants under Zn-deficient conditions (Xin et al. 2020). Moreover, nanosensors are pivotal for real-time surveillance of Zn levels in both plants and soil. These sensors provide valuable insights into plant nutrient status, enabling precise adjustments in fertilizer application for optimal growth (Mahmoud and Fawzy 2021). Nanotechnology also supports the development of smart delivery systems for Zn ions, ensuring they reach specific plant tissues where they are most needed, such as grains in cereals (Xin et al. 2020).

Moreover, ZnO-nanoparticles can assist in soil remediation by facilitating the solubilization of insoluble Zn compounds, making them more accessible for plant uptake (Saleem et al. 2023). This capability is crucial in contaminated soils where conventional methods may be ineffective. Overall, nanotechnology offers innovative solutions to enhance Zn uptake efficiency in plants, contributing to sustainable agriculture and improved food security.

17 Conclusion and Future Perspective

Zinc is critical for plant metabolism, enzyme function, and ion transport, making its availability in soil vital for optimal crop growth. However, low Zn availability leads to decreased productivity and reduced grain nutritional value, posing significant challenges to global food security. Enhancing Zn uptake efficiency in plants is essential for addressing global nutritional deficiencies and improving agricultural productivity. Recent advancements in molecular breeding and biotechnological methods, including strategies like Zn solubilization, PS and organic acid complex formation, and enhanced Zn uptake via TRPs, hold promise for addressing dietary Zn deficiencies in plants. Understanding the regulatory mechanisms of Zn TRPs under deficiency conditions is essential for developing effective strategies to improve Zn acquisition in diverse crops. Moreover, the identification and characterization of genes associated with ZUE offer valuable insights into the genetic and molecular foundations of Zn uptake, facilitating the development of genetically enhanced crop varieties with superior Zn uptake and utilization capabilities. Future research should prioritize unraveling the molecular and genetic mechanisms governing Zn uptake across different crops, integrating omics technologies to comprehensively understand Zn nutrition, and creating biofortified crops with heightened Zn accumulation in edible tissues. Promoting sustainable agricultural practices that optimize Zn availability in soil and encouraging global collaboration and knowledge sharing among researchers, policymakers, and agricultural stakeholders are also crucial. These efforts will accelerate the adoption of Zn-efficient crop varieties and sustainable agricultural practices, ultimately improving food security, enhancing nutritional quality, and contributing to the global goal of ensuring access to safe and nutritious food for all.

Acknowledgments We acknowledge https://www.vectorstock.com/ and https://www.biorender.com/ for designing the figures.

Declaration of Interests All authors have reviewed and confirmed no conflicts of interest.

Author Contribution VM: conceptualization, writing original draft, editing, table, and figure drafting. PS: writing original draft and editing. KS: review, discussion, and editing. VV: review, discussion, and editing. JS: review, discussion, and editing. SS: scientific feedback, edited the manuscript (MS), and provided valuable insights to improve the MS.

References

Aiqing Z, Zhang L, Ning P, Chen Q, Wang B, Zhang F, Yang X, Zhang Y (2022) Zinc in cereal grains: concentration, distribution, speciation, bioavailability, and barriers to transport from roots to grains in wheat. Crit Rev Food Sci Nutr 62:7917–7928

Ajeesh Krishna TP, Maharajan T, Victor Roch G, Ignacimuthu S, Antony Ceasar S (2020) Structure, Function. Regulation and phylogenetic relationship of ZIP family transporters of plants, Front Plant Sci, p 11

Alloway BJ (2008) Zinc in soils and crop nutrition

Amini S, Arsova B, Hanikenne M (2022) The molecular basis of zinc homeostasis in cereals. Plant Cell Environ 45:1339–1361

Arrivault S, Senger T, Krämer U (2006) The Arabidopsis metal tolerance protein AtMTP3 maintains metal homeostasis by mediating Zn exclusion from the shoot under Fe deficiency and Zn oversupply. Plant J 46:861–879

Ayyar S, Appavoo S, N M (2020) Role of zinc nutrition for increasing zinc availability, uptake, yield, and quality of maize (Zea Mays L.) grains: an overview role of zinc nutrition for increasing zinc availability, uptake, yield, and quality of maize (Zea mays L.) grains: an overview. Commun Soil Sci Plant Anal 51: 2001–2021

Bamrungsetthapong T, Vichukit V, Nakasathien S (2010) Cloning and molecular characterization of the zinc transporter (ZIP) gene from Cassava. Nat Sci 44

Banerjee S, Roy P, Nandi S, Roy S (2023) Advanced biotechnological strategies towards the development of crops with enhanced micronutrient content. Plant Growth Regul 100:355–371

Bastakoti S (2023) Role of zinc in management of plant diseases: a review. Cogent Food Agric 9:2194483

Blaudez D, Kohler A, Martin F, Sanders D, Chalot M (2003) Poplar metal tolerance protein 1 confers zinc tolerance and is an Oligomeric vacuolar zinc transporter with an essential leucine zipper motif. Plant Cell 15:2911–2928

Burleigh SH, Kristensen BK, Bechmann IE (2003) A plasma membrane zinc transporter from Medicago truncatula is up-regulated in roots by Zn fertilization, yet down-regulated by arbuscular mycorrhizal colonization. Plant Mol Biol 52

Cai H, Huang S, Che J, Yamaji N, Ma JF (2019) The tonoplast-localized transporter OsHMA3 plays an important role in maintaining Zn homeostasis in rice. J Exp Bot 70:2717–2725

Chakraborty S, Mishra AK (2020) Mitigation of zinc toxicity through differential strategies in two species of the cyanobacterium Anabaena isolated from zinc polluted paddy field. Environ Pollut 263:114375

Cheng N-H, Pittman JK, Shigaki T, Lachmansingh J, LeClere S, Lahner B, Salt DE, Hirschi KD (2005) Functional association of arabidopsis CAX1 and CAX3 is required for normal growth and ion homeostasis. Plant Physiol 138:2048–2060

Clemens S (2022) The cell biology of zinc. J Exp Bot 73:1688–1698

de Valença AW, Bake A (2016) Micronutrient management for improving harvests, human nutrition, and the environment. Scientific project, assigned by food & business knowledge platform Netherlands 24

Dong Y, Ogawa T, Lin D, Koh H-J, Kamiunten H, Matsuo M, Cheng S (2006) Molecular mapping of quantitative trait loci for zinc toxicity tolerance in rice seedling (Oryza sativa L.). Field Crop Res 95:420–425

Fahsi N, Mahdi I, Mesfioui A, Biskri L, Allaoui A (2021) Plant growth-promoting Rhizobacteria Isolated from the Jujube (Ziziphus lotus) plant enhance wheat growth, Zn uptake, and heavy metal tolerance. Agriculture 11

Fariduddin Q, Saleem M, Khan TA, Hayat S (2022) Zinc as a versatile element in plants: an overview on its uptake, translocation, assimilatory roles, deficiency and toxicity symptoms. In: Khan ST, Malik A (eds) Microbial biofertilizers and micronutrient availability: the role of zinc in agriculture and human health. Springer, Cham, pp 137–158

Fiol DF, Terrile MC, Frik J, Mesas FA, Álvarez VA, Casalongué CA (2021) Nanotechnology in plants: recent advances and challenges. J Chem Technol Biotechnol 96:2095–2108
Gainza-Cortés F, Pérez-Dïaz R, Pérez-Castro R, Tapia J, Casaretto JA, González S, Peña-Cortés H, Ruiz-Lara S, González E (2012) Characterization of a putative grapevine Zn transporter, VvZIP3, suggests its involvement in early reproductive development in Vitis vinifera L. BMC Plant Biol 12:111
Gao S, Xiao Y, Xu F, Gao X, Cao S, Zhang F, Wang G, Sanders D, Chu C (2019) Cytokinin-dependent regulatory module underlies the maintenance of zinc nutrition in rice. New Phytol 224:202–215
Ghosh S, Jha SK, Pandey GK (2021) Chapter 6 - Cation/H+ exchanger in plants: roles in development and stress response. In: Upadhyay SK (ed) Calcium transport elements in plants. Academic Press, pp 89–101
Gong F, Qi T, Hu Y, Jin Y, Liu J, Wang W, He J, Tu B, Zhang T, Jiang B, Wang Y, Zhang L, Zheng Y, Liu D, Huang L, Wu B (2022) Genome-wide investigation and functional verification of the ZIP family transporters in wild emmer wheat. Int J Mol Sci 23
Goodarzi A, Namdjoyan S, Soorki AA (2020) Effects of exogenous melatonin and glutathione on zinc toxicity in safflower (Carthamus tinctorius L.) seedlings. Ecotoxicol Environ Saf 201:110853
Grotz N, Guerinot ML (2006) Molecular aspects of Cu, Fe and Zn homeostasis in plants. Biochimica et Biophysica Acta (BBA) Mol Cell Res 1763:595–608
Guerinot ML (2000) The ZIP family of metal transporters. Biochimica et Biophysica Acta Biomembr 1465:190–198
Gupta N, Ram H, Kumar B (2016) Mechanism of Zinc absorption in plants: uptake, transport, translocation and accumulation. Rev Environ Sci Biotechnol 15:89–109
Gupta OP, Singh AK, Singh A, Singh GP, Bansal KC, Datta SK (2022) Wheat biofortification: utilizing natural genetic diversity, genome-wide association mapping, genomic selection, and genome editing technologies. Front Nutr 9
Hafeez BM, Khanif Y, Saleem M (2013) Role of zinc in plant nutrition- a review. Journal of Experimental Agriculture International 3:374–391
Hall BP, Guerinot ML (2006) The role of ZIP family members in iron transport. In: Barton LL, Abadia J (eds) Iron nutrition in plants and rhizospheric microorganisms. Springer, Dordrecht, pp 311–326
Hamzah Saleem M, Usman K, Rizwan M, Al Jabri H, Alsafran M (2022) Functions and strategies for enhancing zinc availability in plants for sustainable agriculture. Front Plant Sci 13
Hao P, Lv X, Fu M, Xu Z, Tian J, Wang Y, Zhang X, Xu X, Wu T, Han Z (2022) Long distance mobile mRNA *CAX3* modulates iron uptake and zinc compartmentalization. EMBO Rep 23:e53698
Haroon M, Khan ST, Malik A (2021) Zinc-solubilizing bacteria: an option to increase zinc uptake by plants. In: Microbial biofertilizers and micronutrient availability: the role of zinc in agriculture and human health. Springer, pp 207–238
Hefferon K (2019) Biotechnological approaches for generating zinc-enriched crops to combat malnutrition. Nutrients 11
Hermans C, Johnson GN, Strasser RJ, Verbruggen N (2004) Physiological characterisation of magnesium deficiency in sugar beet: acclimation to low magnesium differentially affects photosystems I and II. Planta 220:344–355
Huang S, Yamaji N (2024) Metal transport systems in plants. Annu Rev Plant Biol 75
Huang S, Sasaki A, Yamaji N, Okada H, Mitani-Ueno N, Ma JF (2020) The ZIP transporter family member OsZIP9 contributes to root zinc uptake in rice under zinc-limited conditions1 [OPEN]. Plant Physiol 183:1224–1234
Hussain D, Haydon MJ, Wang Y, Wong E, Sherson SM, Young J, Camakaris J, Harper JF, Cobbett CS (2004) P-type ATPase heavy metal transporters with roles in essential zinc homeostasis in Arabidopsis. Plant Cell 16:1327–1339

Hussain A, Ali S, Rizwan M, Ziaurrehman M, Javed MR, Imran M, Chatha SAS, Nazir R (2018) Zinc oxide nanoparticles alter the wheat physiological response and reduce the cadmium uptake by plants. Environ Pollut 242:1518–1526

Ishimaru Y, Suzuki M, Kobayashi T, Takahashi M, Nakanishi H, Mori S, Nishizawa NK (2005) OsZIP4, a novel zinc-regulated zinc transporter in rice. J Exp Bot 56:3207–3214

Ji C, Li J, Jiang C, Zhang L, Shi L, Xu F, Cai H (2022) Zinc and nitrogen synergistic act on root-to-shoot translocation and preferential distribution in rice. J Adv Res 35:187–198

Jiang Y, Chen X, Chai S, Sheng H, Sha L, Fan X, Zeng J, Kang H, Zhang H, Xiao X, Zhou Y, Vatamaniuk OK, Wang Y (2021) TpIRT1 from Polish wheat (Triticum polonicum L.) enhances the accumulation of Fe, Mn, Co, and Cd in Arabidopsis. Plant Sci 312:111058

Kawachi M, Kobae Y, Tomioka R, Maeshima M (2011) The role of membrane transport in the detoxification and accumulation of zinc in plants. In: Sherameti I, Varma A (eds) Detoxification of heavy metals. Springer, Berlin/Heidelberg, pp 129–142

Keeran NS, Usha B, Ganesan G (2021) 2 - P-type ATPases and their role in metal homeostasis in plants. In: Roychoudhury A, Tripathi DK, Deshmukh R (eds) Metal and nutrient transporters in abiotic stress. Academic Press, pp 33–54

Khan GA, Bouraine S, Wege S, Li Y, de Carbonnel M, Berthomieu P, Poirier Y, Rouached H (2014) Coordination between zinc and phosphate homeostasis involves the transcription factor PHR1, the phosphate exporter PHO1, and its homologue PHO1;H3 in Arabidopsis. J Exp Bot 65:871–884

Kobae Y, Uemura T, Sato MH, Ohnishi M, Mimura T, Nakagawa T, Maeshima M (2004) Zinc transporter of Arabidopsis thaliana AtMTP1 is localized to vacuolar membranes and implicated in zinc homeostasis. Plant Cell Physiol 45:1749–1758

Kraemer U (2009) The dilemma of controlling heavy metal accumulation in plants. New Phytol 181:3–5

Krishna TPA, Ceasar SA, Maharajan T, Ramakrishnan M, Duraipandiyan V, Al-Dhabi NA, Ignacimuthu S (2017) Improving the zinc-use efficiency in plants: a review. SABRAO J Breed Genet 49

Krishna TPA, Maharajan T, Ceasar SA (2023) The role of membrane transporters in the biofortification of zinc and iron in plants. Biol Trace Elem Res 201:464–478

Kumar L, Meena NL, Singh U (2016) Zinc transporter: mechanism for improving Zn availability. In: Singh U, Praharaj CS, Singh SS, Singh NP (eds) Biofortification of food crops. Springer, New Delhi, pp 129–146

Lay-Pruitt KS, Wang W, Prom-u-thai C, Pandey A, Zheng L, Rouached H (2022) A tale of two players: the role of phosphate in iron and zinc homeostatic interactions. Planta 256:23

Lee S, Rahman MM, Nakanishi H, Nishizawa NK, An G, Nam HG, Jeon J-S (2023) Concomitant activation of OsNAS2 and OsNAS3 contributes to the enhanced accumulation of iron and zinc in rice. Int J Mol Sci 24

Lešková A, Javot H, Giehl RFH (2022) Metal crossroads in plants: modulation of nutrient acquisition and root development by essential trace metals. J Exp Bot 73:1751–1765

Li S, Zhou X, Huang Y, Zhu L, Zhang S, Zhao Y, Guo J, Chen J, Chen R (2013) Identification and characterization of the zinc-regulated transporters, iron-regulated transporter-like protein (ZIP) gene family in maize. BMC Plant Biol 13:114

Li P, Chen F, Cai H, Liu J, Pan Q, Liu Z, Gu R, Mi G, Zhang F, Yuan L (2015a) A genetic relationship between nitrogen use efficiency and seedling root traits in maize as revealed by QTL analysis. J Exp Bot 66:3175–3188

Li S, Zhou X, Li H, Liu Y, Zhu L, Guo J, Liu X, Fan Y, Chen J, Chen R (2015b) Overexpression of ZmIRT1 and ZmZIP3 enhances iron and zinc accumulation in transgenic Arabidopsis. PLoS One 10

Li X, Zeng R, Liao H (2016) Improving crop nutrient efficiency through root architecture modifications. J Integr Plant Biol 58:193–202

Li S, Liu X, Zhou X, Li Y, Yang W, Chen R (2019) Improving zinc and iron accumulation in maize grains using the zinc and iron transporter ZmZIP5. Plant Cell Physiol 60:2077–2085

Lin Y-F, Aarts MGM (2012) The molecular mechanism of zinc and cadmium stress response in plants. Cell Mol Life Sci 69:3187–3206

Lin Y-F, Liang H-M, Yang S-Y, Boch A, Clemens S, Chen C-C, Wu J-F, Huang J-L, Yeh K-C (2009) Arabidopsis IRT3 is a zinc-regulated and plasma membrane localized zinc/iron transporter. New Phytol 182:392–404

Liu Y, Kong L, Gong C, Yang G, Xu E, Chen W, Zhang W, Chen X (2023) Identification of plant cadmium resistance gene family in Brassica napus and functional analysis of BnPCR10.1 involved in cadmium and copper tolerance. Plant Physiol Biochem 202:107989

López-Millán A-F, Ellis DR, Grusak MA (2004) Identification and characterization of several new members of the ZIP family of metal ion transporters in Medicago truncatula. Plant Mol Biol 54:583–596

Mahmoud AED, Fawzy M (2021) Nanosensors and nanobiosensors for monitoring the environmental pollutants. In: Makhlouf ASH, Ali GAM (eds) Waste recycling technologies for nanomaterials manufacturing. Springer, Cham, pp 229–246

Martín-Barranco A, Spielmann J, Dubeaux G, Vert G, Zelazny E (2020) Dynamic control of the high-affinity iron uptake complex in root epidermal cells1. Plant Physiol 184:1236–1250

Martins V, Carneiro F, Conde C, Sottomayor M, Gerós H (2017) The grapevine VvCAX3 is a cation/H+ exchanger involved in vacuolar Ca2+ homeostasis. Planta 246:1083–1096

Mills RF, Peaston KA, Runions J, Williams LE (2012) HvHMA2, a P1B-ATPase from barley, is highly conserved among cereals and functions in Zn and Cd transport. PLoS One 7

Milner MJ, Seamon J, Craft E, Kochian LV (2013) Transport properties of members of the ZIP family in plants and their role in Zn and Mn homeostasis. J Exp Bot 64:369–381

Mitra GN (2015) Zinc (Zn) uptake. In: Mitra GN (ed) Regulation of nutrient uptake by plants: a biochemical and molecular approach. Springer, New Delhi, pp 127–133

Mondal TK, Ganie SA, Rana MK, Sharma TR (2014) Genome-wide analysis of zinc transporter genes of maize (Zea mays). Plant Mol Biol Report 32:605–616

Moreau S, Thomson RM, Kaiser BN, Trevaskis B, Guerinot ML, Udvardi MK, Puppo A, Day DA (2002) GmZIP1 encodes a symbiosis-specific zinc transporter in soybean. J Biol Chem 277:4738–4746

Mousavi SR, Galavi M, Rezaei M (2013) Zinc (Zn) importance for crop production – a review. International Journal of Agronomy and Plant Production 4:64–68

Munir T, Rizwan M, Kashif M, Shahzad A, Ali S, Amin N, Zahid R, Fakhar-e-Alam M, Muhammad Imran D (2018) Effect of zinc oxide nanoparticles on the growth and Zn uptake in wheat (Triticum aestivum L.) by seed priming method. Dig J Nanomater Biostruct 13:315–323

Nandal V, Solanki M (2021) ZN as a vital micronutrient in plants. J Microbiol Biotechnol Food Sci 11:e4026

Natasha N, Shahid M, Bibi I, Iqbal J, Khalid S, Murtaza B, Bakhat HF, Farooq ABU, Amjad M, Hammad HM, Niazi NK, Arshad M (2022) Zinc in soil-plant-human system: a data-analysis review. Sci Total Environ 808:152024

Nishiyama R, Kato M, Nagata S, Yanagisawa S, Yoneyama T (2012) Identification of Zn–Nicotianamine and Fe–2′-deoxymugineic acid in the phloem sap from rice plants (Oryza sativa L.). Plant Cell Physiol 53:381–390

Northover GHR, Garcia-España E, Weiss DJ (2021) Unravelling the modus operandi of phytosiderophores during zinc uptake in rice: the importance of geochemical gradients and accurate stability constants. J Exp Bot 72:1517–1526

Northover GHR, Mao Y, Blasco S, Vilar R, Garcia-España E, Rocco C, Hanif M, Weiss DJ (2022) Synergistic use of siderophores and weak organic ligands during zinc transport in the rhizosphere controlled by pH and ion strength gradients. Sci Rep 12:6774

Palusińska M, Barabasz A, Kozak K, Papierniak A, Maślińska K, Antosiewicz DM (2020) Zn/Cd status-dependent accumulation of Zn and Cd in root parts in tobacco is accompanied by specific expression of ZIP genes. BMC Plant Biol 20:37

Pedas P, Husted S (2009) Zinc transport mediated by barley ZIP proteins are induced by low pH. Plant Signal Behav 4:842–845

Peramaiyan P, Craufurd P, Kumar V, Seelan LP, McDonald AJ, Balwinder-Singh KA, Singh S (2022) Agronomic biofortification of zinc in rice for diminishing malnutrition in South Asia. Sustain For 14

Podar D (2010) Plant transporters involved in heavy metal homeostasis. Extreme Life Biospeology Astrobiol 2

Ramegowda Y, Venkategowda R, Jagadish P et al (2013) Expression of a rice Zn transporter, OsZIP1, increases Zn concentration in tobacco and finger millet transgenic plants. Plant Biotechnol Rep 7, 309–319. https://doi.org/10.1007/s11816-012-0264-x

Ramesh SA, Choimes S, Schachtman DP (2004) Over-expression of an Arabidopsis zinc transporter in Hordeum vulgare increases short-term zinc uptake after zinc deprivation and seed zinc content. Plant Mol Biol 54:373–385

Rani N, Chauhan A, Kaur S, Solanki MK, Tripathi M, Jain D, Singh S, Upadhyay SK, Kaur G (2023) Molecular mechanistic of Zn-solubilizing bacteria for agronomic eminence: recent updates and futuristic development. J Plant Growth Regul. https://doi.org/10.1007/s00344-023-11111-y

Rehman R, Asif M, Cakmak I, Ozturk L (2021) Differences in uptake and translocation of foliar-applied Zn in maize and wheat. Plant Soil 462:235–244

Rodrigues WFC, Lisboa ABP, Lima JE, Ricachenevsky FK, Del-Bem L-E (2023) Ferrous iron uptake via IRT1/ZIP evolved at least twice in green plants. New Phytol 237:1951–1961

Sadeghzadeh B (2013) A review of zinc nutrition and plant breeding. J Soil Sci Plant Nutr 13

Saleem S, Malik A, Khan ST (2023) Prospects of ZnO-nanoparticles for Zn biofortification of Triticum aestivum in a zinc-solubilizing bacteria environment. J Soil Sci Plant Nutr 23:4350–4360

Saxena V, Bharti MK, Kumar P, Singh J, Patel VB (2023) Effect of zinc uptake on alcohol dehydrogenase, protein and mineral contents of hydroponically grown chickpea (Cicer arietinum). J Plant Nutr 46:867–876

Shimizu A, Yanagihara S, Kawasaki S, Ikehashi H (2004) Phosphorus deficiency-induced root elongation and its QTL in rice (Oryza sativa L.). Theor Appl Genet 109:1361–1368

Shukla P, Kidwai M, Narayan S, Shirke PA, Pandey KD, Misra P, Chakrabarty D (2023) Phytoremediation potential of Solanum viarum Dunal and functional aspects of their capitate glandular trichomes in lead, cadmium, and zinc detoxification. Environ Sci Pollut Res 30:41878–41899

Shukla K, Mishra V, Singh J, Varshney V, Verma R, Srivastava S (2024) Nanotechnology in sustainable agriculture: A double-edged sword. J Sci Food Agric 104:5675–5688

Sinclair SA, Krämer U (2012) The zinc homeostasis network of land plants. Biochimica et Biophysica Acta (BBA) – Molecular Cell Research 1823:1553 1567

Singh B, Natesan SKA, Singh BK, Usha K (2005) Improving zinc efficiency of cereals under zinc deficiency. Curr Sci 88:36–44

Socha AL, Guerinot ML (2014) Mn-euvering manganese: the role of transporter gene family members in manganese uptake and mobilization in plants. Front. Plant Sci 5

Song W-Y, Choi KS, Kim DY, Geisler M, Park J, Vincenzetti V, Schellenberg M, Kim SH, Lim YP, Noh EW, Lee Y, Martinoia E (2010) Arabidopsis PCR2 is a zinc exporter involved in both zinc extrusion and long-distance zinc transport. Plant Cell 22:2237–2252

Song W-Y, Hörtensteiner S, Tomioka R, Lee Y, Martinoia E (2011) Common functions or only phylogenetically related? The large family of PLAC8 motif-containing/PCR genes. Mol Cells 31:1–7

Song W-Y, Lee H-S, Jin S-R, Ko D, Martinoia E, Lee Y, An G, Ahn S-N (2015) Rice PCR1 influences grain weight and Zn accumulation in grains. Plant Cell Environ 38:2327–2339

Song Z, Li S, Li Y, Zhou X, Liu X, Yang W, Chen R (2024) Identification and characterization of yellow stripe-like genes in maize suggest their roles in the uptake and transport of zinc and iron. BMC Plant Biol 24:3

Srivastav A, Ganjewala D, Singhal RK, Rajput VD, Minkina T, Voloshina M, Srivastava S, Shrivastava M (2021) Effect of ZnO nanoparticles on growth and biochemical responses of wheat and maize. Plan Theory 10

Srivastava PC, Rawat D, Pachauri SP, Shrivastava M (2015) Strategies for enhancing zinc efficiency in crop plants. In: Rakshit A, Singh HB, Sen A (eds) Nutrient use efficiency: from basics to advances. Springer, New Delhi, pp 87–101

Stanton C, Sanders D, Krämer U, Podar D (2022) Zinc in plants: Integrating homeostasis and biofortification. Mol Plant 15:65–85

Steele KA, Price AH, Shashidhar HE, Witcombe JR (2006) Marker-assisted selection to introgress rice QTLs controlling root traits into an Indian upland rice variety. Theor Appl Genet 112:208–221

Takahashi R, Ishimaru Y, Shimo H, Ogo Y, Senoura T, Nishizawa NK, Nakanishi H (2012) The OsHMA2 transporter is involved in root-to-shoot translocation of Zn and Cd in rice. Plant Cell Environ 35:1948–1957

Tan L, Qu M, Zhu Y, Peng C, Wang J, Gao D, Chen C (2020) Zinc transporter5 and zinc transporter9 function synergistically in zinc/cadmium uptake1. Plant Physiol 183:1235–1249

Tian P, Feng Y-X, Li C-Z, Zhang P, Yu X-Z (2023) Transcriptional analysis of heavy metal P1B-ATPases (HMAs) elucidates competitive interaction in metal transport between cadmium and mineral elements in rice plants. Environ Sci Pollut Res 30:287–297

Tiong J, McDonald GK, Genc Y, Pedas P, Hayes JE, Toubia J, Langridge P, Huang CY (2014) HvZIP7 mediates zinc accumulation in barley (Hordeum vulgare) at moderately high zinc supply. New Phytol 201:131–143

Verma PK, Verma S, Chakrabarty D, Pandey N (2021) Biotechnological approaches to enhance zinc uptake and utilization efficiency in cereal crops. J Soil Sci Plant Nutr 21:2412–2424

Wairich A, Ricachenevsky FK, Lee S (2022) A tale of two metals: biofortification of rice grains with iron and zinc. Front Plant Sci 13

Wang CY, Jenkitkasemwong S, Duarte S, Sparkman BK, Shawki A, Mackenzie B, Knutson MD (2012) ZIP8 is an iron and zinc transporter whose cell-surface expression is up-regulated by cellular iron loading. J Biol Chem 287:34032–34043

Wang F-H, Qiao K, Shen Y-H, Wang H, Chai T-Y (2021) Characterization of the gene family encoding metal tolerance proteins in Triticum urartu: Phylogenetic, transcriptional, and functional analyses. Metallomics 13:mfab038

Xin X, Judy J, Sumerlin B, He Z (2020) Nano-enabled agriculture: from nanoparticles to smart nanodelivery systems. Environ Chem 17

Yağız AK, Yavuz C, Naeem M, Das DS, Aksoy E (2022) Genome editing for nutrient use efficiency in crops. In: Prakash CS, Fiaz S, Fahad S (eds) Principles and practices of OMICS and genome editing for crop improvement. Springer, Cham, pp 347–383

Yang Z, Yang F, Liu J-L, Wu H-T, Yang H, Shi Y, Liu J, Zhang Y-F, Luo Y-R, Chen K-M (2022) Heavy metal transporters: Functional mechanisms, regulation, and application in phytoremediation. Sci Total Environ 809:151099

Yano K, Ohmori Y, Fujiwara T (2017) Introduction to GWAS and MutMap for identification of genes/QTL using next-generation sequencing. In: Plant macronutrient use efficiency: molecular and genomic perspectives in crop plants. Elsevier, pp 307–321

Yin S, Duan M, Fang B, Zhao G, Leng X, Zhang T (2023) Zinc homeostasis and regulation: Zinc transmembrane transport through transporters. Crit Rev Food Sci Nutr 63:7627–7637

Yoneyama T, Ishikawa S, Fujimaki S (2015) Route and regulation of zinc, cadmium, and iron transport in rice plants (Oryza sativa L.) during vegetative growth and grain filling: metal transporters, metal speciation, grain cd reduction and Zn and Fe biofortification. Int J Mol Sci 16:19111–19129

Exploitation of Microbes as a Means for Biofortification of Zinc in Crop Plants

J. R. Shubha and M. Divyashree

Abstract

Zinc is an important micronutrient that is vital to human and plant health. Zinc deficiency is also referred to as hidden hunger, and the worldwide prevalence of zinc deficiency causes nutritional insecurities in a large population. About 50% of Indian soils lack the required zinc in an available form for plants, affecting crop plants' growth and productivity. This, in turn, causes zinc deficiency and subsequent health complications among the people consuming such foods. Applying chemical zinc fertilizers in the soil will not ensure the fortification of zinc in staple food crops as they are transformed into insoluble forms, hindering uptake by crop plants. As the deficiency in food crops is mainly attributed to the accessibility of zinc by plants but not their presence, supplementing zinc in more accessible forms to plants can serve as a means to curb the deficiency. This problem can be resolved by using microbe-mediated biofortification approaches and ensuring nutrient-rich quality food for all. Here, zinc-solubilizing properties of rhizosphere microflora, mainly plant growth-promoting rhizobacteria, and endophytes can be harnessed to enhance zinc concentration in edible crop plants. These bacteria use various mechanisms, including producing organic acids and metallophores and reducing soil pH to biotransform insoluble zinc to soluble forms, facilitating the uptake of zinc by plants, and applying zinc-solubilizing

J. R. Shubha (✉)
Department of Food Safety and Nutrition, Nitte University Centre for Science Education and Research (NUCSER), Paneer Campus, Nitte Deemed to be University, Mangaluru, India
e-mail: shubha.jr@nitte.edu.in

M. Divyashree
Department of Environmental Health and Toxicology, Nitte University Centre for Science Education and Research (NUCSER), Paneer Campus, Nitte (Deemed to be University), Mangaluru, Karnataka, India

S. Srivastava, H. Kaur (eds.), *Zinc in Soil-Plant Continuum*,
https://doi.org/10.1007/978-981-96-4253-3_7

bacteria as alternative bioinoculants can serve as an efficient, sustainable agricultural approach for zinc biofortification.

Keywords

Agronomic approaches · Biofortification · Microbial inoculation · Transgenics · Zinc solubilization

1 Zinc and Human Health

Micronutrients, including zinc, iron, and selenium, play a vital role in the metabolic activity of animal, plant, and microbial kingdoms. Their deficiency, also referred to as "hidden hunger," signifies the importance of these micronutrients on health (de Valença et al. 2017). Zinc is the second most trace element distributed in the human body. Zinc is required for the optimum functioning of numerous biochemical processes in the human body. It is an indispensable part of at least 300 different enzymes that facilitate normal cell growth and division and protein folding and is involved in the regulation of various gene expression systems (Chasapis et al. 2012; Shyamal et al. 2016; Suganya et al. 2020; Frassinetti et al. 2006; Mohamad et al. 2023; Saper and Rash 2009). Healthy body development during infancy, childhood, adolescence, and pregnancy requires optimum zinc intake and ICMR has proposed RDA (Recommended dietary allowance) of zinc required at different growth stages for the Indian population. According to ICMR, the RDA of zinc in men is 17 mg; for women, it is 13.2 mg; for pregnant women, it is 14.5 mg; and for lactating women, it is 14 mg. Whereas, infants 6–12 months require 2.5 mg; children 1–3y, 4–6y, and 7–9y need 3.3 mg, 4.5 mg/d, and 5.9 mg/d, respectively; boys/girls 10–12y need 8.5 mg/d; boys 13–15y and 16–18y need 14.3 mg/d and 17.6 mg/d, respectively; and girls 13–15y and 16–18y need 12.8 mg/d and 14.2 mg/d, respectively.

Zinc is an essential nutrient for health; it is often in low doses in the human diet. Low levels of zinc lead to health impairments in immunological functions, retardation of a child's growth, and also complications during pregnancy (Bhardwaj et al. 2022; Hess and King 2009). Low levels of zinc are also known to be associated with an increased risk of Alzheimer's disease, autism spectrum disorder (ASD), depression, cancer, diabetes, and cardiovascular diseases (Khan et al. 2016; Little et al. 2010). Zinc, along with other micronutrients, including iron, iodine, and vitamin A, is known to account for 20% of deaths in children below 5 years of age (Prentice et al. 2008). Thus, it is important to critically assess zinc status, ascertain the deficiency levels, and ensure adequate supply of zinc for good health and improve global health status.

Numerous methods have been implemented to create awareness of the importance of zinc in health, encouraging the inclusion of zinc-rich or zinc-fortified dietary components in daily life and some extreme conditions providing zinc supplements (Basharat et al. 2019; Maret and Sandstead 2006). Though zinc is a most essential micronutrient, it cannot be stored in the human body for long time

(Maxfield et al. 2023), and the challenges associated with its bioavailability from ingested dietary sources have led to the need for a constant supply of this micronutrient in adequate amounts to prevent zinc deficiency (Frassinetti et al. 2006). In this direction, the production of agricultural products with improved nutritional quality, i.e., fortification of zinc, can serve as a potential alternative that can serve as a long-term measure ensuring an adequate supply of zinc to ensure food security in the society (Cakmak 2008; Gómez-Galera et al. 2010).

2 Plants and Zinc

On the other hand, zinc is an important micronutrient essential for plant growth, overall development, and protection against disease, and it is also known to play an imperative role in stress resistance (Cabot et al. 2019; Hassan et al. 2020). Among yield-limiting micronutrients, zinc is placed in fourth place in Indian soil, and its deficiency is known to increase to 63% from 49% by 2025 (Arunachalam et al. 2013). Zinc is a vital component of various biomolecules including proteins, lipids, and biosynthesis of auxins. Zinc plays a key role in carbohydrate metabolism, protein, and chlorophyll synthesis. Zinc also aids in protecting lipid membranes against oxidative damage caused by reactive oxygen species (Srivastava and Shrivastava 2017). A lower concentration of zinc affects plant growth, resulting in low levels of indole-3-acetic acid (IAA) in the shoot apical meristems and young leaves (Cakmak et al. 1989). Smaller leaves with interveinal chlorosis in young leaves, spots on older, and retarded development are signs of zinc deficiency in plants (Marschner 2011). Food crops were found to contain low quantities of zinc in the grains when grown in zinc-deficient soils. Thus, the presence of an adequate amount of available zinc that can be absorbed by crop plants is necessary for optimum growth and yield. Further, ideal zinc levels in crop plants are essential to ensure an adequate supply of micronutrient zinc to the population.

Zinc-deficient soil leads to the production of zinc-deficient food crops and causes zinc malnutrition in the population, affecting the nutritional security of the world. This also leads to hindered crop growth, productivity, and low-quality food commodities, causing economic loss to the growers (Kamran et al. 2017; Khan et al. 2007; Upadhayay et al. 2018). Plant uptake of zinc depends on numerous factors, like environmental conditions, form, and availability of zinc, that affect crop plant growth, yield, and nutrient concentration. These factors, along with the solubility of zinc in soil, result in food crops with low quantities of zinc and zinc micronutrient deficiency among the population, threatening the nutritional security of the nation. The presence of ideal zinc levels in crop plants is essential to warrant an adequate supply of zinc to the population and ensure nutrition security. Thus, global research aims to improve plant growth with increased yield and zinc concentration, i.e., nutritionally enriching food crops, commercializing them, and curbing micronutrient deficiency-related diseases. This can be achieved by providing adequate zinc to crop plants in a soluble or absorbable form.

3 Fortification: Approaches to Overcome Zinc Deficiency in Food Crops

The increasing population has led to an increased demand for staple food globally. Different crop improvement programs have been able to increase the quantity of food while compromising the quality, as these staple foods are majorly known to meet calorific but not micronutrient requirements. However, recently, to meet the increasing demand and its associated food securities concerning different micronutrients, many stakeholders, including the government, different agencies, and researchers, are looking forward to food fortification as an efficient strategy for promoting micronutrient deficiency. Literature shows that growing food crops in zinc-deficient soils leads to low concentrations of zinc in food grains and is one of the reasons for zinc deficiency among consumers. Hence, food fortification acts as a means for addressing malnutrition without having to change one's food system. For example, fortifying commonly consumed food crops such as maize and wheat with micronutrients can serve as a potential intervention in securing nutritional values and socioeconomic benefits (Cakmak 2008; Welch and Graham 2004).

4 Fortification

Food fortification is the practice of deliberately adding micronutrients to combat micronutrient deficiency and increase their nutritional value. Though high-income countries have made food fortification mandatory, i.e., iodized salt that dates back to the 1920s in North America and Europe, low-middle-income countries are still picking up the pace in the last few decades because of urbanization and increased economic status (Spohrer et al. 2013). Fortification or the thoughtful increase in the essential nutrient concentration in staple foods has great potential in improving the nutritional quality with public health benefits. Fortification effectiveness in a given country depends on various factors, including the country's socioeconomic status, the extent of the prevalence of the deficiency, affected population group, food production systems available infrastructure, and government regulations (De-Regil et al. 2011). To address the challenges of micronutrient deficiency in a population, food fortification can be mainly divided into three types:

1. *Large-Scale Food Fortification (LSFF)*

   This method also referred to as industrial food fortification involves the addition of required micronutrients to regularly consumed food such as flour, salt, sugar, oil, and other condiments during processing. This approach can be either initiated and regulated by the government bodies as they are mandatory or by food product industries voluntarily adding different micronutrients during processing within regulatory limits. With this large target population can be reached and micronutrient deficiency can be curtailed. For example, the use of iodized salt has increased drastically from 20% to 70% between 1990 and 2008 (De-Regil et al. 2011).

2. *Point of Use*

   Also termed as home fortification, point of use refers to the addition of minerals and vitamins to cooked food before consumption. The term point of use was adopted by the WHO in 2012 to be used in places such as schools and refugee camps to fight micronutrient deficiency. However, it came forefront in 2016 when it was recommended as a strategic intervention to fight micronutrient deficiency in children of age groups 6–24 months (De-Regil et al. 2017). In this direction, micronutrient packets (MPNs) containing essential vitamins and nutrients in powder form can be used to sprinkle on cooked food without altering the organoleptic properties of the food (De-Regil et al. 2011).

3. *Biofortification*

   The third but most interesting aspect in fighting against micronutrient deficiency is biofortification or the purposeful addition of essential nutrients in food crops during their cultivation to enhance nutritional quality and attain health benefits using biological means. This has gained significant momentum in recent years. Biofortification involves agronomic or modern biotechnological techniques to increase the nutrient concentration in foods without affecting the characteristics of the crop. This technique is an important tool in ameliorating nutrient deficiency diseases, hence called nutrition-sensitive agricultural intervention (Talsma and Pachon 2017). The difference between conventional fortification and biofortification is that the former technique focuses on enhancing the nutritional quality during crop plant growth rather than during the processing of the crops. Thus, biofortification is advantageous where implementation of conventional fortification is difficult or limited (Nestel et al. 2006).

Biofortified crops are said to be nutrient-dense, with enhanced retention of nutrients post-processing, bioavailability, and absorption, thus improving the micronutrient status after consuming biofortified foods (La Frano et al. 2014). Successful projects include zinc fortified wheat, beans, rice, corn, and potato production. Promising pieces of evidence have been accumulating with other essential nutrient fortification that includes iron-fortified crops, which have been shown to reverse iron deficiency in school children in India (Finkelstein et al. 2015) and in pregnant women in Rwanda (Haas et al. 2016). A few other works include the biofortification of carotenoids, proteins, and amino acids; thus, biofortification can serve as an efficient method for increasing the desired nutrient density in food crops. As biofortification can be employed for common stable crops, added advantages include cost-efficiently targeting poor populations that do not have access to supplements or fortified processed foods (Birol et al. 2021). Biofortification of required nutrients in crop plants can be achieved by following three main techniques:

Conventional biofortification: It involves a conventional method where plants with high nutrient concentrations are naturally selected and crossbred with other local varieties to obtain suitable, efficient crops. This involves the selection of desired genetic traits within natural boundaries (Talsma and Pachon 2017).

Transgenic biofortification involves directly introducing desired genes or related traits into the host plant, modifying their genetic components and increasing nutrient density (Shahzad et al. 2021).

Agronomic biofortification: This involves enhancing the nutrient concentration in edible parts of the crop plant with the use of fertilizer directly on the soil or application of sprays enriched with desired nutrients to foliar parts or by irrigation with added minerals. Examples include NPK8 fertilizer or pesticides rich in zinc. This is the primary method that has been adopted for many years to achieve a higher zinc concentration in crop plants. The use of inorganic zinc sulfates to increase zinc concentration in soil is also known to have a negative effect on human health and the environment (Guo et al. 2021; Joy et al. 2015; White and Broadley 2005).

However, the problem is not resolved to the expected extent, limiting the success rate of this approach. This is due to the inadequate availability of zinc for absorption by the plant, as it mainly depends on the solubility of the supplied zinc but not the total zinc present in the soil (Cakmak 2008). The availability of zinc also depends on soil type, solubility of applied fertilizers, and the chemical reactions in the soil environment, making it difficult to absorb by the plants. Within 7 days of application, 96–99% of applied inorganic zinc is converted into different insoluble counterparts depending on the soil type and physicochemical nature (Saravanan et al. 2004). Soil pH and moisture content directly affect the solubility of zinc, thereby affecting its bioavailability.

However, extensive use of these agrochemicals resulted in soil fertility loss, an imbalanced environment, and damaged soil ecology, posing health risks to humans and other animals. Hence, switching to alternative practices that can reduce the abovementioned detrimental effects is of great scope. In this direction, bio-based approaches involving biofertilizers, additives, and bio-inoculants are important in providing an eco-friendly, sustainable agricultural approach. Applying microbial inoculants involving zinc-solubilizing endophytes, plant growth-promoting rhizobacteria, or altering rhizosphere microflora is a promising alternative. These microbial inoculants augment zinc solubility by transforming the unavailable zinc into the most available form. The approach of using microbes as potential bio-inoculants in the biofortification of plants can improve zinc uptake by various mechanisms as the plants share a complex relationship with surrounding microflora. Microbes play a specific role in improving the uptake of nutrients and protecting against pathogens and stress by altering their metabolism according to the requirements of the niche (Mitter et al. 2013). Thus, manipulating the microbiome may aid in improving crop productivity and nutritional aspects, thereby contributing to the sustainable approach. Piling up evidence suggests the ability of microbes to enhance the absorption by the plants, enhancing their growth, yield, and accumulation of zinc in the produce (Saravanan et al. 2007; Hussain et al. 2015).

5 Microbial Biofortification

Soil fertility status affects the biofortification levels in crop plants. However, supplementation of soil chemically using fertilizers will only increase Zn concentration in soil but not guarantee its uptake by plants (Majeed et al. 2022; Sharma et al. 2013). Microorganisms play an essential role in the biotransformation of zinc (Fig. 1) and its bioavailability, which can serve as an alternative to other biofortification techniques. Zinc-solubilizing bacteria present a great choice as they possess diverse capabilities to increase crop growth and productivity to ensure food security (Hussain et al. 2018). Microbial biofortification offers an efficient intervention that is both economical and scalable in agricultural systems.

Enhancement of zinc concentration in edible parts of crop plants is one of the important needs of the present agri-sector that can combat zinc deficiency in larger populations. A complex yet important signaling network exists between microbes and plants in the environment. Several studies have highlighted the ability of zinc-solubilizing bacteria to enhance the zinc concentration in edible crop plants. This can serve as a tool to achieve micronutrient deficiency problems in an environmentally friendly, sustainable manner and avoid chemical fertilizers alone. Plant growth stimulation by zinc-solubilizing bacteria can be tested in in vitro settings, making it a suitable mediator in zinc biofortification (Kamran et al. 2017; Upadhayay et al. 2022; Vaid et al. 2019). Owing to these advantages, the cost-effective approach of microbe-mediated zinc biofortification in food crops is gaining interest as these beneficial agents act as facilitators of nutrient uptake (Shakeel et al. 2015) (Fig. 1).

6 Mechanisms Involved in Zinc Solubilization

The risk associated with zinc malnutrition is the poor supply of zinc from the soil to the plant. The soil is well-off with an abundance of zinc, but its dearth in crops prevails because of an inaccessible form of zinc in the soil (Sunithakumari et al. 2016). Rhizosphere microbes are known to support plant growth by providing required micronutrients, including zinc, through various mechanisms. Several mechanisms are involved in the solubilization of zinc including the efflux of organic acids, chelating agent production, extrusion of protons, and also by the use of an oxidoreduction system of the cell membrane and cell surfaces (Crane et al. 1985; Hughes and Poole 1991; Wakatsuki 1995). Organic acids bind to zinc cations, lowering soil pH and facilitating solubilization. Similarly, anions chelate zinc, thus improving bioavailability in soil. Production of extracellular organic acids, including acetic, formic, lactic, succinic, isovaleric, and isobutyric acids, is also associated with the solubilization of zinc (Mumtaz et al. 2017). 2-ketogluconic acids and gluconic acid were detected in *Pseudomonas fluorescens* culture broth that aided in zinc solubilization (Di Simine et al. 1998). Gluconic acid and its derivatives produced by bacteria are known to play a role in biotransforming insoluble zinc compounds to their soluble counterparts (Fasim et al. 2002; Saravanan et al. 2007). This prospect was also confirmed by strains of *Pseudomonas, Acinetobacter*, and *Gluconacetobacter*

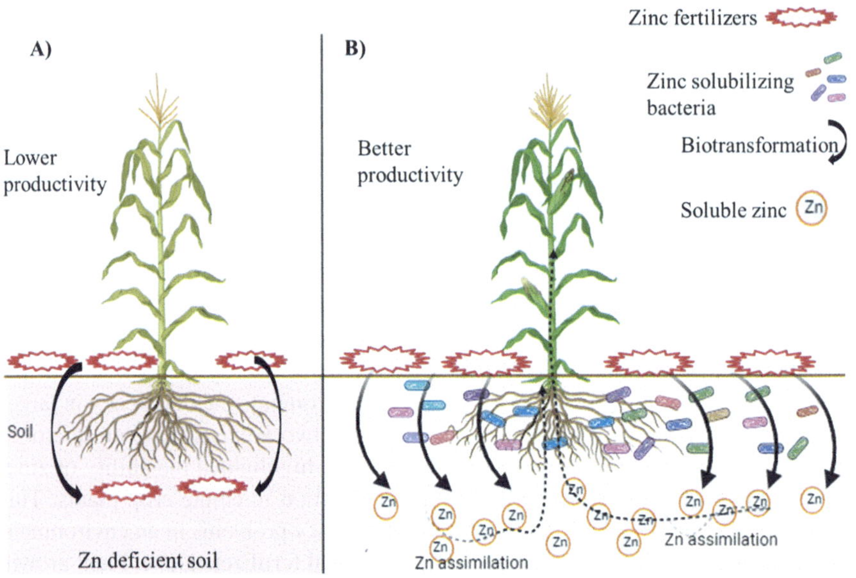

Fig. 1 Zinc-solubilizing microbes as mediators of zinc biofortification: (**a**) without zinc-solubilizing bacterial, applied zinc fertilizers become unavailable; (**b**) in the presence of zinc-solubilizing bacteria, insoluble zinc is biotransformed into soluble counterparts and readily available for uptake by plants

that produced excess gluconic acid and resulted in subsequent zinc solubilization (Di Simine et al. 1998; Fasim et al. 2002; Saravanan et al. 2007; Vaid et al. 2019). Few others like *Burkholderia cepacian* are known to solubilize zinc by means of producing other organic acids such as acetic, formic, oxalic, and tartaric acid (Li et al. 2010). Along with this, inorganic acids such as carbonic acid and nitric acids produced by microbes also aid in the solubilization of zinc (Nahas 1996; Sayer and Gadd 1997).

Lowering pH is also one of the means of zinc solubilization by bacteria as observed in many studies. A positive correlation was observed between lowered pH and zinc solubilization in the culture medium of rhizosphere isolates *Ralstonia pickettii, Klebsiella pneumoniae, Pseudomonas aeruginosa, and Burkholderia cepacia* and improved biofortification of zinc in rice seedling and overall growth (Gontia-Mishra et al. 2017). It was observed by Goteti et al. (2013) in a pot trial inoculated with *Pseudomonas* sp. and *Bacillus* sp. Increased uptake of zinc, manganese, potassium, and nitrogen in plants directly correlates with decreased pH. Metallophores, or zinc chelators, are another group of compounds that support in zinc solubilization (Hussain et al. 2018; Upadhayay et al. 2018). Hence, incorporating metallophores-producing microbes can serve as a promising approach towards zinc biofortification. It was observed that *Enterobacter cancerogenus*, *Pseudomonas monteilii*, and *Microbacterium saperdae* when inoculated to soil system secreted zinc chelators assisting increased zinc solubility, bioavailability, and enhanced uptake of zinc by plant root system (Whiting et al. 2001). Other chelating agents such as ethylenediamine-tetra acetic acid produced by *Pseudomonas* sp., *Agrobacterium* Sp., and *Azospirillum lipoferum* when incorporated as biofertilizers were able to augment zinc solubilization from its salts (Tariq and Hameed 2007).

Few studies indicated that the bacterial approach for zinc solubilization is similar to that of phosphate solubilization. The environmental conditions, mode of nutrition, and the bacterial metabolism type will also determine the effectiveness of zinc solubilization (Costerousse et al. 2017; Upadhayay et al. 2022). However, mostly the heterotrophs have been studied in the case of zinc solubilization and its application to improved zinc accumulation in edible crop plants (Costerousse et al. 2017; Gontia-Mishra et al. 2017; Kamran et al. 2017; Ramesh et al. 2014; Vaid et al. 2019). Along with the phytohormones, siderophores, antibiotics, antimycotics, and hydrogen cyanide produced by microbes are also known to have a beneficial effect on plant growth (Rodríguez and Fraga 1999; Upadhayay et al. 2018). Thus, microbes with beneficial properties serve as mediators to enhance the efficiency of biofortification in crop plants.

7 Microbes as Mediators for Zinc Solubilization in Soil

Studies are still in the early stages as reports available in zinc-solubilizing endophytes and rhizosphere bacteria are limited. Most of the studies attribute the zinc-solubilizing effect of bacteria to its ability to produce organic acids. Genus *Bacillus* is among the prominent group studied and reported for their influence in solubilizing zinc in soil systems (Fasim et al. 2002; Mumtaz et al. 2017; Saravanan et al. 2007; Upadhayay et al. 2022). Reported cultures include *Bacillus subtilis* QST713 (Moreno-Lora et al. 2019), *B. aryabhattai* (Ramesh et al. 2014), and co-inoculants *B. aryabhattai* ZM31 and *B. subtilis* ZM63 (Mumtaz et al. 2017). Environment from which bacterial culture is isolated may affect the zinc-solubilizing capacity of the bacteria. For example, *Bacillus* sp. isolated from zinc ore sphalerite was found to solubilize ZnCO3, ZnO, and ZnS more efficiently (Saravanan et al. 2004; Sharma et al. 2013).

Backing up evidence suggests the significance of microbes in the biofortification of food crops with micronutrients, including zinc, iron, etc. Desai et al. (2012) reported that reduction in soil pH to acidic conditions was indirectly proportional to increased zinc availability in the culture broth. Work by Di Simine et al. (1998) demonstrated that *the Pseudomonas fluorescens* strain successfully solubilized zinc phosphate with the help of gluconic acid production, which in turn reduced soil pH, aiding solubilization. Similarly, Goteti et al. (2013) suggested that the presence of zinc in soil alone will not assure the uptake by plants, but elemental mineral mobility matters and microbes such as PGPR bacteria play an important role in enhancing the zinc uptake. PGPR bacteria are linked to plant development as they are associated with nitrogen fixation and solubilization of essential nutrients, including zinc, phosphates, iron-chelating siderophores, extracellular enzymes, hormones, exopolysaccharides, and ammonia (Ahemad and Kibret 2014; Zablotowicz et al. 1991).

The soil microbiome profile is one of the parameters used to determine the health of the soil. Establishing a strong root system also plays an important role in plant growth. Zinc-solubilizing bacteria, along with other bacteria that can aid in making micronutrients accessible and helping overall growth and productivity, are also termed plant probiotics. Efficient colonization of the plant probiotic bacterial cultures in the rhizosphere region of crop plants is necessary for providing a suitable environment for bacterial metabolism. This provides appropriate conditions for the production of necessary metabolites that aid in zinc solubilization and uptake by plants. Microorganisms, specifically PGPR bacteria, when incorporated into the soil along with fertilizers of either chemical or organic origin, are known to enhance the uptake of micronutrients, which in turn results in increased zinc concentration in edible plant parts (Mishra et al. 2005; Shivay et al. 2010). These PGPR microbes colonize plant root systems and aid in the overall development of plants through various mechanisms as discussed above (Glick 1995), and their application in the agri-sector is continuously expanding (Jalal et al. 2021; Rana et al. 2012; Shahane and Shivay 2022).

The role of PGPR bacteria alone or in consortium in zinc solubilization is widely recognized by researchers. PGPR bacteria are known to expand root surface area by producing auxin and help enhance the uptake of soluble nutrients from the soil (Rana et al. 2012; Hussain et al. 2015). Introducing a consortium of PGPR into the soil is known to increase the soil richness, which influences plant micronutrient uptake (Nuttall et al. 2017). Thus, PRGR is not only useful in enriching micronutrients in crop plants but also plays an additional role in improving crop yield, maintaining fertility and the biodiversity of the soil (Ditta et al. 2022; Jalal et al. 2022; Nestel et al. 2006; White and Broadley 2005).

Piling up evidence has indicated increased popularity of PGPR greatly owing to their involvement in enhancing the micronutrient composition in food grains. Species of *Acinetobacter, Bacillus, Brevibacterium, Enterobacter, and Pseudomonas genera have been reported as potential candidates for biofortification as they were found to solubilize insoluble micronutrients and enrich edible parts of crop plants* (Gopalakrishnan et al. 2016). Work by Ahmad et al. (2023) indicated that the application of biofertilizer consisting PGPR bacterial species *Bacillus subtilis* ZM63, *B. aryabhattai* ZM31, *B. aryabhattai* S10, and *Paenibacillus polymyxa* ZM27 solely, in combination or in consortium, enhanced the zinc status of maize along with growth and productivity, the consortium being most promising (Ahmad et al. 2023).

This was further supported by the work of Burd et al. (2011), Shaukat et al. (2010), and Siddiqui and Shaukat (2002), where the use of zinc-solubilizing PGPR *Azospirillum*, *Pseudomonas*, and *Azotobacter* strains for seed bacterization was found to enhance plant growth and nutrient availability and reduce metal toxicity. *Rhizobium radiobacter* (LB2), a zinc- and phosphate-solubilizing bacteria, was found to increase nutritional quality in terms of zinc concentration increase in lettuce crops. Other bacteria of interest include *Pantoea rodasii* and *Burkholderia cepacian* used in rice biofortification (Upadhayay et al. 2022).

Using endophytes presents an alternative opportunity for zinc biofortification in crop plants. They are known to enhance zinc localization in edible parts of rice (Tariq and Hameed 2007; Wang et al. 2014), wheat (Rehman et al. 2018), and chickpea (Ullah et al. 2020). It was demonstrated by Krithika and Balachandar (2016) and Upadhayay et al. (2022) that the presence of zinc-solubilizing bacteria influences genes related to transporter families responsible for zinc transportation in plants. Singh et al. (2021) reported an increase in transporter gene-ZIP up to 4.96-fold with subsequent increase in zinc transportation when inoculated with *Bacillus amyloliquefaciens* and *Trichoderma harzianum* as co-inoculants in wheat plants. Two endophytic bacterial strains, *Enterobacter* sp., SaCS20 and Sphingomonas sp., SaMR12, were found to augment zinc bioaccessibility in the rhizosphere, which led to increased zinc density-rich rice grain yield (Wang et al. 2014).

Other documented strains with abilities to improve zinc density in food grains include *Exiguobacterium aurantiacum* (Shaikh and Saraf 2017), *B. subtilis* (Moreno-Lora et al. 2019), *Pseudomonas fragi* (Kamran et al. 2017), *Anabaena* sp., *Providencia* sp. (Rana et al. 2012), and *Bacillus tequilensis* CRS38 (Yadav et al. 2022) in wheat. Similarly *B. altitudinis* SRI-178 enhanced zinc concentration in chickpea (Gopalakrishnan et al. 2016), *Pseudomonas* spp. VZ74 in tomato (Karnwal 2021), and *B. aryabhattai* MDSR14 in soybean (Ramesh et al. 2014). Thus, it can be proposed that a bacteria-mediated biofortification strategy can offer a long-term eco-friendly solution and help in curbing micronutrient malnutrition to a greater extent (Hussain et al. 2018).

8 Biofortification and Sustainable Agricultural Approach

Bioeconomy or green economy is harnessing the knowledge of biotechnology in the development of agriculture science to produce better-yielding nutrient-rich crops and crop-derived products. Microbe-based biofertilizers, biopesticides, and bioinoculants and their application can efficiently build a sustainable bioeconomy (Uzoh and Babalola 2018). Bio-based pesticides and fertilizers are of great importance as the means to increase soil fertility, better crop plants, and conserve environmental health. They are essential for the long-term sustainable development of the agricultural sector as they can minimize the use of chemical fertilizers and pesticides to a greater extent. In this regard, the application of zinc-solubilizing bacteria as a biofortifying agent can serve as a green strategy for improved crop plant growth and production of zinc-rich food products, ensuring nutritional safety in the population (Rengalakshmi et al. 2018; Truu et al. 2008).

9 Conclusion

Improving zinc concentration in crop plants and thus combating zinc deficiency-related health problems have great scope in the modern world. It is essential to maintain a healthy population and ensure food security among a larger population worldwide. This can be achieved by exploiting microbe-assisted biofortification approaches to ensure the presence of required zinc concentrations in edible food crops. As the work is still in progress, there is a need for the isolation and identification of a greater number of microbial cohorts capable of solubilizing zinc and making it accessible to plants. The works can greatly improve plant nutrition by providing zinc in more available form, enhancing growth, and thus guaranteeing zinc supply through food crops to a larger population. Further in-depth work on zinc-solubilizing mechanisms, optimizing their formulations, and field application strategies is necessary to provide deeper insights and ensure their applicability in large areas in a sustainable, eco-friendly manner and combat zinc deficiency problems worldwide.

References

Ahemad M, Kibret M (2014) Mechanisms and applications of plant growth promoting rhizobacteria: current perspective. J King Saud Univ Sci 26(1):1–20. https://doi.org/10.1016/J.JKSUS.2013.05.001

Ahmad M, Hussain A, Dar A, Luqman M, Ditta A, Iqbal Z, Ahmad HT, Nazli F, Soufan W, Almutairi K, Sabagh AE (2023) Combating iron and zinc malnutrition through mineral biofortification in maize through plant growth promoting Bacillus and Paenibacillus species. Front Plant Sci 13:1094551. https://doi.org/10.3389/FPLS.2022.1094551/FULL

Arunachalam P, Kannan P, Prabukumar G, Govindaraj M (2013) Zinc deficiency in Indian soils with special focus to enrich zinc in peanut. Afr J Agric Res 8(50):6681–6688. https://doi.org/10.5897/AJARx12.015

Basharat S, Ahsan F, Sikander S, Noreen S, Hamna H (2019) Role of ketogenic diet in pediatric epilepsy. J Psychol Clin Psychiatry 10(3):118–122. https://doi.org/10.15406/JPCPY.2019.10.00639

Bhardwaj AK, Chejara S, Malik K, Kumar R, Kumar A, Yadav RK (2022) Agronomic biofortification of food crops: an emerging opportunity for global food and nutritional security. Front Plant Sci 13:1055278. https://doi.org/10.3389/FPLS.2022.1055278/BIBTEX

Birol E, Foley J, Aytekin D (2021) Biofortification: the evidence: a summary of multi-disciplinary research to help inform transformation of food systems through the adoption and scaling of nutrient-enriched staple food crops. HarvestPlus, Washington, DC

Burd GI, Dixon DG, Glick BR (2011) Plant growth-promoting bacteria that decrease heavy metal toxicity in plants. Can J Microbiol 46(3):237–245. https://doi.org/10.1139/W99-143

Cabot C, Martos S, Llugany M, Gallego B, Tolrà R, Poschenrieder C (2019) A role for zinc in plant defense against pathogens and herbivores. Front Plant Sci 10. https://doi.org/10.3389/FPLS.2019.01171

Cakmak I (2008) Enrichment of cereal grains with zinc: agronomic or genetic biofortification? Plant Soil 302(1–2):1–17. https://doi.org/10.1007/S11104-007-9466-3

Cakmak I, Marschner H, Bangerth F (1989) Effect of zinc nutritional status on growth, protein metabolism and levels of Indole-3-acetic acid and other phytohormones in bean (Phaseolus vulgaris L.). J Exp Bot 40(3):405–412. https://doi.org/10.1093/JXB/40.3.405

Chasapis CT, Spiliopoulou CA, Loutsidou AC, Stefanidou ME (2012) Zinc and human health: an update. Arch Toxicol 86(4):521–534. https://doi.org/10.1007/S00204-011-0775-1/METRICS

Costerousse B, Schönholzer-Mauclaire L, Frossard E, Thonar C (2017) Identification of heterotrophic zinc mobilization processes among bacterial strains isolated from wheat rhizosphere (Triticum aestivum L.). Appl Environ Microbiol 84(1). https://doi.org/10.1128/AEM.01715-17

Crane FL, Sun IL, Clark MG, Grebing C, Löw H (1985) Transplasma-membrane redox systems in growth and development. BBA Rev Bioenerg 811(3):233–264. https://doi.org/10.1016/0304-4173(85)90013-8

de Valença AW, Bake A, Brouwer ID, Giller KE (2017) Agronomic biofortification of crops to fight hidden hunger in sub-Saharan Africa. Glob Food Sec 12:8–14. https://doi.org/10.1016/J.GFS.2016.12.001

De-Regil LM, Suchdev PS, Vist GE, Walleser S, Peña-Rosas JP (2011) Home fortification of foods with multiple micronutrient powders for health and nutrition in children under two years of age. Cochrane Database Syst Rev 2011(9). https://doi.org/10.1002/14651858.CD008959.PUB2

De-Regil LM, Jefferds MED, Peña-Rosas JP (2017) Point-of-use fortification of foods with micronutrient powders containing iron in children of preschool and school-age. Cochrane Database Syst Rev 2017(11). https://doi.org/10.1002/14651858.CD009666.PUB2

Desai S, Kumar PG, Sultana U, Pinisetty S, Hassan Ahmed MS, Daniel Amalraj LE, Reddy G (2012) Potential microbial candidate strains for management of nutrient requirements of crops. Afr J Microbiol Res 6(17):3924–3931. https://doi.org/10.5897/AJMR12.224

Di Simine CD, Sayer JA, Gadd GM (1998) Solubilization of zinc phosphate by a strain of Pseudomonas fluorescens isolated from a forest soil. Biol Fertil Soils 28(1):87–94. https://doi.org/10.1007/S003740050467/METRICS

Ditta A, Ullah N, Imtiaz M, Li X, Jan AU, Mehmood S, Rizwan MS, Rizwan M (2022) Zn biofortification in crops through Zn-solubilizing plant growth-promoting Rhizobacteria. 115–133. https://doi.org/10.1007/978-3-030-91499-8_7

Fasim F, Ahmed N, Parsons R, Gadd GM (2002) Solubilization of zinc salts by a bacterium isolated from the air environment of a tannery. FEMS Microbiol Lett 213(1):1–6. https://doi.org/10.1111/J.1574-6968.2002.TB11277.X

Finkelstein JL, Mehta S, Udipi SA, Ghugre PS, Luna SV, Wenger MJ, Murray-Kolb LE, Przybyszewski EM, Haas JD (2015) A randomized trial of iron-biofortified pearl millet in school children in India. J Nut 145(7):1576–1581. https://doi.org/10.3945/JN.114.208009

Frassinetti S, Bronzetti GL, Caltavuturo L, Cini M, Croce CD (2006) The role of zinc in life: a review. J Environ Pathol Toxicol Oncol 25(3):597–610. https://doi.org/10.1615/JENVIRONPATHOLTOXICOLONCOL.V25.I3.40

Glick BR (1995) The enhancement of plant growth by free-living bacteria. Canadian J Microbio 41(2):109–117. https://doi.org/10.1139/M95-015

Gómez-Galera S, Rojas E, Sudhakar D, Zhu C, Pelacho AM, Capell T, Christou P (2010) Critical evaluation of strategies for mineral fortification of staple food crops. Transgenic Res 19(2):165–180. https://doi.org/10.1007/S11248-009-9311-Y/METRICS

Gontia-Mishra I, Sapre S, Tiwari S (2017) Zinc solubilizing bacteria from the rhizosphere of rice as prospective modulator of zinc biofortification in rice. Rhizosphere 3:185–190. https://doi.org/10.1016/J.RHISPH.2017.04.013

Gopalakrishnan S, Vadlamudi S, Samineni S, Sameer Kumar CV (2016) Plant growth-promotion and biofortification of chickpea and pigeonpea through inoculation of biocontrol potential bacteria, isolated from organic soils. Springerplus 5(1). https://doi.org/10.1186/S40064-016-3590-6

Goteti PK, Emmanuel LDA, Desai S, Shaik MHA (2013) Prospective zinc solubilising bacteria for enhanced nutrient uptake and growth promotion in maize (Zea mays L.). Int J Microbiol 2013. https://doi.org/10.1155/2013/869697

Guo XX, Zhao D, Zhuang MH, Wang C, Zhang FS (2021) Fertilizer and pesticide reduction in cherry tomato production to achieve multiple environmental benefits in Guangxi, China. Sci Total Environ 793. https://doi.org/10.1016/J.SCITOTENV.2021.148527

Haas JD, Luna SV, Lung'aho MG, Wenger MJ, Murray-Kolb LE, Beebe S, Gahutu JB, Egli IM (2016) Consuming iron biofortified beans increases iron status in Rwandan women after 128 days in a randomized controlled feeding trial. J Nutr 146(8):1586–1592. https://doi.org/10.3945/JN.115.224741

Hassan MU, Aamer M, Chattha MU, Haiying T, Shahzad B, Barbanti L, Nawaz M, Rasheed A, Afzal A, Liu Y, Guoqin H (2020) The critical role of zinc in plants facing the drought stress. Agriculture 10(9):396. https://doi.org/10.3390/AGRICULTURE10090396

Hess SY, King JC (2009) Effects of maternal zinc supplementation on pregnancy and lactation outcomes. Food Nutr Bull 30(1 Suppl):S60. https://doi.org/10.1177/15648265090301S105

Hughes MN, Poole RK (1991) Metal speciation and microbial growth – the hard (and soft) facts. J Gen Microbiol 137(4):725–734. https://doi.org/10.1099/00221287-137-4-725/CITE/REFWORKS

Hussain A, Arshad M, Zahir Z, Asghar M (2015) Prospects of zinc solubilizing bacteria for enhancing growth of maize. Pakistan J Agri Sci.

Hussain A, Zahir ZA, Asghar HN, Ahmad M, Jamil M, Naveed M, Zaman Akhtar MFU (2018) Zinc solubilizing bacteria for zinc biofortification in cereals: a step toward sustainable nutritional security. In: Role of rhizospheric microbes in soil: volume 2: nutrient management and crop improvement, pp 203–227. https://doi.org/10.1007/978-981-13-0044-8_7

Jalal A, Galindo FS, Boleta EHM, da Silva Oliveira CE, Dos Reis AR, Nogueira TAR, Moretti Neto MJ, Mortinho ES, Fernandes GC, Teixeira Filho MCM (2021) Common bean yield and zinc use efficiency in association with diazotrophic bacteria co-inoculations. Agronomy 11(5):959. https://doi.org/10.3390/AGRONOMY11050959

Jalal A, Oliveira CEDS, Freitas LA, Galindo FS, Lima BH, Boleta EHM, Da Silva EC, Do Nascimento V, Nogueira TAR, Buzetti S, Teixeira Filho MCM (2022) Agronomic biofortification and productivity of wheat with soil zinc and diazotrophic bacteria in tropical savannah. Crop Pasture Sci 73(8):817–830. https://doi.org/10.1071/CP21457

Joy EJM, Stein AJ, Young SD, Ander EL, Watts MJ, Broadley MR (2015) Zinc-enriched fertilisers as a potential public health intervention in Africa. Plant Soil 389(1–2):1–24. https://doi.org/10.1007/S11104-015-2430-8/FIGURES/2

Kamran S, Shahid I, Baig DN, Rizwan M, Malik KA, Mehnaz S (2017) Contribution of zinc solubilizing bacteria in growth promotion and zinc content of wheat. Front Microbiol 8. https://doi.org/10.3389/FMICB.2017.02593

Karnwal A (2021) Pseudomonas spp., a zinc-solubilizing vermicompost bacteria with plant growth-promoting activity moderates zinc biofortification in tomato. Int J Veg Sci 27(4):398–412. https://doi.org/10.1080/19315260.2020.1812143

Khan R, Gurmani AR, Khan MS, Gurmani AH (2007) Effect of zinc application on rice yield under wheat rice system. Pak J Biol Sci 10(2):235–239. https://doi.org/10.3923/PJBS.2007.235.239

Khan SA, Khan SA, Narendra AR, Mushtaq G, Zahran SA, Khan S, Kamal MA (2016) Alzheimer's disease and autistic Spectrum disorder: is there any association? CNS Neurol Disord Drug Targets 15(4):390–402. https://doi.org/10.2174/1871527315666160321104303

Krithika S, Balachandar D (2016) Expression of zinc transporter genes in rice as influenced by zinc-solubilizing enterobacter cloacae strain ZSB14. Front Plant Sci 7(APR2016). https://doi.org/10.3389/FPLS.2016.00446

La Frano MR, de Moura FF, Boy E, Lönnerdal B, Burri BJ (2014) Bioavailability of iron, zinc, and provitamin A carotenoids in biofortified staple crops. Nutr Rev 72(5):289–307. https://doi.org/10.1111/NURE.12108

Li WC, Ye ZH, Wong MH (2010) Metal mobilization and production of short-chain organic acids by rhizosphere bacteria associated with a Cd/Zn hyperaccumulating plant, Sedum alfredii. Plant Soil 326(1–2):453–467. https://doi.org/10.1007/S11104-009-0025-Y

Little PJ, Bhattacharya R, Moreyra AE, Korichneva IL (2010) Zinc and cardiovascular disease. Nutrition 26(11–12):1050–1057. https://doi.org/10.1016/J.NUT.2010.03.007

Majeed A, Rashid I, Niaz A, Ditta A, Sameen A, Al-Huqail AA, Siddiqui MH (2022) Balanced use of Zn, Cu, Fe, and B improves the yield and sucrose contents of sugarcane juice cultivated in sandy clay loam soil. Agronomy 12(3):696. https://doi.org/10.3390/AGRONOMY12030696

Maret W, Sandstead HH (2006) Zinc requirements and the risks and benefits of zinc supplementation. J Trace Elem Med Biol 20(1):3–18. https://doi.org/10.1016/J.JTEMB.2006.01.006

Marschner H (2011) Mineral nutrition of higher plants, 3rd edn. eBook ISBN: 9780123849069. EBook ISBN: 9780123849069

Maxfield L, Shukla S, Crane JS (2023) Zinc deficiency. Causes and Management of Nutritional Deficiency Disorders, 245–260. https://doi.org/10.4018/979-8-3693-2947-4.ch014

Mishra BN, Prasad R, Gangaiah B (2005) Organic manures for increased rice productivity and sustained supply of Fe to rice. Acta Agron Hung 52(4):391–397. https://doi.org/10.1556/AAGR.52.2004.4.8

Mitter B, Brader G, Afzal M, Compant S, Naveed M, Trognitz F, Sessitsch A (2013) Advances in elucidating beneficial interactions between plants, soil, and bacteria. Adv Agron 121:381–445. https://doi.org/10.1016/B978-0-12-407685-3.00007-4

Mohamad NS, Tan LL, Ali NIM, Mazlan NF, Sage EE, Hassan NI, Goh CT (2023) Zinc status in public health: exploring emerging research trends through bibliometric analysis of the historical context from 1978 to 2022. Env Sci Pollution Res 30(11):28422–28445. https://doi.org/10.1007/S11356-023-25257-5

Moreno-Lora A, Recena R, Delgado A (2019) Bacillus subtilis QST713 and cellulose amendment enhance phosphorus uptake while improving zinc biofortification in wheat. Appl Soil Ecol 142:81–89. https://doi.org/10.1016/J.APSOIL.2019.04.013

Mumtaz MZ, Ahmad M, Jamil M, Hussain T (2017) Zinc solubilizing bacillus spp. potential candidates for biofortification in maize. Microbiol Res 202:51–60. https://doi.org/10.1016/J.MICRES.2017.06.001

Nahas E (1996) Factors determining rock phosphate solubilization by microorganisms isolated from soil. World J Microbiol Biotechnol 12(6):567–572. https://doi.org/10.1007/BF00327716

Nestel P, Bouis HE, Meenakshi JV, Pfeiffer W (2006) Biofortification of staple food crops. J Nutr 136(4):1064–1067. https://doi.org/10.1093/JN/136.4.1064

Nuttall JG, O'Leary GJ, Panozzo JF, Walker CK, Barlow KM, Fitzgerald GJ (2017) Models of grain quality in wheat—a review. Field Crop Res 202:136–145. https://doi.org/10.1016/J.FCR.2015.12.011

Prentice AM, Gershwin ME, Schaible UE, Keusch GT, Victora CG, Gordon JI (2008) New challenges in studying nutrition-disease interactions in the developing world. J Clin Invest 118(4):1322–1329. https://doi.org/10.1172/JCI34034

Ramesh A, Sharma SK, Sharma MP, Yadav N, Joshi OP (2014) Inoculation of zinc solubilizing Bacillus aryabhattai strains for improved growth, mobilization and biofortification of zinc in soybean and wheat cultivated in Vertisols of Central India. Appl Soil Ecol 73:87–96. https://doi.org/10.1016/J.APSOIL.2013.08.009

Rana A, Joshi M, Prasanna R, Shivay YS, Nain L (2012) Biofortification of wheat through inoculation of plant growth promoting rhizobacteria and cyanobacteria. Eur J Soil Biol 50:118–126. https://doi.org/10.1016/J.EJSOBI.2012.01.005

Rehman A, Farooq M, Naveed M, Ozturk L, Nawaz A (2018) Pseudomonas-aided zinc application improves the productivity and biofortification of bread wheat. Crop Pasture Sci 69(7):659–672. https://doi.org/10.1071/CP17441

Rengalakshmi R, Manjula M, Prabavathy VR, Jegan S, Selvamukilan B (2018) Building bioeconomy in agriculture: harnessing soil microbes for sustaining ecosystem services. In: Towards a sustainable bioeconomy: principles, challenges and perspectives, World sustainability series, pp 261–277. https://doi.org/10.1007/978-3-319-73028-8_14

Rodríguez H, Fraga R (1999) Phosphate solubilizing bacteria and their role in plant growth promotion. Biotechnol Adv 17(4–5):319–339. https://doi.org/10.1016/S0734-9750(99)00014-2

Saper RB, Rash R (2009) Zinc: an essential micronutrient. Am Fam Physician 79(9):768

Saravanan VS, Subramoniam SR, Raj SA (2004) Assessing in vitro solubilization potential of different zinc solubilizing bacterial (ZSB) isolates. Braz J Microbiol 35(1–2):121–125. https://doi.org/10.1590/S1517-83822004000100020

Saravanan VS, Madhaiyan M, Thangaraju M (2007) Solubilization of zinc compounds by the diazotrophic, plant growth promoting bacterium gluconacetobacter diazotrophicus. Chemosphere 66(9):1794–1798. https://doi.org/10.1016/J.CHEMOSPHERE.2006.07.067

Sayer JA, Gadd GM (1997) Solubilization and transformation of insoluble inorganic metal compounds to insoluble metal oxalates by aspergillus Niger. Mycol Res 101(6):653–661. https://doi.org/10.1017/S0953756296003140

Shahane AA, Shivay YS (2022) Soil health management in organic production system—a review. Int J Bio-Resour Stress Manag 13:1186–1200. https://doi.org/10.23910/1.2022.3141a

Shahzad R, Jamil S, Ahmad S, Nisar A, Khan S, Amina Z, Kanwal S, Aslam HMU, Gill RA, Zhou W (2021) Biofortification of cereals and pulses using new breeding techniques: current and future perspectives. Front Nutr 8. https://doi.org/10.3389/FNUT.2021.721728/FULL

Shaikh S, Saraf M (2017) Biofortification of Triticum aestivum through the inoculation of zinc solubilizing plant growth promoting rhizobacteria in field experiment. Biocatal Agric Biotechnol 9:120–126. https://doi.org/10.1016/J.BCAB.2016.12.008

Shakeel M, Rais A, Hassan MN, Hafeez FY (2015) Root associated Bacillus sp. improves growth, yield and zinc translocation for basmati rice (Oryza sativa) varieties. Front Microbiol 6. https://doi.org/10.3389/FMICB.2015.01286

Sharma A, Patni B, Shankhdhar D, Shankhdhar SC (2013) Zinc – an indispensable micronutrient. Physiol Mol Biol Plants 19(1):11–20. https://doi.org/10.1007/S12298-012-0139-1

Shaukat K, Affrasayab S, Hasnain S (2010) Growth responses of Triticwn aestivum to plant growth promoting rhizobacteria used as a bio fertilizer. Res J Microbiol 5(10):1022–1030. https://doi.org/10.3923/JM.2006.330.338

Shivay Y, Prasad R, Rahal A (2010) Studies on some nutritional quality parameters of organically or conventionally grown wheat. Cereal Res Commun 38(3):345–352. https://doi.org/10.1556/CRC.38.2010.3.5

Shyamal M, Mazumdar P, Maity S, Samanta S, Sahoo GP, Misra A (2016) Highly selective turn-on fluorogenic chemosensor for robust quantification of Zn(II) based on aggregation induced emission enhancement feature. ACS Sens 1(6):739–747. https://doi.org/10.1021/ACSSENSORS.6B00289/SUPPL_FILE/SE6B00289_SI_001.PDF

Siddiqui IA, Shaukat SS (2002) Zinc and glycerol enhance the production of nematicidal compounds in vitro and improve the biocontrol of Meloidogyne javanica in tomato by fluorescent pseudomonads. Lett Appl Microbiol 35(3):212–217. https://doi.org/10.1046/J.1472-765X.2002.01162.X

Singh UB, Malviya D, Singh S, Singh P, Ghatak A, Imran M, Rai JP, Singh RK, Manna MC, Sharma AK, Saxena AK (2021) Salt-tolerant compatible microbial inoculants modulate physio-biochemical responses enhance plant growth, Zn biofortification and yield of wheat grown in saline-sodic soil. Int J Environ Res Public Health 18(18). https://doi.org/10.3390/IJERPH18189936

Spohrer R, Larson M, Maurin C, Laillou A, Capanzana M, Garrett GS (2013) The growing importance of staple foods and condiments used as ingredients in the food industry and implications for large-scale food fortification programs in southeast Asia. 34(2 Suppl). https://doi.org/10.1177/15648265130342S107

Srivastava S, Shrivastava M (2017) Zinc supplementation imparts tolerance to arsenite stress in *Hydrilla verticillata* (L.f.) Royle. Int J Phytoremediation 19:353–359

Sunithakumari K, Padma Devi SN, Vasandha S (2016) Zinc solubilizing bacterial isolates from the agricultural fields of Coimbatore, Tamil Nadu, India on JSTOR. Curr Sci 110(2):196–205. https://www.jstor.org/stable/24906745

Suganya A, Saravanan A, Manivannan N (2020) Role of zinc nutrition for increasing zinc availability, uptake, yield, and quality of maize (Zea Mays L.) grains: an overview. Commun Soil Sci Plant Anal 51(15):2001–2021. https://doi.org/10.1080/00103624.2020.1820030

Talsma EF, Pachon H (2017) Biofortification of crops with minerals and vitamins. E-Library of Evidence for Nutrition Actions (ELENA). https://www.who.int/tools/elena/bbc/biofortification

Tariq M, Hameed S (2007) Plant root associated bacteria for zinc mobilization in rice. https://www.researchgate.net/publication/279695421

Truu M, Truu J, Ivask M (2008) Soil microbiological and biochemical properties for assessing the effect of agricultural management practices in Estonian cultivated soils. Eur J Soil Biol 44(2):231–237. https://doi.org/10.1016/J.EJSOBI.2007.12.003

Ullah A, Farooq M, Hussain M (2020) Improving the productivity, profitability and grain quality of kabuli chickpea with co-application of zinc and endophyte bacteria Enterobacter sp. MN17. Arch Agron Soil Sci 66(7):897–912. https://doi.org/10.1080/03650340.2019.1644501

Upadhayay VK, Singh AV, Pareek N (2018) An insight in decoding the multifarious and splendid role of microorganisms in crop biofortification. Int J Curr Microbiol App Sci 7(06):2407–2418. https://doi.org/10.20546/IJCMAS.2018.706.286

Upadhayay VK, Singh AV, Khan A, Sharma A (2022) Contemplating the role of zinc-solubilizing bacteria in crop biofortification: an approach for sustainable bioeconomy. Front Agron 4:903321. https://doi.org/10.3389/FAGRO.2022.903321/BIBTEX

Uzoh IM, Babalola OO (2018) Rhizosphere biodiversity as a premise for application in bio-economy. Agric Ecosyst Environ 265:524–534. https://doi.org/10.1016/J.AGEE.2018.07.003

Vaid SK, Srivastava PC, Pachauri SP, Sharma A, Rawat D, Mathpal B, Shankhadhar SC, Shukla AK (2019) Residual effect of zinc applied to rice on zinc nutrition of succeeding wheat crop inoculated with zinc solubilizing microbial consortium. Israel J Plant Sci 66(3–4):227–237. https://doi.org/10.1163/22238980-00001019

Wakatsuki T (1995) Metal oxidoreduction by microbial cells. J Ind Microbiol 14(2):169–177. https://doi.org/10.1007/BF01569900/METRICS

Wang Y, Yang X, Zhang X, Dong L, Zhang J, Wei Y, Feng Y, Lu L (2014) Improved plant growth and Zn accumulation in grains of rice (Oryza sativa L.) by inoculation of endophytic microbes isolated from a Zn Hyperaccumulator, Sedum alfredii H. J Agric Food Chem 62(8):1783–1791. https://doi.org/10.1021/JF404152U

Welch RM, Graham RD (2004) Breeding for micronutrients in staple food crops from a human nutrition perspective. J Exp Bot 55(396):353–364. https://doi.org/10.1093/JXB/ERH064

White PJ, Broadley MR (2005) Biofortifying crops with essential mineral elements. Trends Plant Sci 10(12):586–593. https://doi.org/10.1016/J.TPLANTS.2005.10.001

Whiting SN, De Souza MP, Terry N (2001) Rhizosphere bacteria mobilize Zn for hyperaccumulation by Thlaspi caerulescens. Environ Sci Technol 35(15):3144–3150. https://doi.org/10.1021/ES001938V

Yadav RC, Sharma SK, Varma A, Rajawat MVS, Khan MS, Sharma PK, Malviya D, Singh UB, Rai JP, Saxena AK (2022) Modulation in biofertilization and biofortification of wheat crop by inoculation of zinc-solubilizing rhizobacteria. Front Plant Sci 13:777771. https://doi.org/10.3389/FPLS.2022.777771/BIBTEX

Zablotowicz RM, Tipping EM, Lifshitz R, Kloepper JW (1991) Plant growth promotion mediated by bacterial rhizosphere colonizers. In: The rhizosphere and plant growth, pp 315–326. https://doi.org/10.1007/978-94-011-3336-4_70

Breeding Approaches and Modern Tools to Fight Zinc Deficiency in Plants

Vívian Ebeling Viana, Viviane Kopp da Luz, and Camila Pegoraro

Abstract

Zinc (Zn) is an essential micronutrient to plants as it is involved in different processes, influencing growth, development, and yield. Consequently, Zn deficiency in agricultural soils restricts crop yield, in addition to affecting the nutritional quality of plant-based foods. Zn deficiency also causes serious human health problems. Therefore, elucidating the chemical reactions of Zn in the soil as well as the uptake and transport by the plants is necessary. Different approaches to mitigate Zn deficiency effects can be outlined. Application of fertilizers, biostimulators, and nanoparticles and the development of new genotypes that improved Zn use efficiency are some important approach to deal with Zn deficiency. The creation of new cultivars can be achieved using conventional or molecular breeding techniques. In conventional breeding, the available genetic diversity is explored, using hybridization and artificial selection tools, while in molecular breeding, different tools are applied to identify candidate genes, such as molecular markers, genetic mapping, and gene expression analyses. The identified genes can be manipulated through transgenics and genome editing. Considering the impact of Zn deficiency on the productivity of cultivated plants and on human health, this chapter aims to provide information about genes involved in Zn uptake, breeding strategies, and modern tools applied to fight against Zn deficiency.

V. E. Viana
Universidade Federal de Pelotas, Fitossanidade, Pelotas, Brazil
e-mail: vivian.viana@ufpel.edu.br

V. K. da Luz · C. Pegoraro (✉)
Universidade Federal de Pelotas, Fitotecnia, Pelotas, Brazil
e-mail: viviane.kopp@ufpel.edu.br; cpegoraro.faem@ufpel.edu.br

S. Srivastava, H. Kaur (eds.), *Zinc in Soil-Plant Continuum*,
https://doi.org/10.1007/978-981-96-4253-3_8

Keywords

Zinc uptake · Crop improvement · Plant nutrition · Genome editing · Transgenic

1 Introduction

Zinc (Zn) is an essential mineral nutrient involved in plant growth due to its function in different processes and defense systems. Zn is crucial for plant yield and quality. Plants uptake Zn from the soil, and under deficiency, there is growth restriction, reduction in the size of internodes and leaves, chlorosis and necrosis, and delayed maturation (Hacisalihoglu 2020).

Chemical interactions that occur in the soil can result in a limited amount of arable agricultural soils in different regions of the world (Saleem et al. 2022). On the other hand, Zn excess in the soil causes plant phytotoxicity, reducing yield and quality. Increased plant Zn concentration promotes interruption of Zn-dependent physiological processes. However, Zn deficiency is more widespread than Zn toxicity (Kaur et al. 2024). Plants have (intraspecific and interspecific) genetic differences among them regarding Zn requirement, Zn efficiency, and tolerance to high Zn concentrations (Gupta et al. 2016).

Plants are the main source of Zn into the food chain. Therefore, Zn deficiency in the soil, in addition to affecting plants, has an impact on human health. Zn deficiency in humans occurs due to a diet of foods poor in Zn and negatively affects cell growth and the reproductive and immune systems. In this condition, respiratory diseases, diarrhea, skin problems, memory problems, hair loss, and birth problems are common (Younas et al. 2023).

The damage caused by Zn deficiency on plants and humans is constantly studied. Many physiological mechanisms such as root uptake, transport, leaf sequestration, and biochemical use of Zn in the plant have been investigated. Furthermore, studies on the manipulation of genes involved in these processes have also been performed and contribute to understanding Zn homeostasis in crops (Henriques et al. 2012).

One of the strategies used to mitigate the effects of Zn deficiency in plants is the application of fertilizers. However, in this case there is an associated cost, and the efficiency of application depends on the physical and chemical properties of the soil. Another more favorable and lower cost approach is ameliorate plant zinc use efficiency and grain zinc content through plant breeding (Henriques et al. 2012).

According to the methods applied and tools available, plant breeding can be classified into two main types, conventional breeding and molecular or modern breeding. Conventional breeding is supported by the amplification of genetic variability through artificial hybridization, followed by selection, evaluation, release, and multiplication of the new cultivar. Modern or molecular breeding emerged when molecular biology techniques were integrated into conventional breeding to obtain greater genetic gain (Anand et al. 2023). Among these techniques are molecular markers, genetic mapping, marker-assisted selection, and genomic selection (Anand et al. 2023). Gene expression analysis is another fundamental tool (Biswas 2023). Genetic mapping and transcript analysis enable the recognition of target genes that control traits of agronomic importance, which can be manipulated using transgenics or genome editing aiming to develop improved plants (Permyakova and Deineko 2024).

Within this context, this chapter aims to bring information about the mechanisms of Zn uptake and transport by plants and breeding approaches and modern tools to fight zinc deficiency in plants.

2 Importance of Zn

Zinc displays many important roles in the plant physiology. It plays a crucial function as a constituent of hundreds of enzymes: lyases, transferases, hydrolases, isomerases, oxidoreductases, and ligases. Zinc is the unique element that is part of all of these classes of enzymes and influences the properties of these enzymes (Hamzah Saleem et al. 2022).

Zinc-binding proteins act in many biological processes such as RNA transcription, protein translation, photosynthesis, and reactive oxygen species (ROS) metabolism (Natasha et al. 2022). As a key component of antioxidant enzymes such as superoxide dismutase, Zn is essential for cellular defense mechanisms that mitigate stress-induced free radical levels (Castillo-González et al. 2018).

The other essential biophysical-chemical functions of Zn in plants are control of gene expression, protein synthesis, participation in phosphate and carbohydrate metabolism, maintenance of the structural integrity of ribosomes, role in the structure and function of biomembranes (Noman et al. 2019; Castillo-González et al. 2018). Thus, Zn deficiency in plants can impair biophysical-chemical mechanisms essential under normal and stress conditions (Natasha et al. 2022).

Approximately 1230 proteins from *Arabidopsis thaliana* contain, bind, or are involved in Zn transport. Zinc has different functions in all individuals. A significant number of people suffer from moderate zinc deficiency. Among the effects of Zn deficiency are respiratory illnesses, diarrheal disorders, malaria, and about 800,000 deaths per year in the world's poor population (Sinclair and Krämer 2012; Hussain et al. 2022; Younas et al. 2023). Skin problems, blindness, cognitive losses, and hair loss in humans are caused by Zn deficiency (Hussain et al. 2022). In general, Zn deficiency affects different parts of the world where the people consume especially cereal-based diets, which are poor in bioavailable Zn (Sinclair and Krämer 2012).

In plants, Zn deficiency causes disturbances in physiological processes, reduced stress tolerance, and symptoms such as chlorosis, growth inhibition, necrosis, and small new leaves, which often lead to negative effects on productivity (Younas et al. 2023). Generally, plants show symptoms of Zn deficiency when the concentration of this element is below 15 to 20 mg Zn per kg of shoot dry biomass. Among the symptoms are restricted biomass production, lower fertility, chlorosis in the leaves, increased shoot branching, and early senescence of older leaves. Zn deficiency leads to an increase in the ROS synthesis resulting from reduced Cu/Zn superoxide dismutase (CuZn-SOD) activity, impaired protein synthesis, and elevated Fe accumulation (Sinclair and Krämer 2012).

Although Zn is important for the plant, excess of this element causes damage and causes chlorosis. Thus, Zn homeostasis in the plant is essential for its growth. The regulation of Zn homeostasis in the plant encompasses the uptake control, intracellular binding to metal chelators, efflux from the cell, and retention in vacuoles (Song et al. 2010).

2.1 Zn in Soil and Zn Deficiency in Plants

The lack of Zn in the soil is one of the main abiotic conditions that impairs to plants, impacting significantly more than 49% of the world's cultivable areas with occurrence in many countries (Hacisalihoglu 2020). Total Zn content in soil is dependent on bedrock, weathering, organic matter, texture, and pH (Sharma et al. 2013).

Zinc is found in three primary fractions in soil: first—water-soluble Zn (including Zn^{2+} and soluble organic fractions); second—Zn adsorbed and exchangeable in the colloidal fraction (associated with clay particles, humic compounds, and Al and Fe hydroxides); and third—insoluble Zn complexes and minerals (Broadley et al. 2007). Typically, Zn concentration in soils varies from 10 to 300 mg kg^{-1} (average of 50 mg kg^{-1}) (Sharma et al. 2013), and concentration greater than 150 mg kg^{-1} is being considered high (Landon 2014; Noulas et al. 2018). Not all Zn in the soil is available to plants; it depends on the physicochemical characteristics of the soil; interaction with other nutrients, plant roots, and microorganisms in the rhizosphere; and in addition the ability of each genotype to absorb Zn.

Certain soils have a low total Zn concentration, such as sandy soils. Likewise, there are soils with a low level of Zn available to plants because of its strong adsorption, as occurs in calcareous soils, saline soils, vertisols, and highly weathered soils (Noulas et al. 2018; Mossa et al. 2020). In calcareous and alkaline soils, with high pH, the Zn is the most limiting nutrient, followed by nitrogen (N) and phosphorus (P). In arid and semi-arid soils, Zn deficiency is occasioned by high calcium carbonate and low organic matter content. Other factors of low Zn in soil comprehend mineral poverty of the original soil material, high pH (>7), high salt content, formation of complexes with soil, and high phosphorus content (Younas et al. 2023).

Phosphorus application reduces the Zn concentration in grains by 17% to 56%, while Fe, Cu, and Mn are maintained or even raised. In soils with low Zn content, excess P aggravates Zn disturbance and crop growth (Akhtar et al. 2019). Regarding the interaction between Zn and P, their relationships have a negative correlation. Increasing inorganic phosphate (Pi) level in soil enlarges Zn adsorption on soil particles (Verma et al. 2021). In wheat, the increase in P results in a significant reduction in Zn levels in shoots and grains. The negative effect of P, however, can be mitigated with the concomitant application of Zn (Ova et al. 2015).

When soil temperatures are below recommended, Zn deficiency is more pronounced. Low temperatures cause a reduction in the microbial decomposition of organic matter, which would release Zn to the crop. in contrast, increasing soil temperature increases Zn supply, the diffusion rate of soil colloids to plant roots, and organic matter mineralization rates (Noulas et al. 2018).

Lack of Zn causes restriction of growth, stress tolerance, and chlorophyll synthesis (Sharma et al. 2013). Zn deficiency induces many physiological changes in addition to making the plant more susceptible to disease (Bastakoti 2023). At Zn concentrations below 15–20 mg kg^{-1} dry weight (Sinclair and Krämer 2012), disturbances in physiological processes occur, resulting in effects such as chlorosis, growth inhibition, delayed maturity, necrosis and new leaves small, and biomass reduction, which often lead to negative effects on productivity (Malavolta 2006). An important reduction in growth and grain yield under Zn deficiency conditions was observed in wheat, oats, barley, and triticale showed (Ajeesh Krishna et al. 2020).

The creation of cultivars with greater efficiency in absorbing Zn, adapted to soils with low Zn concentrations, has a significant function in sustainable agricultural production. In this way, not only the dependence on fertilizers will be reduced but also losses in crop yield and quality are minimized (Hacisalihoglu 2020).

2.2 Strategies for Zn Uptake and Transport by Plants

Zinc is absorbed by plants through their roots and leaves, predominantly in the divalent form of zinc (Zn^{2+}). However, since Zn is part of the soil solution and under conditions of adequate organic matter content, it can be chelated with organic radicals or in complex forms; consequently, it can also be absorbed by plants in the form of soluble organic chelates (Hassan et al. 2020). In soils with high pH, it is assumed that Zn is also absorbed as a monovalent cation ($ZnOH^{+}$) (Noulas et al. 2018).

The mechanism of Zn absorption through leaves is not yet completely understood. Nutrient absorption by leaves is affected by their surface properties, which include the thickness of the protective wax layer and the composition and structure of the cuticle, beyond the number of trichomes (Sturikova et al. 2018). As well as the physiological state, including the phenological phase, incidence of stress, and plant health, environmental conditions such as humidity, temperature, and lighting are equally important. The different Zn forms can be absorbed by the cuticle, and generally, nonpolar lipophilic compounds diffuse through the cuticle, although the principle of penetration of hydrophilic polar compounds is not yet known (Fernández and Brown 2013).

The transport of nutrients among cellular spaces, the apoplasts, is also import for Zn distribution in the plant. Zinc, like iron and calcium, is one of the nutrients that is found in the form of positively charged cations. Because of the composition pectins of the middle lamella and primary cell walls, the apoplast is predominantly negatively charged, which results in the binding and consequent accumulation of cations in this compartment. As a result, translocation of these cations to other plant organs becomes a difficult process (Sturikova et al. 2018).

During the absorption, Zn^{2+} are transported by the root epidermis, cortex, endodermis, pericycle, and xylem and are then moved to the stem, leaves, phloem, and seeds. Several studies were performed with the aim of elucidating the mechanisms of plants that demonstrate effectiveness in absorbing Zn in the face of stress caused by low Zn availability, aiming to establish effective plant breeding strategies.

Although studies exist on these mechanisms, considerable experimental evidence arises mainly from root absorption studies (Hacisalihoglu 2020).

In mineral absorption, roots play a dynamic role, releasing various compounds into the rhizosphere, thus enabling proper functioning and growth. Zinc is absorbed by roots as Zn^{2+} through mass flow and diffusion processes (Gupta et al. 2016). These passive absorption processes include the water molecules (solvent) and the difference in Zn concentrations across the root cell plasma membrane (RCPM). The main driving force of Zn^{2+} absorption (cation capture) is the hyperpolarization of the RCPM, mediated by the activity of the RCPM H+-ATPase system (Gupta et al. 2016; Hassan et al. 2020). However, unlike water, charged Zn ions cannot freely cross cell membranes (Alberts et al. 2007), being transported by specific transport proteins.

Similar to the observed in other mineral elements, the process of Zn accumulation in plants is controlled at each stage, from uptake from the soil, radial transport through the root tissues, transport from the roots to the shoot, until the distribution and redistribution among tissues. Each phase requires the expression of Zn transporter proteins specific to the plant cell and tissue type (Yang et al. 2020). Zinc transporters are all present in all parts of the cell, encompassing the plasma membrane, chloroplasts, and vacuoles. The objective is to separate Zn into different compartments so that it remains at adequate physiological concentrations in the cell and toxicity does not occur (Sinha et al. 2022).

Several genes that encode Zn transport proteins or regulate their expression have been identified. Some important families of Zn transport proteins include zinc-iron permease (ZIP) family/ZRT-, IRT-like proteins, heavy metal ATPases (HMA), metal tolerance protein (MTP), YSL and OPT proteins, PCR (plant cadmium resistance), VIT (vacuolar iron transporter), and cation exchanger (CAX). In addition to Zn, many of these transporters are also act in the transport of Mn, Fe, Cd, Co, and Cu (Ajeesh Krishna et al. 2017).

Besides to Zn transporters, some miRNA families have been characterized as their involvement in Zn status signaling. In *Sorghum bicolor*, the miR166, miR171, miR172, and miR398, miR399, and miR319 are upregulated modulating the expression of transporter genes in response to Zn deficiency (Li et al. 2013b). Zn deficiency reduces copper/zinc superoxide dismutase (CuZn-SOD) activity and the upregulation of miR398 represses *CuZn-SOD* expression in roots, while the reduced expression of miR528 in leafs leads to a relatively stable level of CuZn-SOD in Zn deficiency.

2.2.1 Zinc-Iron Permease (ZIP)/ZRT-, IRT-Like Proteins Family

Iron (Fe^{2+}) and zinc (Zn^{2+}) are often transported by proteins belonging to the ZIP (ZRT/IRT-like protein) family. These proteins are found in all organisms that represent different levels of evolution, from bacteria to fungi, plants, and mammals. ZIPs participate in the absorption and transport of metals, maintaining homeostasis by transporting them to the cytoplasm (Li et al. 2013a). ZIPs are the main transporters for the absorption and translocation of Zn and Fe in plants.

Iron-regulate transporter 1 (AtIRT1) is the first ZIP gene identified, encoding a protein with the function of transporting Fe on the root surface of *Arabidopsis* plants. Later in rice and several other species, members of the ZIP family were also studied, such as maize (*Zea mays*), wheat (*Triticum aestivum*), barley (*Hordeum vulgare*), grape (*Vitis vinifera*), clover (*Medicago truncatula*), alfalfa (*Medicago sativa*), cowpea (*Phaseolus vulgaris*), and palm (*Areca catechu*) (Verma et al. 2021; An et al. 2021; Ajeesh Krishna et al. 2017; Cardini et al. 2021). For example, overexpression of rice *OsZIP4*, *OsZIP5*, and *OsZIP8* increases root Zn level but reduces its quantity in shoots and seeds. OsZIP3 is also involved in Zn unloading and regulates Zn distribution in developing tissues (Verma et al. 2021).

In rice, the transporters OsZIP1, OsZIP5, and OsZIP9 are strongly expressed in exo- and endodermal root cells under Zn-deficient soils and function in cellular uptake of Zn across the plasma membrane. *OsZIP1* is upregulated in plants under Zn deficiency and Zn, Cu, and Cd excess and in this case functioning as an exporter and contributing to detoxification by decreasing the concentration of these metals in roots (Stanton et al. 2022). OsZIP1 may be involved in the retransfer of Zn ions during leaf aging under normal and Zn deficiency conditions, but overexpression of *ZIP1* can reduce the concentrations of Zn, Cu, and Cd in rice.

In rice, OsZIP4 has two distinct major functions in Zn translocation. First, OsZIP4 located in the phloem of the differentiated vascular bundle (DVB) is associated with loading Zn into the phloem for distribution to transition zones (TBs) and other tissues, encompassing new leaves and panicles. Second, OsZIP4 is highly expressed in TBs and transports Zn to meristem cells to meet active growth requirements (Mu et al. 2021). OsZIP4 and OsZIP8 are actively associated with Zn grain filling in rice (Mu et al. 2021; Ajeesh Krishna et al. 2020). OsZIP4 is involved in the transport and redistribution of Zn in rice and is mainly expressed in the vascular bundle and mesophyll cells of leaves, in the phloem of stems and roots, and in the meristem and roots. Studies suggest that *OsZIP4* expression is more likely to occur in young leaves during Zn-deficient environments (Ma et al. 2023).

OsZIP9 transporter is mainly expressed in lateral roots that act in environmental adaptation. Its transcription is increased under Zn deficiency conditions, while it is reduced under Zn excess conditions. These patterns suggest that OsZIP9 acts as a regulator, maintaining Zn homeostasis and adjusting its expression according to plant uptake needs at different Zn concentrations (Yang et al. 2020).

In rice, when *OsZIP4*, *OsZIP5*, and *OsZIP8* were overexpressed, there was an elevation in Zn level in the root; however, decreases Zn concentration in shoots and seeds (Lee et al. 2010). OsZIP3 participates in Zn transport and regulates Zn distribution to developing organs (Sasaki et al. 2015). In barley, the overexpression of HvZIP7, a protein that functions as a specific low-affinity Zn transporter, led to an increase in Zn levels in seeds and leaves when plants were exposed to a high, non-toxic, or toxic Zn source (Viana et al. 2021). Suppression of OsZIP7 caused drastic decrease in grain yield, which was related to Zn deficiency in plants (Tang et al. 2021)

Zinc-containing proteins act in DNA replication and controlling gene transcription. Consequently, developing organs need a high amount of Zn for active growth, as in the case of rice shoot meristem, which has the highest levels of Zn among organs and tissues (Mu et al. 2021). Two transporters, OsHMA2 and OsZIP3, have been recognized for their important function in Zn transport to developing tissues in rice (Huang et al. 2020).

The activity of several proteins of the ZIP family is regulated by transcription factors of the basic leucine zipper (bZIP) family, which are associated to the zinc deficiency response element (ZDRE) of many genes susceptible to Zn deficiency. In *Arabidopsis*, ZIP1, ZIP3, ZIP4, ZIP5, ZIP9, ZIP10, ZIP12, and IRT3 genes contain ZDRE, and their transcription is controlled by bZIP19 and bZIP23 under Zn deficiency (Verma et al. 2021). Under conditions of Zn deficiency, bZIP19 and bZIP23 transcription factors bind to the Zn deficiency response element (RTGTCGACAY) in the promoter region of target genes and activate their expression, contributing to Zn uptake and distribution (Ajeesh Krishna et al. 2022). The AtbZIP19 and AtbZIP23 transcription factors regulate the *AtZIP4* and *nicotianamine synthase 2* (*NAS2*) transcription under Zn deficiency. These genes act in Zn uptake and nicotianamine (NA)-mediated Zn distribution in *Arabidopsis* under Zn deficiency (Lilay et al. 2021).

Twelve *ZmZIP* genes (*ZmZIP1–ZmZIP12*) were identified in maize. Ten *ZmZIP* genes (*ZmZIP1*, *ZmZIP2*, and *ZmZIP4–ZmZIP11*) are tissue specific with high transcription in the flag leaf under Zn deficiency conditions. *ZmZIP2*, *ZmZIP5*, *ZmZIP6*, *ZmZIP8*, and *ZmZIP11* are expressed in grain under Zn deficiency environment. The proteins ZmZIP4, ZmZIP5, ZmZIP7, and ZmZIP9 are localized in the cytoplasm and chloroplast (Mondal et al. 2014; Ajeesh Krishna et al. 2020). *ZmZIP5* and *ZmZIP11* have strong transcription in the flag leaf, which may have a determining function in transferring Zn from the flag leaf to the developing grain, resulting in significant accumulation of Zn in the grain (Mondal et al. 2014; Ajeesh Krishna et al. 2020).

In wheat genome, a total of 16 ZIP genes have been mentioned. In shoots, *TaZIP3*, *TaZIP5*, *TaZIP6*, *TaZIP7*, and *TaZIP13* showed greater transcription under Zn-deficient environment; nevertheless, the timing varied among each gene. *TaZIP3*, *TaZIP5*, *TaZIP7*, and *TaZIP13* genes showed elevation of transcription in roots under Zn deficiency, while transcription of only the *TaZIP6* gene remained relatively constant (Evens et al. 2017). Another study revealed that in the flag leaf, the transcription of *TdZIP1*, *TdZIP3*, and *TdZIP7* genes decreased and transcription of *TdZIP10* and *TdZIP15* increased during grain development (Deshpande et al. 2018).

2.2.2 Heavy Metal ATPases (HMA) Family

HMA is a particular membrane transporter with occurrence in the xylem parenchymal cell and is associated in Zn transport in the xylem. In *Arabidopsis*, AtHMA2 and AtHMA4 are present in vascular vessels of root tissue and assists in Zn transport in the xylem (Pasricha et al. 2021). Zinc efflux from root to shoot especially depends on HMA2 and HMA4 transporters (Claus et al. 2013). AtHMA3 occurs in the vacuolar membrane and assists in Zn detoxification through vacuolar sequestration. AtHMA3 is also suggested to play a role in the storage of Cd, Pb, and Co in the vacuole (Gupta et al. 2016). Furthermore, AtHMA1 is located within the chloroplast membrane and has a key function in the detoxification of Zn ions by removing the

Zn^{2+} ion from plastids to the cytoplasm of *Arabidopsis*. The translocation of Zn^{2+} and Cd^{2+} ions is selectively driven by AtHMA2 through coordination with conserved cysteine and glutamine residues (Hasan et al. 2022).

Knockout of both, *AtHMA4* and *AtHMA2*, or only of the *AtHMA4* caused the reduction in Zn translocation from roots to shoots. In rice, mutation of *OsHMA2* also resulted in a decrease in Zn and Cd translocation from roots to shoots at the vegetative stage (Yamaji et al. 2013). OsHMA2 has an important function in the preferential distribution of Zn from roots and nodes to developing organs, which need high Zn content (Yamaji et al. 2013).

2.2.3 Cation Diffusion Facilitator (CDF) or Metal Tolerance Protein (MTP) Family

Members of the cation diffusion facilitator (CDF) family associated with micronutrient distribution and homeostasis have a determining function in all organisms. In plants, these transporters, also called Metal Tolerance Proteins (MTPs), have several divalent metal cationic substrates, such as Mn^{2+}, Fe^{2+}, and Zn^{2+}, which in general transport out of the cytoplasm, either to subcellular compartments or to the apoplast.

Strategies to maintain metal balance have been improved by plants. Membrane transporters have a determining function in maintaining an appropriate quantity in plant cells by compartmentalizing or effluxing metal ions. Proteins of the cation diffusion facilitator (CDF) family have been shown to contribute to the control of metal ion balance (Wang et al. 2021a). The compartmentalization of metals in the vacuole, endoplasmic reticulum, or Golgi apparatus by transporters located in endomembranes is decisive for plant tolerance to stress caused by excess metal (Tang et al. 2021).

MTP family members across kingdoms were divided into three clusters taking into account their metal specificities: Zn-MTP, Mn/Fe-MTP, and Fe/Zn-MTP. In *Arabidopsis*, Zn-MTPs AtMTP1 and AtMTP3 occur in the tonoplast and function in the vacuolar sequestration of excess Zn. Therefore, AtMTP3 is necessary to detoxify Zn in Fe deficiency (Alejandro et al. 2023). AtMTP1, also known as ZAT, was the first identified protein of the cation diffusion facilitator (CDF) family. AtMTP1 is located in the vacuolar membrane of root cells and is associated with vacuolar sequestration (storage of excess Zn in the cytoplasm in vacuoles) to maintain fine Zn homeostasis (Kobae et al. 2004).

Many CDF proteins present six transmembrane domains (TMDs) and a C-terminal domain preserved in the cytoplasm (Gustin et al. 2011). In general, MTP genes are not essential for Zn transport activity but, however, can make easy vacuolar sequestration of excess Zn in the cytoplasm. Most MTPs are located in the tonoplast and function as Zn and Cd antiporters involved in the sequestration or efflux of these ions to minimize metal toxicity. Studies have demonstrated that AtMTP1 and AtMTP3 are associated with Zn transport in *Arabidopsis*. Additionally, both proteins function in the vacuolar sequestration of excess Zn (Das et al. 2021). However, AtMTP1 and AtMTP3 show divergent results in Zn accumulation. AtMTP3 reduces the concentration of released Zn, while AtMTP1 increases Zn levels in roots (Kaur and Garg 2021). Furthermore, AtMTP5 and AtMTP12 are responsible for the accumulation of Zn in the Golgi complex (El-Sappah et al. 2023).

According to the exclusivity in the special metals transported, MTP transporters can be grouped into seven phylogenetic clusters: Zn-CDFs in groups 1–4, 5, and 12 function as MTP1–MTP4, MTP5, and MTP12, groups 6 and 7 function as Fe/Zn-CDFs containing MTP6 and MTP7, and Mn-CDFs are placed in groups 8 (MTP8) and 9 (MTP9-MTP11) (Gustin et al. 2011; Shirazi et al. 2023). MTP proteins have previously been studied in rice, wheat, soybean, peanut, tobacco, grape, bean, *Populus trichocarpa*, *Medicago truncatula*, and *Nicotiana tomentosiformis* (Shirazi et al. 2023; Yilmaz et al. 2023).

Sequence homology and phylogenetic studies showed the closest evolutionary relationship of AtMTP1 from *Arabidopsis* with soybean and *Medicago truncatula*. Furthermore, it was observed the coexpression of *AtMTP1* with a number of closely related genes involved in Cd/Zn transport in plants. Moreover, *AtMTP1* has high transcription in apical roots, especially during senescence and seed germination (Das et al. 2021). In soybean, *GmaMTP7.1* and *GmaMTP11.1* were more expressed in flowers, indicating that they may be involved in the early development of leaves and flowers, besides to the expected role of GmaMTP7.1 in development and maturation (El-Sappah et al. 2023).

2.2.4 Yellow Stripe-Like (YSL) and Oligopeptide Transporter (OPT) Families

Yellow stripe-like (YSL) proteins are homologous proteins designated after the maize membrane-bound iron phytosiderophores transporter, yellow stripe 1 (YS1). YSLs are part of the oligopeptide transport (OPT) family of proteins, which are classified into two subfamilies, namely, peptide transporters (PT) and YSL (Chowdhury et al. 2022). YSL1 and YSL3 are associated with the movement of Zn-NA, Fe-NA, and Cu-NA complexes between xylem and phloem (Waters et al. 2006).

In several phases of Fe and Zn transport, the same protein families are responsible; nevertheless, the two metals are handled separately by plants, often through the participation of different proteins of multiple gene families (Borrill et al. 2014). Nicotianamine (NA) and other metal chelators are fundamentals for the radial dispersion of Fe and Zn through the root, and the movement of Zn to the vacuole affects the overall transport of Zn through the roots to the shoot (Haydon et al. 2012). Fe and Zn are carried in the xylem, and Zn can move as a cation or in a complex with citrate and other organic, and Fe is chelated by citrate (Lu et al. 2013; Rellán-Álvarez et al. 2010). In wheat, as xylem is discontinuous, all nutrients enter the grain through the phloem. Fe and Zn are transported as complexes with NA or small proteins in the phloem (Borrill et al. 2014).

Translocation of Zn ions through the stem to the organ of assimilation (such as seeds) occurs in the phloem, Zn in the symplast is bound to NA chelates. Phloem cell protein transporters have not yet been precisely identified; until now, they are part of the YSL (yellow stripe-like) class, and these are the oligopeptide transporters (Sturikova et al. 2018).

AtYSL1 and *AtYSL3* have high transcription in the leaf in the vascular bundle of senescent leaves and is necessary for Zn allocation to developing seeds, although neither shows Zn-dependent expression. Consequently, it is believed that AtYSL1 and AtYSL3 may function in the transport of Zn-NA complexes in phloem, which is the main Zn forms found in phloem sap (Stanton et al. 2022). YSL proteins are considered to transport Zn-NA and Zn-DMA in rice. The combination of overexpression of *ferritin* and *NAS* with *OsYSL2* caused a 1.6-fold increase in Zn levels in rice grains (Verma et al. 2021).

YSL transporters are associated with the transport and discharge of Zn in the xylem. In barley, the endospermatic aleurone layer, as well as in the embryo of grains, Zn^{2+} uptake occurs by members of the YSL transporter family (YSL 6, 9 and 12), with the help of Zn complexes (NA-Zn) (Kaur and Garg 2021). When Zn-dependent expression control is absent, Zn transport via YSL transporters depends on the level of Zn-NA complexes, which can increase when NA concentration increases under Zn excess, Zn deficiency, or Fe deficiency (Haydon et al. 2012). In *Arabidopsis*, YSL1 and YSL3 are associated with Zn remobilization from aged leaves as well as pollen development (Kaur and Garg 2021).

In *Arabidopsis*, AtYSL1 has a important function in long-distance transport of the NA-Zn complex and in seed Zn transport. The YSL1 gene, which encodes a transporter of NA-Zn complexes, is involved in Zn transport in the phloem (Cardini et al. 2021). YSL1 and YSL3 from *Arabidopsis* are vital in Fe transport and responsible for the transport of Fe, Cu, and Zn from the source to the sink, that is, the grain filling site. Nonetheless, YSL4 and YSL6 appear to be associated with iron transport and metal mobilization to *Arabidopsis* seeds (Mahendrakar et al. 2020).

2.2.5 Plant Cadmium Resistance (PCR) Family

The PCR (Plant Cadmium Resistance) family receives its name from its specific function of restricting the transport of cadmium ions (Cd^{2+}) across the cell membrane, contributing to increasing plant resistance to Cd. Several members of this family have been identified in many plant species. In *Arabidopsis*, for example, the PCR family is multigenic, composed of 12 members, while in *Oryza sativa*, 21 members of PCR family were found (Gupta et al. 2016). Structural studies of AtPCR1 proteins show that they are small plasma membrane-bound proteins with a cysteine-rich domain and are associated with the efflux of heavy metals (Song et al. 2004). Previous research demonstrates that AtPCR2 helps in the transport of Zn and Fe and also exhibits resistance to Cd (Song et al. 2010).

These proteins function as secondary transporters especially in epidermal cells and in the xylem of young roots. *PCR2* was expressed in both roots and shoots of *Arabidopsis* and is associated with plant survival under Zn excess and deficiency. PCR2 is also decisive for Zn redistribution and detoxification in plants such as *Arabidopsis*. The main role of PCR2 is Zn transport and maintenance of optimal Zn concentration in *Arabidopsis* roots (Ajeesh Krishna et al. 2017).

The membrane protein PCR2 is associated with Zn detoxification in the presence of high concentrations and Zn transfer from the root to the shoot. This dual function is possible due to the dual expression pattern of *PCR2* in the xylem parenchyma of young root parts and in the root epidermis in the root differentiation zone, where root hairs develop (Song et al. 2010; Guan et al. 2023). In rice, OsPCR1 influences both Zn accumulation and grain weight (Song et al. 2015).

2.2.6 Vacuolar Iron Transporter (VIT) Family

Proper cell function requires an optimal concentration of Zn in the cytoplasm. Membrane transporters have a determining function in the transport, storage, and redistribution of Zn to intracellular organelles (Ajeesh Krishna et al. 2022). The VIT family mediates the transport of heavy metals through the tonoplast and thus assists in their vacuolar sequestration. Previous discoveries about VITs addressed their function in iron trafficking from the cytosol to the vacuole, but later their role in the transport of other micronutrients (Zn and Mn) was highlighted.

Two members of the rice VIT family, namely, OsVIT1 and OsVIT2, function to mediate the uptake of Fe and Zn into flag leaves through vacuolar storage and thereby regulate metal translocation from the flag leaf to developing grains (Zhang et al. 2012). In barley, *VIT1_1* is highly transcribed in leaf transfer cells, where it assists in the mobilization of Zn to vacuoles, while *VIT1_2* is strongly transcribed in aleurone cells (Kaur and Garg 2021).

2.2.7 Cation Exchanger (CAX) Family

Members of the CAX family have been found in many species (Emery et al. 2012). CAX transporters are divalent cation/H+ antiporters and are localized in vacuoles, which are associated with cation transport in plants (Neeraj et al. 2009). They have been predominantly involved in the transport of Ca^{2+} ions; complementary studies have highlighted its ability to transport a diversity of ions, including Zn.

MHX, a member of the CAX family, located in the vacuolar membrane is involved in the transport of Mg and Zn through the tonoplast in *Arabidopsis*. CAX transporters are crucial for the redistribution of cations including Zn in *Arabidopsis* (Ajeesh Krishna et al. 2017). The vacuolar transporters *MTP1*, *MTPc4*, *VIT1–1*, and *CAX1a,b* are strongly transcribed in transfer cells and are believed to promote the flow of Zn^{2+} into the vacuole (Caldelas and Weiss 2017).

3 Strategies to Attenuate the Effects of Zn Deficiency

The application of fertilizers with Zn is among the main approaches used to prevent Zn deficiency in plants. Fertilizers can be spread on topsoil, foliar sprayed, or used to treat seeds (Alloway 2009). Another tool to increase Zn uptake by crops is the use of biostimulants. Biostimulators can be classified into natural substances, artificial compounds, inorganic elements (Si and Se), and fungi (arbuscular mycorrhizal) and rhizobacteria (beneficial plant growth-promoting bacteria) (Saleem et al. 2022; Jalal et al. 2023).

Nanotechnology is another approach applied to ameliorate Zn uptake by plants. Nano-fertilizers can be attached to nanomaterials to regulate nutrient delivery to meet the plant nutritional needs. The application of Zn oxide (ZnO) nanoparticles had a great effect on plant growth and yield. Nanoparticles can be applied to leaves or roots (Saleem et al. 2022; Jalal et al. 2023).

The creation of new cultivars able to deal Zn deficiency and improved Zn use efficiency is another approach to mitigate the effects of lack of this mineral in plants (Zeng et al. 2021). In this case, conventional breeding and/or molecular breeding techniques and biotechnology tools can be used (Rose et al. 2013). Crop improvement to better uptake micronutrients such as Zn from soil or fertilizer sources is one of the most economical solutions (Rose et al. 2013).

As a genetic trait, tolerance to Zn deficiency is defined as Zn efficiency. A tolerant genotype or Zn efficiency is one capable of growing and producing well in areas lacking Zn (Sadeghzadeh 2013). Understanding the morphological and physiological aspects, the genetic and molecular architecture, and the mechanisms associated with Zn efficiency is necessary to assist the breeding process (Rose et al. 2013; Sadeghzadeh 2013).

3.1 Conventional Breeding

The presence of genetic variability is one of the prerequisites for the development of genotypes with high Zn efficiency (Nagarathna et al. 2010). Previous studies found great variation in tolerance to Zn deficiency (Zn efficiency) among species and among genotypes within the same species (Graham and Rengel 1993; Hacisalihoglu 2020). Therefore, to explore this variability, one of the first steps of breeding is the screening of genotypes with high Zn efficiency in segregating populations (Sadeghzadeh 2013) or in the germplasm of the species of interest (Abdoli and Esfandiari 2017; Hacisalihoglu 2020).

The preferred genetic resources for crossing are modern elite cultivars, as they have a high frequency of good alleles. The creation of genetic variability can result in changes in gene expression, appearance of new alleles, crossing-over between nonhomologous chromosomes, transposable element, epigenetic modifications in DNA, and modifications in epistatic inheritance patterns (Behl et al. 2003). However, due to the narrow genetic variability elite cultivars, wild relatives and different domesticated species can be used in crosses.

In addition to improving Zn efficiency, grain yield potential must be considered. The selected parents must present high yield and efficient absorption of Zn in soil deficient in this element and be able to take advantage of the applied Zn in normal soils. The crossing between these genotypes will generate segregating populations with high genetic variability for yield and Zn efficiency (Behl et al. 2003).

Conduction and selection in segregating populations varies according to the type of reproduction of the species and the type of parents used. In autogamous plants such as wheat, due to sampling variations and environmental effects, it is recommended to use lower selection pressure in F2. To maintain adequate genetic variability it is essential to select numerous individual plants in F2, as done in the SSD and pedigree methods (Behl et al. 2003). When desired to include local accessions or exotic species, crossing with F2 plants or backcrossing with both parents is recommended. This procedure must be done to promote the breaking of bonds and/or the restoration of chromosomal balance in the resulting populations. At F3 high selection pressure for yield and Zn efficiency can be applied. In the F4 generation, plants must be grown in environments with deficiency and sufficient amounts of Zn. Subsequently, large-scale trials must be carried out in different locations, together with control cultivars. Seeds from selected plants will be multiplied and then released as new cultivars (Behl et al. 2003).

3.2 Molecular Breeding

3.2.1 Molecular Markers

Molecular markers are nucleotide sequences that present polymorphism between individuals. The polymorphism of a given sequence within a population is due to deletions, insertions, duplications, and nucleotide translocations. Most of the time, these mutations do not alter gene function, as they do not occur within genes, but rather close to/linked to them (Hasan et al. 2021).

Molecular markers have numerous uses, among which are genetic mapping, investigation of genetic diversity, and phylogenetic studies (Grover and Sharma 2014). Genetic mapping and genetic diversity studies play an important role in the genetic improvement of plants, increasing the efficiency of the process. With advances in hybridization, polymerase chain reaction (PCR), and DNA sequencing techniques, the resolution and usability of markers have improved considerably (Grover and Sharma 2014).

Several molecular markers have been developed over the years as follows:

DNA Markers: Restriction fragment length polymorphism (RFLP), random amplified polymorphic DNA (RAPD), amplified fragment length polymorphism (AFLP), simple sequence repeat (SSR) or microsatellite, single nucleotide polymorphism (SNP), random amplified microsatellite polymorphism (RAMP), sequence-related amplified polymorphism (SRAP), inter simple sequence repeats (ISSR), sequence characterized amplified region (SCAR), retrotransposons microsatellite amplification polymorphism (REMAP), retrotransposon-based insertion polymorphism (RBIP), inter-primer binding site (iPBS), and inter-SINE amplified polymorphism (ISAP).

Resistance Gene Markers: Resistance gene analogue polymorphism (RGAP).

RNA Markers: Inter small RNA polymorphism (iSNAP) and expressed sequence tag-SSR (EST-SSR).

Targeted Fingerprinting Markers: Promoter anchored amplified polymorphism (PAAP), direct amplification of length polymorphisms (DALP), target region amplified polymorphism (TRAP), start codon targeted (SCoT), conserved region amplification polymorphism (CoRAP) (Hasan et al. 2021).

Furthermore, other markers such as targeted fingerprinting markers (TFM), DALP, chloroplast simple sequence repeat (cpSSRs), selectively amplified microsatellite polymorphic loci (SAMPL), single-strand conformational polymorphism (SSCP), cleaved amplified polymorphic sequences (CAPS), diversity array technology (DArT), sequence tagged site (STS), inter-retrotransposon amplified polymorphism (IRAP) (Amiteye 2021), genotyping by sequencing (GBS), restriction site associated DNA Tags (RAD) (Grover and Sharma 2014), and sequence specific amplified polymorphism (SSAP) are available. Finally, gene expression can also be considered a marker (Pegoraro et al. 2016; Campos et al. 2017).

Different researchers have used molecular markers for studies of tolerance to Zn deficiency in plants (Table 1). The vast majority of research aims to map genes and QTLs associated with Zn efficiency. After confirmation, molecular markers can be applied for marker-assisted selection (MAS) as an indirect selection criterion for this characteristic (Sadeghzadeh 2013). The use of MAS allows selection for Zn efficiency regardless of environmental conditions and growth stage (Sadeghzadeh 2013). However, MAS has some limitations for quantitative traits, as it cannot identify genes with minor effects, and few markers are used, only the most significant ones. To overcome this limitation, genomic selection (GS) can be used (Jeon et al. 2023).

Table 1 Application of molecular markers in studies for tolerance to Zn deficiency in plants

Marcador	Espécie	Tool	Objetivo	Referência
SSR, SRAP, RAPD	*Capsicum annuum*	Linkage mapping	Zn deficiency	Pinar et al. (2022)
SNP	*Zea mays*	Linkage mapping	Zn deficiency	Xu et al. (2022a)
SNP	*Zea mays*	Linkage mapping, association mapping	Zn deficiency	Xu et al. (2022b)
AFLP	*Arabidopsis thaliana*	Linkage mapping	Zn deficiency	Ghandilyan et al. (2012)
SNPs	*Arabidopsis thaliana*	Association mapping	Zn deficiency	Chen et al. (2016)
Gene	*Arabidopsis thaliana*	Gene expression	Zn deficiency	Campos et al. (2017)
RFLP, SSR	*Hordeum vulgare*	Linkage mapping	Zn deficiency	Lonergan et al. (2009)
GBS	*Phaseolus vulgaris*	Linkage mapping	Zn deficiency	Diaz et al. (2022)
SNP	*Triticum* sp.	Association mapping	Zn biofortification	Velu et al. (2018)
SNP	*Triticum aestivum*	Linkage mapping	Zn biofortification	Wang et al. (2021b)
GBS	*Triticum* sp.	Association mapping	Zn biofortification	Juliana et al. (2022)
SSR	*Oryza sativa*	Association mapping	Zn biofortification	Pradhan et al. (2020)
SSR, SNP	*Oryza sativa*	Meta-QTLs	Zn biofortification	Joshi et al. (2023)

In addition to research on tolerance to Zn deficiency, markers have been applied to mapping genes involved in Zn accumulation in grains (biofortification), aiming to minimize the impacts of hidden hunger (Table 1).

3.2.2 Genetic Mapping

The creation of molecular markers and sophisticated statistical tools has enabled the location of genomic regions that control important quantitative traits such as disease resistance, tolerance to abiotic stresses, yield, and quality (Gupta et al. 2019). Quantitative traits are controlled by polygenes or minor genes, have a complex segregation pattern, are strongly influenced by the environment, and are difficult to be accurately mapped. Genetic loci linked to a quantitative trait are called quantitative trait locus/loci (QTL) (Zia et al. 2022). Linkage mapping and association mapping are the methods applied to map QTLs (Xu et al. 2017; Gupta et al. 2019). Both methods detect QTLs by marker-trait association, and among the variations found between them is the type of mapping population used (Xu et al. 2017).

Linkage mapping (QTL interval mapping; meiotic mapping or genetic mapping) uses molecular markers to identify genomic regions that control traits of interest, in order to determine the relative position of genes on chromosomes, as well as the distance between them. In this technique, it is necessary to create a structured population that presents segregation for the markers and the target trait. The distance between markers present on the same chromosome is determined through recombination. High recombination indicates that the markers are more distant from each other (Zia et al. 2022). Biparental populations from parents contrasting for the trait of interest are constructed. In this case, F2 populations, backcrosses (BC), recombinant inbred lines (RILs), near-isogenic lines (NILs), or doubled haploids (DH) can be evaluated. The process comprises genotyping (molecular markers) and phenotyping (for trait of interest) of the mapping population and the application of statistical analyzes to detect QTLs. Among the limitations of linkage mapping are the low recombination rate that occur during the creation of the mapping population and the dependence on the genetic diversity of the parents, which in many cases differ for few characteristics (Xu et al. 2017).

The construction of multiparental populations was proposed to mitigate the limitations of linkage mapping from biparental populations, providing high-resolution QTL mapping. Multiparental populations include nested association mapping (NAM) and multiparent advanced generation intercrosses (MAGIC). Analyzes of multiparental populations are similar to those of biparental populations, but statistical methods created to analyze biparental populations cannot be readily applied to multiparental populations (Xu et al. 2017).

Association mapping (genome-wide association study (GWAS) or linkage disequilibrium (LD) mapping) uses historical recombination over many of generations, which results in high mapping resolution and a large number of alleles. In this approach, natural populations are used, with no need to construct biparental or multiparental populations. The mapping population will be composed of elite cultivars, landraces, wild relatives, and exotic accessions. Genotyping (molecular markers) and phenotyping (for characteristics of interest) of this population are necessary.

Next, heritability in a broad sense, LD, population structure, and kinship must be determined. Finally, association analyses are carried out using appropriate statistical methods for identifying QTLs (Xu et al. 2017).

Different studies to detect QTLs linked to Zn efficiency using linkage mapping and association mapping were developed (Table 1). The genes and/or sequences identified, as well as the markers linked to the QTLs, can help breeders in screening for superior genotypes (MAS) (Sadeghzadeh 2013). Furthermore, the detected genes can be used in the development of cultivars through genetic engineering tools. The identification of genes and/or chromosomal loci that control Zn efficiency in a plant species can be used in different crops due to homology, synteny, and collinearity of genomes (Sadeghzadeh 2013).

Linkage mapping and association mapping are also widely used to identify QTLs that control Zn accumulation in grains (biofortification) (Table 1). Genomic regions that control Zn accumulation in grains are also involved in the efficiency of uptake of this element from the soil.

3.2.3 Gene Expression Analysis

The selection of genes involved in Zn efficiency can be done from the study of the transcriptome, using RNA-Seq and microarray tools. In this case, the expression profile of all genes expressed in a certain tissue, condition, or genotype is obtained. Likewise, expression studies of specific genes using the RT-qPCR tool are also employed.

RNA-Seq

High-throughput DNA sequencing technologies have been adjusted for RNA sequencing (RNA-Seq), making it possible to determine the sequence and quantify mRNAs or total RNA in an economical way (Deshpande et al. 2023). The implementation of the experiment to collect samples for RNA-Seq studies defines the quality and reliability of the results obtained. Initially, the treatments must be defined and what will make up each replicate, whether a plant or a set of plants. Regarding the number of replicates, a minimum of three are recommended (for the application of statistical analysis). However, for more in-depth studies, in which you want to detect poorly expressed genes, 6–12 replicates per treatment are recommended (Upton et al. 2023).

RNA-Seq analysis is highly sensitive, capable of detecting variation between replicates subjected to the same treatment under highly controlled conditions. In this case, intergenerational effects must be considered, that is, the effect of the environment on the plant that produced the seed that will be used in RNA-Seq experiments. Therefore, it is important that the seeds used in the experiment were produced in the same environment. Another factor is the use of bulk fabrics, which should be avoided. This is because genes have differential expression between tissues, and some important genes can be masked. Finally, it is recommended to collect tissues at the same time of day to avoid the effects of the circadian cycle (Upton et al. 2023).

After sampling, fixing, and freezing the tissues, RNA is extracted. To prepare an RNA-Seq library with short-read sequencing, the RNA is copied into cDNA, multiplied via polymerase chain reaction (PCR), and cleaved. The RNA-Seq library will enter the sequencing platform, which will generate the reads (Deshpande et al. 2023).

Computational or bioinformatics analysis is responsible for decoding the biological complexity of transcriptomes. For this purpose, a large number of software are available. The first step consists of filtering the raw data to remove sequencing errors, removing adapters and readings with low precision. The second step is the alignment of the reads against a reference genome, to establish where the reads originate. Furthermore, the alignment provides coverage. Alignment can also identify which transcripts are expressed and measure their levels of expression. The third step is the quantification of transcripts by counting the reads that align with each reference transcript. Adaptations are made to quantify the expression of different gene isoforms. The fourth step is to estimate differences in the expression of a gene among treatments using statistical tests (Deshpande et al. 2023).

RNA-Seq studies to obtain the transcriptional profile related to Zn deficiency were developed in *Zea mays* (Mager et al. 2018), *Glycine max* (Zeng et al. 2019a), *Oryza sativa* (Bandyopadhyay et al. 2017; Nanda et al. 2017; Zeng et al. 2019b; Lu et al. 2021), and *Triticum aestivum* (Gupta et al. 2021).

RT-qPCR

Reverse transcription followed by quantitative polymerase chain reaction (RT-qPCR) allows measuring the quantity of a target sequence in a certain sample. The technique combines PCR and fluorescent dyes (SYBR Green) to monitor target amplification. There are two main RT-qPCR methods, absolute and relative quantifications. Relative quantification is mainly used for gene expression analysis (Chen et al. 2023).

To apply the technique, the use of the MIQE manual (Bustin et al. 2009) is recommended. In general, experiments should be conducted in a similar way to RNA-Seq experiments, with treated samples and control samples (calibrator). Total RNA is extracted from the samples of interest, and after eliminating contaminating genomic DNA, first-strand cDNA is produced. The cDNA will be diluted for relative quantification analysis.

Primers for the genes of interest/target and for the reference genes (housekeeping or normalizers), with small amplicons, must be synthesized. To validate the primers, a pool of cDNA from the treatments is created, which will be subjected to five serial dilutions. In validation, the specificity of the primers is evaluated based on the occurrence of a single peak in the melt curve. The efficiency of the primers is also determined, which should preferably be between 90% and 100%.

Next, the transcriptional profile of target and reference genes is analyzed under different conditions. The relative expression of a target gene is made in relation to reference genes (normalized by reference genes). Therefore, it is required to check which reference genes are more stable, that is, which do not show changes in expression in the face of different treatments. Different programs can be used to analyze the stability of reference genes, such as geNorm (Vandesompele et al. 2002), NormFinder (Andersen et al. 2004), and BestKeeper (Pfaffl et al. 2004). The most stable genes are used to calculate relative quantification of the target gene. One of the most used formula is $2^{-\Delta\Delta Ct}$ (Livak and Schmittgen 2001).

The RT-qPCR technique has been applied to identify target genes involved in tolerance to Zn deficiency in different plant species, testing different genes. Furthermore, this tool is used to validate transcriptomics results (RNASeq and Microarray) and in studies with mutant and transgenic plants. Among the species evaluated are *Medicago truncatula* (Liao et al. 2022), *Medicago sativa* (Cardini et al. 2021), *Sorghum bicolor* (Li et al. 2013b), *Oryza sativa* (Begum et al. 2016; Lilay et al. 2020; Zhan et al. 2022), *Arabidopsis thaliana* (Campos et al. 2017), *Areca catechu* (An et al. 2021), *Zea mays* (Xu et al. 2021), *Lycopersicon esculentum* (Kendziorek et al. 2016), *Hordeum vulgare* (Tiong et al. 2015), and *Chenopodium quinoa* (Lv et al. 2023).

3.3 Biotechnological Tools

The importance of obtaining food with greater production efficiency in terms of productivity, associated with the growing need for food with high quality in nutritional terms, has boosted the use of biotechnological tools in plants. To use biotechnological tools, a vast interdisciplinary knowledge involving cell biology, molecular biology, and biochemistry, among other areas, is required. Understanding the tools of molecular markers, genetic mapping and transcriptomics are fundamental to promoting the advancement of the application of biotechnology in plants.

In addition to developing new genotypes through the introduction of desirable traits (such as genes, promoters, or any nucleotide sequence), plant biotechnology plays a crucial role in identifying the functions of genes involved in traits of interest, thereby serving as a key tool in functional genomics. To achieve this, techniques like directional cloning and insertional mutagenesis are employed to pinpoint major genes that significantly influence phenotypic variation. From this point on, we will discuss transgenic tools, including overexpression and gene silencing as well as the gene editing systems known as the new plant breeding techniques (NPBT).

3.3.1 Basic Aspects of Transgenesis

Transgenics consists of the introduction of genes or any nucleotide sequence from organisms of different sexually incompatible species. On the other hand, there is also cisgenia, the introduction of genes or nucleotide sequences between sexually compatible species. Note that for an organism to be considered cisgenic, it must be free of any sequences originating from sexually incompatible species, such as reporter genes and selection genes.

Transgenesis uses plasmids, as a means of inserting the desired gene or sequence into the target plant, and the best known and used is the tumor inducing (Ti) or root induction (Ri) plasmid from *Agrobacterium tumefaciens* or *Agrobacterium rhizogenes* serving as a vector for the delivery of the sequence of interest. Structurally, Ti plasmids contain the T-DNA (transfer DNA) region, which is transferred to the host plant in response to phenolic signals produced by plant wounds. The T-DNA of Ti plasmids carries "oncogenes" that induce crown gall disease in dicotyledonous plants, leading them to synthesize nutrients that the bacterium can use as a source of carbon and nitrogen (Hwang et al. 2017). Oncogenes are flanked by 25 bp repeat regions known as the left (LB) and right (RB) borders, which surround the DNA. The DNA between these borders is transferred to the plant and integrated randomly in the genome. The LB and RB serve as recognition sites for the virulence proteins VirD1 and VirD2, which create single-strand DNA (ssDNA) breaks necessary for releasing the T-DNA as a single-stranded T-strand molecule (Stachel et al. 1986). To be used in transgenesis, the Ti plasmid is "disarmed" by removing the oncogenes, allowing the insertion of the DNA of interest into the T-DNA. This modified T-DNA is then transferred to the plant through co-cultivation of plant cells with *Agrobacterium* carrying the modified plasmid (vector) (to know more about Ti plasmid, see Hwang et al. 2017). The T-DNA is located on the binary vector, while the non-T-DNA region contains an origin of replication that can be activated in both *Escherichia coli* (used to construct the vector carrying the DNA of interest) and *Agrobacterium*, along with antibiotic resistance genes for bacterial selection. In addition to T-DNA vector or T-binary vector from the Ti-plasmids, a second plasmid is also used, called vir helper vector that process *vir* genes that encode Vir proteins serving as helper and are found in engineered *Agrobacterium* strains for plant transformation (Lee and Gelvin 2008 and Fig. 1). The activity of Vir proteins is essential for the integration of DNA into the plant genome, as they facilitate the cleavage of T-DNA and its transfer from the *Agrobacterium* cell into the plant cells.

Like wild-type *Agrobacterium* T-DNA, modified T-DNA integrates randomly into the genome through illegitimate recombination dependent on DNA polymerase theta through the 3′ end capture at genomic breaks (van Kregten et al. 2016). Therefore, the insertion site and copy number vary in each transformation event, and selection pressure may influence the recovery of T-DNA insertions in gene-rich or transcriptionally active regions of chromatin (Kim et al. 2007).

Due to the fact that Ti/Ri plasmids are large and display a reduced copy number, this makes manipulation in vitro difficult. To this end, binary vectors are currently used in plant transgenics. These vectors present a functional origin of replication for *E. coli* and *Agrobacterium*. Binary vectors present a multiple cloning site (MCS) favoring the insertion of T-DNA border sequences, marker genes selectable for bacteria and plants, bacterial conjugation genes, and origin(s) of replication to allow maintenance in *E. coli* and *Agrobacterium* (Lee and Gelvin 2008).

Briefly, transgenesis can be characterized in eight steps: (1) identification of the gene or sequence of interest, (2) selection of vector and promoter, (3) gene or sequence of interest isolation, (4) cloning, (5) buildup of the expression vector, (6) transformation of target plants, (7) regeneration and selection of transformed plants, and (8) testing of transformation events.

Step 1: The identification of genes of interest can be done in various ways, through reports in the literature and previous experiments, such as a dataset of genes differentially expressed in RNA-seq. Once the gene or sequence of interest has been identified, it is necessary to carefully study the strategy that will be adopted, which is directly associated with the objective of the work.

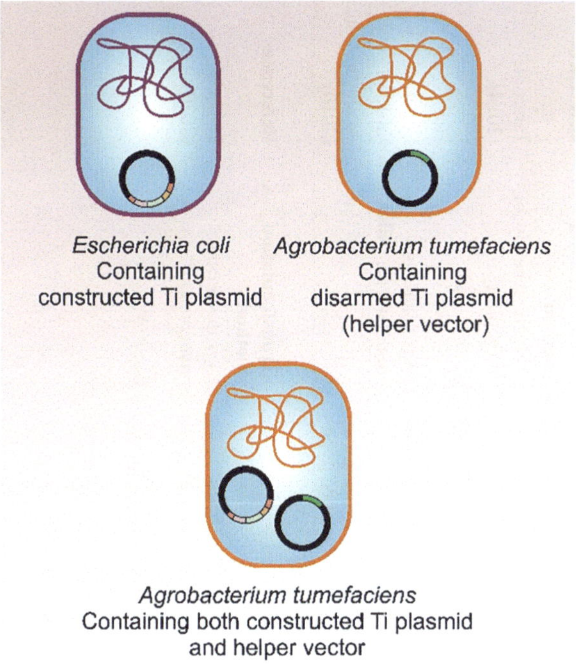

Fig. 1 Schematically view of the two vectors used for plant transformation. Ti plasmid in *Escherichia coli* carrying the constructed vector containing the T-DNA with the desired sequence/gene and in *Agrobacterium tumefaciens* strain carrying the disarmed helper vector used for Vir protein expression. The constructed Ti plasmid form *E. coli* is them transformed into *Agrobacterium* cells that will carry both plasmids, constructed Ti plasmid and helper vector, that is used for plant cell transformation and T-DNA integration

Step 2: Choosing the transformation vector requires some special care. The presence of reporter genes and bacterial and plant selection genes must be analyzed to meet the demands of the laboratory in question. Furthermore, it is necessary to assess the possible size of the sequence of interest that will be inserted to ensure that the plasmid will support the target sequence. Another important aspect is the objective of the transformation; whether overexpression or silencing, we will talk more about these two factors later.

Just like endogenous genes, transgenes require promoters to direct their expression. Therefore, the choice of the appropriate promoter in relation to expression in the target species is extremely important as it depends on the fine-tuned interaction with several proteins as transcription factors. Several promoters have been used to direct transgene expression (Tables 2 and 3). Constitutive promoters such as *Actin*, *Ubiquitin*, and *CaMV 35S* promote constitutive expression of the transgene in all plant tissues and stages. On the other hand, if the objective is targeted expression at a specific tissue or stage, inducible promoters are used. In studies on Zn absorption and translocation in plants, several promoters have already been used (see Tables 2

Table 2 Gene silencing and gene editing reports involving zinc regulators in plants

Silencing						
Gene	**Silencing method**	**Promoter**	**Organism**	**Effect**	**Transformation method**	**References**
MxIRT1	Antisense sequence	*CaMV 35S*	*Oyza sativa L.*	Reduction in shoot and root length	Polyethylene glycol, protoplast	Tan et al. (2014)
				Reduced Zn concentration in grains		
OsZIP1	Antisense sequence; T-DNA insertional mutant	*CaMV 35S*	*Oryza sativa L.*	Reduced Zn concentration in roots	Agrobacterium tumefaciens; callus	Liu et al. (2019)
AtbZIP19/ AtbZIP23	T-DNA insertional mutant	–	*Arabidopsis thalina*	Zinc deficiency hypersensitive phenotype	–	Assunção et al. (2010)
Gene editing						
Gene	**Editing system**	**Strategy**	**Organism**	**Effect**	**Transformation method**	**References**
Oszip5	CRISPR/Cas9 (knockout)	Single-guide RNA	*Oryza sativa* L.	Reduced Zn uptake	*Agrobacterium tumefaciens*	Tan et al. (2020)
Oszip9		Single-guide RNA		Reduced Zn uptake, decrease in growth and in Zn content		
Oszip5/ Oszip9		Double single-guide RNA		Reduced Zn uptake, Zn deficiency		
OsNAS2	CRISPR/Cas9 (knockout a cis acting element in OsNAS2 promoter)	Single-guide RNA	*Oryza sativa* L.	Increased Zn in grains and yield	*Agrobacterium tumefaciens*	Ludwig et al. (2024)

Table 3 Gene overexpression reports involving zinc regulators in plants

Overexpression					
Gene	**Promoter**	**Organism**	**Effect**	**Transformation method**	**References**
AtZIP1	*Ubiquitin*	*Hordeum vulgare*	Increased Zn content in grains Increased Zn uptake	*Agrobacterium tumefaciens*; immature embryos	Ramesh et al. (2004)
OsHMA3	*Ubiquitin*	*Oryza sativa L.*	Increased Zn concentration in roots Upregulation of *OsZIP4, OsZIP5,OsZIP8, OsZIP9*, and *OsZIP10*	*Agrobacterium tumefaciens*; callus	Sasaki et al. (2014)
ZmIRT1 *ZmZIP3*	*CaMV 35S*	*Arabidopsis thaliana*	Increased Zn concentration in roots Reduced Zn concentration in roots	*Agrobacterium tumefaciens*; floral dip	Li et al. (2015)
MxIRT1	*CaMV 35S*	*Oryza sativa L.*	Tolerance to Zn-deficiency Increased Zn content in grains	Polyethylene glycol, protoplast	Tan et al. (2014)
OsZIP1	*CaMV 35S*	*Oryza sativa L.*	Tolerance to Zn-toxicity	*Agrobacterium tumefaciens*; callus	Liu et al. (2019)
OsZIP4	*CaMV 35S*	*Oryza sativa L.*	Increased Zn concentration in roots Reduced Zn concentration in shoots and grains	*Agrobacterium tumefaciens*; callus	Ishimaru et al. (2007)
AtZIP1 *AtMTP1*	*PAT*	*Manihot esculenta*	Increased Zn concentration in roots	*Agrobacterium tumefaciens*; friable embryogenic callus	Gaitán-Solís et al. (2015)
OsIRT1	*Ubiquitin*	*Oryza sativa L.*	Sensitiveness to Zn toxicity Increased Zn content in roots, shoots and grains	*Agrobacterium tumefaciens*; callus	Lee and An (2009)

(continued)

Table 3 (continued)

AhNRAMP1	*CaMV 35S*	*Nicotiana tabacum* L.	Tolerance to Zn deficiency	*Agrobacterium tumefaciens*; standard leaf-disk method	Wang et al. (2019)
	OsiRT1	*Oryza sativa* L.	Increased yield and Zn content in plants; tolerance to Zn deficiency	*Agrobacterium tumefaciens*; callus	
OsZIP9	*Ubiquitin*	*Oryza sativa* L.	Increased Zn content in shoots and brown rice	*Agrobacterium tumefaciens*	Tan et al. (2020)
HvNAS1	*Actin*	*Oryza sativa* L.	Increased Zn content in brown and polished rice grains	*Agrobacterium tumefaciens*	Masuda et al. (2009)
AtbZIP19 (mutated)	*CaMV 35S*	*Arabidopsis thaliana*	Increased Zn content in shoots and seeds	*Agrobacterium tumefaciens*	Lilay et al. (2021)

At Arabidopsis thaliana, *Os Oryza sativa*, *Zm Zea mays*, *Mx* Malus, *Ah Arachis hypogaea* L., *Hv Hordeum vulgare* xiaojinensis
PAT promoter: tuber-specific class I patatin promoter from potato

and 3). Even with adequate selection of promoters, a major challenge is faced in the expression of transgenes. The non-expression of the transgene in the target plant can occur due to cytosolic degradation of the transgenic mRNA or methylation of the promoter regulating transgene expression, resulting in transcriptional gene silencing (Rajeevkumar et al. 2015). The non- or low expression of the transgene can be associated with the high metabolic cost to express the transgene along the entire plant life cycle, which can cause, in some cases, negative effects on plant growth and yield. Therefore, the use of tissue or stage-specific promoters can be a great option to reduce the risks of non-expression of the transgene.

Steps 3 and 4: Once the vector and promoter have been determined, the gene or sequence of interest (hereinafter called an insert) needs to be isolated through excision of DNA fragments from their precursor molecules using restriction endonuclease enzymes, or the isolation and amplification of the target region via PCR. The insert will then be used in cloning reactions for insertion into the expression vector.

Step 5: At this point, two plasmids are commonly used, one for cloning the insert and the other for delivery and expression in the target plant. To transfer the insert from the cloning vector to the expression vector, the co-integration/exchange systems based on the lambda phage integration and excision mechanism have been used (Lee and Gelvin 2008). In this system, the insert is exchanged inside the T-DNA region of a Ti-plasmid via homologous recombination. The Gateway system provided by Invitrogen has been widely used when talking about binary vectors (for more information about Gateway system, see Nakagawa et al. 2009).

Step 6: The vector, now ready for transformation, is inserted into the plant cells. Regarding the type of plant cell to be used, there are some issues that must be taken into consideration. There are two basic types of transgene expression defined by the matrix used for transformation, transient and stable. The transient transformation method does not require the T-DNA to be incorporated into the host genome, it uses differentiated cells and allows a rapid expression of the phenotype caused by the transgene, it is localized, and for a certain time (a few days), that is, it is not inheritable. The most common known methods of transient expression are floral dip and agroinfiltration. Protoplasts have also been used for transient expression as they assist in single cell studies; their regeneration in plants is a laborious process that consumes a lot of time; however, they can be used in stable gene expression (Lörz et al. 1985; Hwang et al. 2017). Floral dip, agroinfiltration and protoplasts have already been used in studies involving Zn translocation absorption (see Tables 2 and 3). On the other hand, for the expression of the transgene to be heritable and expressed in different tissues, it is necessary to transform undifferentiated plant cells, and this method is more laborious. To achieve this, it requires established tissue culture techniques and generally uses embryogenic callus and immature embryos for heritable transformation (see Tables 2 and 3).

The transformation of plant cells requires that the constructed vector enters the plant cell to be trafficked by the cytoplasm, and finally T-DNA should be imported by the nucleus (Hwang et al. 2017). There are some methods for insertion of the construct in plant cells, as electroporation, microinjection, and polyethylene glycol fusion of protoplasts, microprojectile bombardment (biolistics—does not require insertion of the insert into a plasmid), silicon carbide whisker-mediated transformation, pollen tube pathway-mediated transformation, and nanoparticle-mediated gene delivery are all examples of methods used for plant transformation (Su et al. 2023). These methods have limitations such as long time for protoplast regeneration, multiple copies of insertion via biolistics leading to loss of the DNA integrity, low stability, and epigenetic silencing. Due to these limitations, transformation via *Agrobacterium* is the most frequently used (see Tables 2 and 3).

Transformation via *Agrobacterium* requires co-cultivation of plant tissue with transformed bacterial cells for infection and T-DNA transfer. There are established protocols for tissue culture and infection with *Agrobacterium* reported in the literature for different crops, as rice (Nishimura et al. 2006; Sahoo et al. 2011; Slamet-Loedin et al. 2013; Rengasamy et al. 2024), barley (Harwood 2013), wheat (Ishida et al. 2014; Hayta et al. 2021), and maize (Ishida et al. 2003; Vega et al. 2008; Sidorov and Duncan 2009; Kang et al. 2022).

Steps 7 and 8: Following infection, *Agrobacterium* must be eliminated using specific antibiotics. The plants are then placed in culture media containing a selection agent, typically hygromycin. Surviving plants are subsequently moved to regeneration media and, eventually, to root induction media. Potentially transformed plants are then assessed for transformation through PCR and reporter gene analysis.

The plant regeneration step is decisive and often an obstacle in the process of obtaining a transgenic plant. Researchers developed strategies to increase the efficiency of plant regeneration such as the use of some different plant hormones in the regeneration medium and ectopic expression of morphogenetic transcription factors such as *Baby boom* (*Bbm*), *Wuschel* (*Wus2*), *growth-regulating factors,* and its co-factor *GRF-interacting factor1* and *Wuschel-homeobox 5* (for review, see Lee and Wang 2023). The expression of these genes in transforming plants helps in regeneration as well as increasing the possibility of using different species and compatible genotypes.

Although laborious, the plant transformation process ends with the analysis of the transgene phenotype in the host plant. Many interesting findings regarding the increase in Zn accumulation, tolerance to Zn deficiency, and identification of genes have been addressed using transgenic techniques to identify genes involved in the absorption and transport of Zn in plants (Tables 2 and 3).

3.3.2 Gene Silencing

RNA-Mediated Interference (RNAi)

RNAi is a posttranscriptional mechanism that regulates gene expression of natural occurrence in eukaryotic cells (Fire et al. 1998). In brief, the RNAi system is activated by endogenous micro-RNAs (miRNAs) or exogenous double-stranded RNAs (dsRNAs) that are recognized by RNase-III classes of enzymes from the Dicer family and processed (dicing) into approximately 25 nt small interfering RNAs (siRNAs) with overhang ends of two nt. siRNAs are loaded into Argonaute (AGO) proteins to form an RNA-induced silencing complex (RISC). The sense strand of the siRNA, known as the passenger strand, is discarded for degradation, while the antisense strand, referred to as the guide strand, remains. The guide strand, which is complementary to the target mRNA, directs the RISC to the mRNA, leading to its processing and gene silencing (reviewed in Zhao and Guo 2022). The cleaved mRNA fragments serve as a template for the synthesis of dsRNA by an RNA-dependent RNA polymerase (RdRP), initiating a second cycle of RNA interference through the production of secondary siRNAs (Hung and Slotkin 2021).

Transgenic systems aiming at gene silencing by RNAi use a long hairpin RNA (hpRNA) typically constructed by a perfect inverted repeat (IR) complementary to the target gene sequence, forming the dsRNA in the hpRNA stem (reviewed in Zhang et al. 2022). In silencing vectors, the insert isolated from a target gene is introduced in sense and antisense, and between them, there is a functional intron resulting in the expression of an hpRNA. Insertion of the sense and antisense insert can be carried out in a single step when using the Gateway system (reviewed in Ashfaq et al. 2020). Recent studies have shown that hpRNA transgenes are consistently methylated in the inverted repeat (IR) and in the adjacent promoter, leading to transcriptional self-silencing. Zhang and co-workers (2022) showed that nucleotide substitutions in the hpRNA sense sequence disrupts IR structure by preventing methylation and a more uniform and persistent RNAi. Silencing by heterologous 3′ untranslated regions (SHUTR) is another silencing system that uses the cloning of the insert (sense) between a promoter and a 3′ untranslated region (3′-UTR) that forms, in plants, dsRNAs directed to the gene target (reviewed in Ashfaq et al. 2020).

In transgenes, like the promoter, terminators are extremely important in directing the completion of transgene transcription. In the RNAi system, the terminator is the most important structure as determines whether mRNA will be targeted by the RNAi. It is proposed that a weak or mild terminator, when paired with a promoter that overrides the terminator's efficiency, leads to scattered polyadenylation sites and potentially results in the production of transcripts that are not polyadenylated (reviewed in Hung and Slotkin 2021).

Like any transformation system, silencing systems using RNAi have advantages and disadvantages (reviewed in Gilchrist and Haughn 2010). Using silencing via RNAi, it is possible to knockdown multiple homologous genes when using an insert based on a conserved region, a partial loss of function can be obtained in addition to being temporal or targeted to a specific tissue, and it is a heritable transformation. One of the main disadvantages of using RNAi via transgenesis is the off-target effect, the need for plant transformation, and the fact that the silencing potential depends on the construct and the transformants. Currently, there are reports of using RNAi in functional genomics studies to unveil the molecular regulation of Zn transport uptake in plants (Table 2).

It is worth noting that in recent years researchers have studied the exogenous application of dsRNAs and siRNAs in a non-transformative way, called spray-induced gene silencing (SIGS) (Sammons et al. 2011). The first documented instance of exogenous RNA application triggering the RNAi mechanism in plants was reported in 2011, through a Monsanto patent that involved spraying dsRNA onto *Nicotiana benthamiana*, resulting in the production of siRNA targeting the gene encoding phytoene desaturase (Sammons et al. 2011, patent no.: US20110296556A1). Currently, the induction of the RNAi system via SIGS has been used in plant protection against insects, nematodes, and pathogens. SIGS is a biotechnological tool that does not involve transgenics and presents itself as a potential tool for use in studies on the absorption and translocation of Zn to identify gene function in a transient, rapid, and necessary way in cases where silencing the target gene can be lethal.

3.3.3 Virus-Induced Gene Silencing (VIGS)

VIGS constructs involve cloning cDNA fragments into a DNA version of the genome of an RNA virus, which is then used to transfect the plant through *Agrobacterium*. The effects caused by VIGS in the plant are transient in nature, and the dsRNA from the viral genome, carrying the sequence of interest, is generated during virus replication. The dsRNA formed is degraded by the Dicer-like enzymes of the plant cell into small molecules of small interfering RNA (siRNA) activating the siRNA silencing pathway degrading the target gene transcript causing knockout or knockdown phenotype (Rössner et al. 2022).

Like other techniques for functional genomics, the use of VIGS has advantages and disadvantages (reviewed in Gilchrist and Haughn 2010). The use of VIGS does not require the transformation of plants and therefore presents quick results and low cost of use. Furthermore, it is possible to knockdown potentially harmful genes if silenced. The possibility of off-target effects (effects in genes other than the target), variation in the level of silencing and transient expression, specificity for some plant species, and the absence of obtaining missense mutations are considered the main disadvantages of using viral vectors.

Not all viruses can be used for VIGS. Viruses that can be modified to act as VIGS need to have certain traits as they can encode specific proteins that can effect the silencing machinery in the host plant. Furthermore, the viruses used must be capable of infecting the target plant species, which is why the viruses used for the VIGS system are characterized for specific plant species. A list of more than 20 VIGS was provided by Senthil-Kumar and collaborators (2008) and comprises several plant species, including rice, maize, and soybean.

3.3.4 Random Mutations Caused by Insertional Mutagenesis

The insertion of T-DNA via *Agrobacterium* or biolistics and the activation of transposons have been explored to create random gene disruption leading to knockouts. Gene disruption typically leads to a complete loss of function, and T-DNA insertions provide a high-frequency, stable phenotype, allowing for rapid genetic screening of tagged genes. The frequency of T-DNA insertion is higher in euchromatin, mainly in the telomeric and subtelomeric regions in the chromosomes compared to the heterochromatin regions near the centromeric regions (Ram et al. 2019). Insertions in genes of interest can be scanned within a plant population, even in heterozygosity, using PCR and specific primers, one aligned in a region of the gene and the other aligned with the insertion (Gilchrist and Haughn 2010).

With the advent of cost-effective, high-throughput next-generation sequencing (NGS) platforms, it has become possible to characterize mutant collections for the systematic analysis of gene function (Ram et al. 2019). Taking into account the advancement of tools, increasing the possibilities of using transgenics at increasingly specific, stable, and precise levels, target mutations, such as those generated by gene editing tools, will be discussed later in this chapter and their possibilities for use in studies for desirable traits regarding Zn status in plants.

Transposons are small sequences of DNA that can move to different regions within the genome. Its mobility occurs through the insertion of DNA double-strand breaks (DSBs) followed by the random insertion of dsDNA into the genome. The transposon typically produces a single-copy insertion, whereas T-DNA tends to integrate in various configurations. Additionally, T-DNA mutants are considered genetically modified organisms, as they are created through *Agrobacterium*-mediated transformation (Ram et al. 2019).

Insertional mutagenesis is widely applied to generate libraries of loss-of-function mutants that allow identification of the insertion in a manner facilitated by the use of PCR-based techniques (Gilchrist and Haughn 2010). Some banks of insertional mutants are available for rice crops. The Rice Functional Genomic Express Database (http://signal.salk.edu/cgi-bin/RiceGE) and The Rice Tos17 Insertion Mutant Database (https://tos.nias.affrc.go.jp/) are some examples. Insertional mutagenesis has been used in functional genomics studies to understand Zn uptake and translocation in plants (Table 2).

3.4 Gene Overexpression

Overexpression is a strategy used to complement the use of loss-of-function mutants and knockout/knockdown strategies. In some cases, due to redundancy of gene function or when silencing is particularly deleterious, gene silencing strategies fail to provide answers about gene functionality. Drawing a comparison with silencing strategies, overexpression is considered a transformation strategy that leads to gain of function. Gene overexpression is conceptually defined as the use of a strong promoter and upstream activation sequences to drive high-level, typically constitutive transcription of a gene coding sequence (Lloyd 2003). Furthermore, generally in studies involving gain of function, the cDNA corresponding to the target variant or DNA free of non-coding regions is directionally cloned, thus guaranteeing expression in the 5′-3′ direction free from the need for transcriptional processing and removal of non-coding regions.

The most common form of gain of function is using strong constitutive promoters, such as *Ubiquitin* and *CaMV 35S* promoters. Several reports involving studies related to Zn have used constitutive promoters as a strategy for identifying the function of genes and improving the Zn content in grains (Table 3). These studies demonstrate the efficiency of using transgenics and overexpression of genes to assist Zn absorption by the plant and translocation of Zn to the grains, which may be a viable strategy for the development of genotypes showing grains rich in Zn accumulation. Obviously, other traits must be analyzed in these transforming lines, such as possible negative effects on grain productivity and tolerance/resistance to diseases and abiotic stresses.

However, overexpression using the constitutive promoters often creates hypermorphic alleles, in which case the observed phenotype directly implicates the endogenous function of the gene (Zhang 2003). This type of effect is associated with dominant negative (DN) effects that lead to mutant polypeptides able to interrupt the wild-type gene activity when overexpressed (reviewed in Veitia 2008). Furthermore, the use of constitutive promoters can sometimes be considered a failed strategy since the expression of the transgene will occur throughout the entire development cycle of the plant, and as it is constitutively expressed, it is not induced by some environmental stimulus, and the altered phenotype may be interpreted inappropriately (Lloyd 2003).

Therefore, some researchers have used inducible promoters to coordinate transgene directed expression in stages, in tissues, and in response to environmental stimuli. Wang et al. (2019) directed the expression of the *AhNRAMP1* gene, from peanut (*Arachis hypogaea* L.), in rice using the promoter of the rice *OsiRT1* gene. This study is especially important because it was possible to obtain transgenic rice lines with increased productivity, increased Zn amount in the grains, and tolerance to Zn deficiency.

3.5 Targeted Mutation

Targeted mutation is a way of inducing changes in DNA with high precision, that is, in regions of interest. Since their introduction in mid-2013, gene editing technologies have transformed plant breeding, whether through the development of genotypes or functional genomics studies. Due to their important applicability in plants, gene editing tools have been regularly refined to meet the most diverse needs of plant breeding. Four gene editing technologies have been developed: meganucleases (MN), zinc-finger nucleases (ZFNs), transcription activator-like effector nucleases (TALENs), and clustered regulatory interspaced short palindromic repeat (CRISPR) associated with specific nuclease Cas (CRISPR/Cas).

MNs, also known as homing endonucleases (HEases), are site-specific endonucleases that recognize and cleave long DNA sequences ranging from 14 to 40 base pairs with high specificity. These enzymes can be engineered to target a wide variety of sequences. Zinc finger nucleases (ZFNs) are restriction enzymes created by fusing a nonspecific DNA cleavage domain (*FokI*) with site-specific DNA-binding zinc finger proteins (ZFPs). The modular structure of C2H2 zinc finger motifs, each capable of independently recognizing specific DNA sequences, makes them ideal for the development of engineered ZFNs that can target new sequences. Each motif recognizes 3 to 4 nucleotides through its α-helix, and tandem zinc finger modifications enable precise DNA sequence recognition. Transcription activator-like effectors (TALEs) feature a DNA-binding domain composed of monomers with 34 amino acids each, where each domain binds to a unique base pair. The specificity of TALEs is determined by two hypervariable amino acids, known as repeat-variable di-residues (RVDs), located at positions 12 and 13 of each monomer. The TALE is fused with the *Fok I* nuclease (called now as TALEN) promoting the nuclease site specific for the TALEN (reviewed in Viana et al. 2019). All of these early gene editing technologies are primarily protein-based DNA-targeting systems. The breakthrough of the CRISPR system lies in the use of a single-guide RNA (sgRNA) to precisely direct the nuclease (Cas) to its target site for cleavage. Further details about the CRISPR System will be discussed later in a specific section.

All gene editing systems (MNs, ZFNs, TALENs, and CRISPR/Cas) have the common signature of using endonucleases capable of inserting cleavage into the DNA double-strand break (DSB), considered one of the greatest damages to the DNA structure, resulting in the induction of the DNA repair system. DSB repair can occur through two competing and partially overlapping pathways: homologous recombination (HR) and nonhomologous end joining (NHEJ). NHEJ is an error-prone repair mechanism that often results in the insertion or deletion of bases (indels) at DSB sites, leading to mutations such as frameshifts and nonsense mutations. This characteristic is utilized in ZFN, TALEN, and CRISPR-based gene editing platforms. In contrast, HR is a conservative repair mechanism that facilitates the exchange of genetic information between two homologous DNA sequences. Gene editing systems use the insertion of DNA sequences (donor DNA), modified with the variation of interest, to serve as a template and induce a higher frequency of HR repair. However, even with the insertion of a donor DNA, the frequency of HR repair is small in the range of 10–6 in somatic cells, generally used in plant transformation (Barakate et al. 2020). In view of this, heterologous expression of genes coding for recombinase proteins such as RecG, RecA, and RuvC from prokaryotes and Rad52, DMC1, and RAD51 from eukaryotes has been used and has shown an increase in the frequency of HR occurrence (Barakate et al. 2020). In addition to recombinases, transcription factors have been identified to promote HR frequency. The transcription factor SOG1 was identified as the master regulator of DNA damage responses and is ubiquitinated and stabilized due the action of the ubiquitin E3 ligase DDRM1 promoting HR (Wang et al. 2022b).

Undoubtedly, gene editing systems, from the use of MN to the development of the most robust and precise system such as CRISPR, have revolutionized functional genomics. Due to the minor issues regarding the use of these tools in plants, we have been facing a wide range of reports which prompted plant science to a new era generating high amounts of variations. In addition, it helps to improve gene editing platforms related to precision, multiple targets, and the knowledge related to DNA repair.

3.5.1 CRISPR/Cas System

The development of the CRISPR/Cas tool in 2012 innovated science as a whole, marking a new era where it is possible to use an extremely specific and versatile system in the development of lineages presenting specific variations introduced surgically. Nevertheless, the CRISPR/Cas system secured the Nobel Prize in the Field of Chemistry for Dr. Emmanuelle Charpentier and Dr. Jennifer A. Doudna. Given the importance of Zn in plants, some studies have already used genome editing systems aimed at functional genomics (Table 2).

The CRISPR mechanism is an adaptive immune system found in bacteria and archaea, defending against invading nucleic acids. It functions by cleaving the invading DNA at specific sequences, providing immunity by incorporating segments of the invader's DNA (spacers) between two repeated sequences at the proximal end of the CRISPR locus. When a new nucleic acid invades, the CRISPR locus and spacers are transcribed and processed into small interfering CRISPR RNAs (crRNAs), which then combine with transactivating CRISPR RNA (tracrRNA) to activate the Cas nuclease. This complex directs the Cas nuclease to cleave the homologous sequence of the invading DNA (spacer) (reviewed in Bortesi and Fischer 2015).

The CRISPR system used in genetic engineering is a construct formed by fusing crRNA and tracrRNA into a single-guide RNA (sgRNA), which is homologous to the target DNA sequence and specifically directs the Cas nuclease to the target region. Diverse nucleases can be used in the CRISPR system, and the most commonly used are Cas9 and Cpf1/Cas12. Cas9, the most commonly used Cas protein derived from *Streptococcus pyogenes* (SpCas9), requires an adjacent protospacer adjacent motif (PAM) downstream of the target DNA, typically a 5′-NGG-3′ or 5′-NAG-3′ sequence (where N = A, T, C, or G) (Jinek et al. 2012). The Cas9 protein contains two nuclease domains: the HNH domain, which cleaves the sgRNA that is complementary to the target DNA, and the RuvC-like domain, which cleaves the noncomplementary DNA strand containing the PAM sequence. This action results in the generation of a DSB 3 base pairs upstream of the PAM within the protospacer sequence (Jinek et al. 2012). The DSB introduced by the Cas9 is further repaired by NHEJ or HR.

Some reports show the use of Cpf1/Cas12 in plant gene editing system. Cpf1/Cas12 is isolated from *Prevotella* and *Francisella 1* from the family V of CRISPR nucleases that include both endoribonuclease and endodeoxyribonuclease activities. Cpf1/Cas12 has two main differences with respect to Cas9: Cpf1/Cas12 use a thymine-rich PAM motif while Cas9 use a guanine-rich PAM motif which increment the target sequence that can be AT rich for Cpf1/Cas12 or GC rich for Cas9; Cpf1/Cas12 introduces 5′ overhang DNA ends while Cas9 generate blunt DNA ends (Zetsche et al. 2015). In addition, while Cas9 action relies on the need of a crRNA and a tracRNA, Cpf1/Cas12 needs only a crRNA of 44-nucleotide crRNA that can be localized 5′ directing after the repeated sequence and spacer to guide the Cpf1/Cas12 nuclease activity (Zetsche et al. 2015). Recently, CasΦ was introduced in the CRISPR-CasΦ system. It is a hypercompact type V CRISPR system featuring a single CasΦ protein of approximately 70 kilodaltons, about half the size of Cas9 or Cas12a. Like Cpf1/Cas12, CRISPR-CasΦ only requires a crRNA and has a minimal PAM requirement of 5′-TBN-3′ (where B = G, T, or C) (Pausch et al. 2020). Many other different Cas used um CRISPR platform, and a list of gene regulators (promoter, terminator, and intron) is provided by Hassan et al. (2021).

In-depth knowledge about CRISR/Cas system prompted scientists to develop new strategies for gene editing using CRISR/Cas platform. These strategies include dCas9, base editing, prime editing, GRAND editing, and *CRISPR-Combo* among others with the aim of increment specificity, versatility, and robustness of the CRISR/Cas platform.

3.5.2 Multiplex and Ultra-multiplex Genome Editing Systems

Multiplex genome editing systems (MGE) were recently developed to modify two or more specific DNA loci, causing two or more DBSs through a single round of mutagenesis (Van Huffel et al. 2022). MGE is achieved using TALENs and CRISPR platforms, but due the less difficult process in engineering CRISPR plasmids, CRISPR has been the most widely used platform of MGE. Using one Cas protein and multiple sgRNAs to target different DNA regions is the main strategy for MGE, and the aim is to introduce various sgRNA expression cassettes into one binary vector (reviewed in Abdelrahman et al. 2021). Expressing multiple sgRNAs has been proposed in different ways, as reviewed by Abdelrahman and co-authors (2021) as follows:

(a) RNA polymerase III uses transcription factors to recognize promoters for each sgRNA to generate individual sgRNAs.
(b) In the case of the polycistronic CRISPR/Cas system, small gRNAs are cloned to be transcribed as a single transcript, which requires posttranscriptional processing by Csy4, tRNAs, and a hammerhead ribozyme for gRNA separation.

(c) The Cpf1/Cas12 system, which features a self-cleaving gRNA, does not need additional elements for posttranscriptional processing, as it can process and mature crRNAs through its intrinsic ribonuclease activity.

Due the importance in study different genes, its redundancy and interaction by the targeting of multiple in only one transformation round, MGE have been widely used in plant transformation protocols. Van Huffel and collaborators (2022) investigated the number of sgRNAs in a multiplex strategy and its associated phenotype. They developed two main approaches for designing multiplex CRISPR/Cas experiments: the Split–Select–Combine strategy and the Overshoot–Select–Purify strategy. These strategies are based on the number of plants in the plant library, the number of target genes, and the genetic interactions among those target genes. Recently, the ultra-multiplex method was developed by Wu and collaborators (2024) based in CRISPR/Cas platform. The ultra-multiplex system consists of two template vectors, eight donor vectors, four destination vectors, and a primer-design software package. It combines the benefits of Golden Gate cloning for assembling multiple repetitive fragments and Gateway recombination for assembling larger fragments. Additionally, the system modifies the structure of the amplicons used to assemble sgRNA expression cassettes. This plant ultra-multiplex genome editing system can assemble a single binary vector capable of targeting more than 40 genomic loci.

3.5.3 Catalytic Modified Cas9: nCas9 and dCas9

As abovementioned, Cas9 nuclease processes the RuvC and HNH that act as catalytic domains. Each active catalytic domain cleaves one DNA strand introducing a blunt-ended DSBs. Two types of Cas9 nickases (nCas9) were developed introducing a point mutation in RuvC and HNH, making Cas9 with only one active domain (reviewed in Lee et al. 2015). nCas9 is able to insert single-strand breaks (SSBs) in the target DNA. SSBs are breaks in the covalently bonded carbon-phosphate backbone of one strand in the DNA double helix. SSBs do not activate NHEJ pathway and can be repaired via HR or high-fidelity base excision repair (BER) pathway (reviewed in Lee et al. 2015). nCas9 does not introduce uncontrolled indels and is a good option for precise gene editing.

On the other hand, the catalytic deactivated or dead Cas9 (dCas9) does not present RuvC or HNH activity and thus is not able to cleave the target DNA (reviewed in Lee et al. 2015). However, dCas9 can bind to the target, and researchers have utilized dCas9 for various applications. These include its use in multiplexing, in combination with cytosine deaminase (CBE) or adenosine deaminase (ABE) to induce point mutations (reviewed in Abdelrahman et al. 2021), as well as in recruiting proteins for epigenome editing. dCas9 can also be associated with transcriptional activators and repressors to regulate target gene expression (Rao and Wang 2021), fused to fluorescent proteins like GFP for live cell imaging of chromosomal loci (Wang et al. 2017), and employed to modulate RNA methylation and demethylation (Shi et al. 2024).

3.5.4 Modifications of CRISPR/Cas

Advances in the knowledge about Cas nucleases boosted the introduction of some modifications in the CRISPR toolkit. The association of the CRISPR with a nCas9 and a cytosine deaminase (CBE) or adenosine deaminase (ABE) comprises the base editing (BE). BE technology enables precise modifications of genomes (DNA) or transcriptomes (RNA) through the introduction of point mutations. When targeting DNA, BE can be categorized into cytosine base editors (CBEs), adenine base editors (ABEs), C-to-G base editors (CGBEs), dual-base editors, and organellar base editors (for review, see Molla et al. 2021). CBE and ABE systems utilize an nCas9 with the RuvC domain deactivated, enabling the introduction of C:G>T:A and A:T>G:C base transitions, respectively, directly at the DNA sites targeted by the sgRNA. These systems are highly efficient at inducing DNA transitions and serve as alternatives to traditional methods, as they do not require double-strand break (DSB) formation or donor DNA (reviewed in Zhu et al. 2020). CGBE is a new base editor with a uracil DNA N-glycosylase (UNG), and three highly active cytosine deaminases, hAID, hA3A, and Anc689, have been recently applied in rice targeting five genes generating C-to-G conversions (Tian et al. 2022). CBE, ABE, and CGBE are unable to simultaneously convert two different types of nucleotides within the same protospacer sequence. To overcome this limitation, dual-base editors were developed by fusing both cytidine and adenosine deaminases to a single Cas protein, enabling the concurrent introduction of C:T and A:G substitutions in the target genomic region (reviewed in Molla et al. 2021).

New systems are seeking for CRISPR system able to introduce a BE-associated transcriptional activation. The CRISPR-Combo platform, built on a single Cas9 protein, was developed for simultaneous genome editing (including targeted mutagenesis or base editing) and gene activation in plants (Pan et al. 2022). For that, authors used a CBE or ABE associated to a nCas9 into the CRISPR-Act3.0 possessing a sgRNA scaffold that is able to recruit the transcriptional activator.

Prime editing is another important tool that has been applied in plant gene editing. Prime editing utilizes an nCas9 fused to a reverse transcriptase, which is complexed with the prime editing guide RNA (pegRNA) (reviewed in Zhu et al. 2020). The pegRNA directs precise modifications at the nicked site, enabling base editing, including small insertions and deletions (~44 bp), without the need for donor DNA (Anzalone et al. 2019). Different from BEs that caused base transitions, prime editing can introduce transversions. To increase the size of insertions, GRAND editing was developed, enabling the precise insertion of large DNA fragments (ranging from 20 bp to approximately 1 kb) into the target site (Wang et al. 2022a). This method performs gene editing using a reverse transcriptase template (RTT) that is partially aligned with, but nonhomologous to, the target sequences within a pair of pegRNAs.

4 Prospects of Biotechnological Approaches in Zn Studies

Many of the current tools to apply in plants to make then more resilient to Zn deficiency, to improve plant ability to deal with Zn deficiency/toxicity, and to understand the molecular bases behind Zn dynamics in the soil and inside the plant have been part of the toolkit of researchers. Some important studies have revealed key factors that can be used for a precise breeding. The frontier of the knowledge about Zn in plants remains open, and new studies should be conducted in the near future.

From this chapter, we could identify some strategic ways to continue the search for ways to mitigate Zn deficiency. Understanding the panel of genotypes available and their relationship with the Zn dynamics in the soil remains an open way. Analyzing genotypes with respect to their responses facing Zn deficiency and mapping the factor behind this for sure will offer more information mainly combined with MAS, gene expression analysis, and functional genomics. Unveil more about miRNAs and its role in regulating key genes, from roots to seed, and the use of gene editing tools targeting gene promoters will help to go deeply in fight with Zn deficiency.

References

Abdelrahman M, Wei Z, Rohila JS, Zhao K (2021) Multiplex genome-editing technologies for revolutionizing plant biology and crop improvement. Front Plant Sci 22:721203

Abdoli M, Esfandiari E (2017) Assessment of genetic variation and zinc deficient tolerance in spring durum wheat (Triticum durum Desf.) genotypes in calcareous soil with zinc deficiency. J Genet Resour 3(1):7–17

Ajeesh Krishna TP, Ceasar SA, Maharajan T, Ramakrishnan M, Duraipandiyan V, Al-Dhabi NA, Ignacimuthu S (2017) Improving the zinc-use efficiency in plants: a review. SABRAO J Breed Genet 49(3):211–230

Ajeesh Krishna TP, Maharajan T, Victor Roch G, Ignacimuthu S, Antony Ceasar S (2020) Structure, function, regulation and phylogenetic relationship of ZIP family transporters of plants. Front Plant Sci 11:662

Ajeesh Krishna TP, Maharajan T, Ignacimuthu S, Antony Ceasar S (2022) Functional, structural, and transport aspects of ZIP in plants. In: Plant metal and metalloid transporters. Singapore, Springer Nature, pp 207–226

Akhtar M, Yousaf S, Sarwar N, Hussain S (2019) Zinc biofortification of cereals—role of phosphorus and other impediments in alkaline calcareous soils. Environ Geochem Health 41:2365–2379

Alberts B, Johnson A, Walter P, Lewis J, Raff M, Roberts K (2007) Molecular biology of the cell, 5th edn. Taylor & Francis, Garland Science, New York

Alejandro S, Meier B, Hoang MTT, Peiter E (2023) Cation diffusion facilitator proteins of Beta vulgaris reveal diversity of metal handling in dicotyledons. Plant Cell Environ 46(5):1629–1652

Alloway BJ (2009) Soil factors associated with zinc deficiency in crops and humans. Environ Geochem Health 31:537–548. https://doi.org/10.1007/s10653-009-9255-4

Amiteye S (2021) Basic concepts and methodologies of DNA marker systems in plant molecular breeding. Heliyon 7(10):e08093

An Q, Cui C, Khan NM, Zhou G, Wan Y (2021) Genome-wide investigation of zinc-iron permease (ZIP) genes in Areca catechu and potential roles of ZIPs in Fe and Zn uptake and transport. Plant Signal Behav 16(12):1995647

Anand A, Subramanian M, Kar D (2023) Breeding techniques to dispense higher genetic gains. Front Plant Sci 13:1076094

Andersen CL, Jensen JL, Ørntoft TF (2004) Normalization of real-time quantitative reverse transcription-PCR data: a model-based variance estimation approach to identify genes suited for normalization, applied to bladder and colon cancer data sets. Cancer Res 64:5245

Anzalone AV, Randolph PB, Davis JR, Sousa AA, Koblan LW, Levy JM, Chen PJ, Wilson C, Newby GA, Raguram A, Liu DR (2019) Searchand-replace genome editing without double-strand breaks or donor DNA. Nature 576(7785):149–157

Ashfaq MA, Kumar VD, Reddy PSS, Kumar CHA, Kumar KS, Rao NN, Tarakeswari M, Sujatha M (2020) Post-transcriptional gene silencing: basic concepts and applications. J Biosci 45:128

Assunção AG, Herrero E, Lin YF, Huettel B, Talukdar S, Smaczniak C, Immink RG, van Eldik M, Fiers M, Schat H, Aarts MG (2010) Arabidopsis thaliana transcription factors bZIP19 and bZIP23 regulate the adaptation to zinc deficiency. Proc Natl Acad Sci 107(22):10296–10301

Bandyopadhyay T, Mehra P, Hairat S, Giri J (2017) Morpho-physiological and transcriptome profiling reveal novel zinc deficiency-responsive genes in rice. Funct Integr Genomics 17(5):565–581

Barakate A, Keir E, Oakey H, Halpin C (2020) Stimulation of homologous recombination in plants expressing heterologous recombinases. BMC Plant Biol 20:336

Bastakoti S (2023) Role of zinc in management of plant diseases: a review. Cogent Food Agric 9(1):2194483

Begum MC, Islam M, Sarkar MR, Azad MAS, Huda AKMN, Kabir AH (2016) Auxin signaling is closely associated with Zn-efficiency in rice (Oryza sativa L.). J Plant Interact 11(1):124–129

Behl RK, Osaki M, Wasaki J, Watanabe T, Shinano T (2003) Breeding wheat for zinc efficiency improvement in semi-arid climate – a review. Tropics 12(4):295–312

Biswas MK (2023) Emerging horizons in plant genetics and breeding. Int J Mol Sci 24(14):11621

Borrill P, Connorton JM, Balk J, Miller AJ, Sanders D, Uauy C (2014) Biofortification of wheat grain with iron and zinc: integrating novel genomic resources and knowledge from model crops. Front Plant Sci 5:53

Bortesi L, Fischer R (2015) The CRISPR/Cas9 system for plant genome editing and beyond. Biotechnol Adv 33:41–52

Broadley MR, White PJ, Hammond JP, Zelko I, Lux A (2007) Zinc in plants. New Phytol 173(4):677–702

Bustin SA, Benes V, Garson JA, Hellemans J, Huggett J, Kubista M, Mueller R, Nolan T, Pfaffl MW, Shipley GL, Vandesompele J, Wittwer CT (2009) The MIQE guidelines: minimum information for publication of quantitative real-time PCR experiments. Clin Chem 55(4):611–622

Caldelas C, Weiss DJ (2017) Zinc homeostasis and isotopic fractionation in plants: a review. Plant Soil 411:17–46

Campos ACAL, Kruijer W, Alexander R, Akkers RC, Danku J, Salt DE, Aarts MGM (2017) Natural variation in Arabidopsis thaliana reveals shoot ionome, biomass, and gene expression changes as biomarkers for zinc deficiency tolerance. J Exp Bot 68(13):3643–3656

Cardini A, Pellegrino E, White PJ, Mazzolai B, Mascherpa MC, Ercoli L (2021) Transcriptional regulation of genes involved in zinc uptake, sequestration and redistribution following foliar zinc application to Medicago sativa. Plan Theory 10:476

Castillo-González J, Ojeda-Barrios D, Hernández-Rodríguez A, González-Franco AC, Robles-Hernández L, López-Ochoa GR (2018) Zinc metalloenzymes in plants. Interciencia 43(4):242–248

Chen X, Yuan L, Ludewig U (2016) Natural genetic variation of seed micronutrients of Arabidopsis thaliana grown in zinc-deficient and zinc-amended soil. Front Plant Sci 7:1070

Chen Z, Halford NG, Liu C (2023) Real-time quantitative PCR: primer design, reference gene selection, calculations and statistics. Metabolites 13(7):806

Chowdhury R, Nallusamy S, Shanmugam V, Loganathan A, Muthurajan R, Sivathapandian SK et al (2022) Genome-wide understanding of evolutionary and functional relationships of rice yellow stripe-like (YSL) transporter family in comparison with other plant species. Biologia 77:39–53

Claus J, Bohmann A, Chavarría-Krauser A (2013) Zinc uptake and radial transport in roots of Arabidopsis thaliana: a modelling approach to understand accumulation. Ann Bot 112(2):369–380

Das U, Haque AM, Bari MA, Mandal A, Kabir AH (2021) Computational characterization and expression profile of MTP1 gene associated with zinc homeostasis across dicot plant species. Gene Rep 23:101073

Deshpande P, Dapkekar A, Oak M, Paknikar K, Rajwade J (2018) Nanocarrier-mediated foliar zinc fertilization influences expression of metal homeostasis related genes in flag leaves and enhances gluten content in durum wheat. PLoS One 13(1):e0191035

Deshpande D, Chhugani K, Chang Y, Karlsberg A, Loeffler C, Zhang J, Muszyńska A, Munteanu V, Yang H, Rotman J, Tao L, Balliu B, Tseng E, Eskin E, Zhao F, Mohammadi P, Łabaj P, Mangul S (2023) RNA-seq data science: from raw data to effective interpretation. Front Genet 14:997383

Diaz S, Polania J, Ariza-Suarez D, Cajiao C, Grajales M, Raatz B, Beebe SE (2022) Genetic correlation between Fe and Zn biofortification and yield components in a common bean (Phaseolus vulgaris L.). Front Plant Sci 12:739033

El-Sappah AH, Abbas M, Rather SA, Wani SH, Soaud N, Noor Z, Li J (2023) Genome-wide identification and expression analysis of metal tolerance protein (MTP) gene family in soybean (*Glycine max*) under heavy metal stress. Mol Biol Rep 50(4):2975–2990

Emery L, Whelan S, Hirschi KD, Pittman JK (2012) Protein phylogenetic analysis of Ca2+/cation antiporters and insights into their evolution in plants. Front Plant Sci 3:1

Evens NP, Buchner P, Williams LE, Hawkesford MJ (2017) The role of ZIP transporters and group F bZIP transcription factors in the Zn-deficiency response of wheat (*Triticum aestivum*). Plant J 92(2):291–304

Fernández V, Brown PH (2013) From plant surface to plant metabolism: the uncertain fate of foliar-applied nutrients. Front Plant Sci 4:289

Fire A, Xu S, Montgomery MK, Kostas SA, Driver SE, Mello CC (1998) Potent and specific genetic interference by double-stranded RNA in *Caenorhabditis elegans*. Nature 391(6669):806–811

Gaitán-Solís E, Taylor NJ, Siritunga D, Stevens W, Schachtman DP (2015) Overexpression of the transporters AtZIP1 and AtMTP1 in cassava changes zinc accumulation and partitioning. Front Plant Sci 6:492. https://doi.org/10.3389/fpls.2015.00492. PMID: 26217349; PMCID: PMC4496839

Ghandilyan A, Kutman UB, Kutman BY, Cakmak I, Aarts MGM (2012) Genetic analysis of the effect of zinc deficiency on Arabidopsis growth and mineral concentrations. Plant Soil 361:227–239

Gilchrist E, Haughn G (2010) Reverse genetics techniques: engineering loss and gain of gene function in plants. Brief Funct Genomics 9(2):103–110

Graham RD, Rengel Z (1993) Genotypic variation in zinc uptake and utilization by plants. In: Robson AD (ed) Zinc in soils and plants, Developments in plant and soil sciences, vol 55. Springer, Dordrecht

Grover A, Sharma PC (2014) Development and use of molecular markers: past and present. Crit Rev Biotechnol 36(2):290–302

Guan J, Yang Y, Shan Q, Zhang H, Zhou A, Gong S et al (2023) Plant cadmium resistance 10 enhances tolerance to toxic heavy metals in poplar. Plant Physiol Biochem 203:108043

Gupta N, Ram H, Kumar B (2016) Mechanism of zinc absorption in plants: uptake, transport, translocation and accumulation. Rev Environ Sci Biotechnol 15:89–109

Gupta PK, Kulwal PL, Jaiswal V (2019) Association mapping in plants in the post-GWAS genomics era. Adv Genet 104:75–154

Gupta OP, Pandey V, Saini R, Khandale T, Singh A, Malik VK, Narwal S, Ram S, Singh GP (2021) Comparative physiological, biochemical and transcriptomic analysis of hexaploid wheat (T. aestivum L.) roots and shoots identifies potential pathways and their molecular regulatory network during Fe and Zn starvation. Genomics 113(5):3357–3372

Gustin JL, Zanis MJ, Salt DE (2011) Structure and evolution of the plant cation diffusion facilitator family of ion transporters. BMC Evol Biol 11:1–13

Hacisalihoglu G (2020) Zinc (Zn): the last nutrient in the alphabet and shedding light on Zn efficiency for the future of crop production under suboptimal Zn. Plan Theory 9(11):1471

Hamzah Saleem M, Usman K, Rizwan M, Al Jabri H, Alsafran M (2022) Functions and strategies for enhancing zinc availability in plants for sustainable agriculture. Front Plant Sci 13:1033092

Harwood WA (2013) A protocol for high-throughput agrobacterium-mediated barley transformation. In: Cereal genomics: methods and protocols, Methods in molecular biology, vol 1099. Springer, New York

Hasan N, Choudhary S, Naaz N, Sharma N, Laskar RA (2021) Recent advancements in molecular marker-assisted selection and applications in plant breeding programmes. J Genet Eng Biotechnol 19:128

Hasan MN, Islam S, Bhuiyan FH, Arefin S, Hoque H, Jewel NA et al (2022) Genome wide analysis of the heavy-metal-associated (HMA) gene family in tomato and expression profiles under different stresses. Gene 835:146664

Hassan MU, Aamer M, Umer Chattha M, Haiying T, Shahzad B, Barbanti L et al (2020) The critical role of zinc in plants facing the drought stress. Agriculture 10(9):396

Hassan MM, Zhang Y, Yuan G, De K, Chen JG, Muchero W, Tuskan GA, Qi Y, Yang X (2021) Construct design for CRISPR/Cas-based genome editing in plants. Trends Plant Sci 26(11):1133–1152

Haydon MJ, Kawachi M, Wirtz M, Hillmer S, Hell R, Krämer U (2012) Vacuolar nicotianamine has critical and distinct roles under iron deficiency and for zinc sequestration in Arabidopsis. Plant Cell 24(2):724–737

Hayta A, Smedley MA, Clarke M, Forner M, Harwood WA (2021) An efficient agrobacterium-mediated transformation protocol for hexaploid and tetraploid wheat. Curr Protoc 1:e58

Henriques AR, Chalfun-Junior A, Aarts M (2012) Strategies to increase zinc deficiency tolerance and homeostasis in plants. Braz J Plant Physiol 24(1):3–8

Huang S, Wang P, Yamaji N, Ma JF (2020) Plant nutrition for human nutrition: hints from rice research and future perspectives. Mol Plant 13(6):825–835

Hung Y-H, Slotkin RK (2021) The initiation of RNA interference (RNAi) in plants. Curr Opin Plant Biol 61:102014

Hussain A, Jiang W, Wang X, Shahid S, Saba N, Ahmad M et al (2022) Mechanistic impact of zinc deficiency in human development. Front Nutr 9:717064

Hwang H-H, Yua M, Laia E-M (2017) Agrobacterium-mediated plant transformation: biology and applications. Arabidopsis Book e0186:1–31. https://doi.org/10.1199/tab.0186. PMID: 31068763; PMCID: PMC6501860

Ishida Y, Saito H, Hiei Y, Komari T (2003) Improved protocol for transformation of maize (*Zea mays* L.) mediated by *Agrobacterium tumefaciens*. Plant Biotechnol 20(1):57–66

Ishida Y, Tsunashima M, Hiei Y, Komari T (2014) Wheat (Triticum aestivum L.) transformation using immature embryos. In: Agrobacterium protocols: volume 1, Methods in molecular biology, vol 1223. Springer, New York

Ishimaru Y, Masuda H, Suzuki M, Bashir K, Takahashi M, Nakanishi H, Mori S, Nishizawa NK (2007) Overexpression of the OsZIP4 zinc transporter confers disarrangement of zinc distribution in rice plants. J Exp Bot 58(11):2909–2915. https://doi.org/10.1093/jxb/erm147. Epub 2007 Jul 13. PMID: 17630290.

Jalal A, Oliveira CES, Fernandes GC, da Silva EC, da Costa KN, de Souza JS, Leite GS, Biagini ALC, Galindo FS, Teixeira Filho MCM (2023) Integrated use of plant growth-promoting bacteria and nano-zinc foliar spray is a sustainable approach for wheat biofortification, yield, and zinc use efficiency. Front Plant Sci 14:1146808

Jeon D, Kang Y, Lee S, Choi S, Sung Y, Lee TH, Kim C (2023) Digitalizing breeding in plants: a new trend of next-generation breeding based on genomic prediction. Front Plant Sci 14:1092584

Jinek M, Chylinski K, Fonfara I, Hauer M, Doudna JA, Charpentier E (2012) A programmable dual-RNA-guided DNA endonuclease in adaptive bacterial immunity. Science 337:816–821

Joshi G, Soe YP, Palanog A, Hore TK, Nha CT, Calayugan MI, Inabangan-Asilo MA, Amparado A, Pandey ID, Cruz PCS, Hernandez JE, Swamy BPM (2023) Meta-QTL s and haplotypes for efficient zinc biofortification of rice. Plant Genome 16(4):e20315

Juliana P, Govindan V, Crespo-Herrera L, Mondal S, Huerta-Espino J, Shrestha S, Poland J, Singh RP (2022) Genome-wide association mapping identifies key genomic regions for grain zinc and iron biofortification in bread wheat. Front Plant Sci 13:903819

Kang M, Lee K, Finley T, Chappell H, Veena V, Wang K (2022) An improved agrobacterium-mediated transformation and genome-editing method for maize inbred B104 using a ternary vector system and immature embryos. Front Plant Sci 13:860971

Kaur H, Garg N (2021) Zinc toxicity in plants: a review. Planta 253(6):129

Kaur H, Srivastava S, Goyal N, Walia S (2024) Behavior of zinc in soils and recent advances on strategies for ameliorating zinc phyto-toxicity. Environ Exp Bot 2020:105676

Kendziorek M, Klimecka M, Barabasz A, Borg S, Rudzka J, Szczęsny P, Antosiewicz DM (2016) Engineering high Zn in tomato shoots through expression of AtHMA4 involves tissue-specific modification of endogenous genes. BMC Genomics 17:625

Kim S-I, Veena, Gelvin SB (2007) Genome-wide analysis of agrobacterium T-DNA integration sites in the Arabidopsis genome generated under non-selective conditions. Plant J 51:779–791

Kobae Y, Uemura T, Sato MH, Ohnishi M, Mimura T, Nakagawa T, Maeshima M (2004) Zinc transporter of Arabidopsis thaliana AtMTP1 is localized to vacuolar membranes and implicated in zinc homeostasis. Plant Cell Physiol 45(12):1749–1758

Landon JR (2014) Booker tropical soil manual: a handbook for soil survey and agricultural land evaluation in the tropics and subtropics. Routledge, London

Lee S, An G (2009) Over-expression of OsIRT1 leads to increased iron and zinc accumulations in rice. Plant Cell Environ 32(4):408–416. https://doi.org/10.1111/j.1365-3040.2009.01935.x. Epub Jan 14. PMID: 19183299.

Lee L-Y, Gelvin SB (2008) T-DNA binary vectors and systems. Plant Physiol 146:325–332

Lee K, Wang K (2023) Strategies for genotype-flexible plant transformation. Curr Opin Biotechnol 79:102848

Lee S, Jeong HJ, Kim SA, Lee J, Guerinot ML, An G (2010) OsZIP5 is a plasma membrane zinc transporter in rice. Plant Mol Biol 73:507–517

Lee J, Chung JH, Kim HM, Kim DW, Kim H (2015) Designed nucleases for targeted genome editing. Plant Biotechnol J 14(2):448–462

Li S, Zhou X, Huang Y, Zhu L, Zhang S, Zhao Y et al (2013a) Identification and characterization of the zinc-regulated transporters, iron-regulated transporter-like protein (ZIP) gene family in maize. BMC Plant Biol 13(1):1–14

Li Y, Zhang Y, Shi D, Liu X, Qin J, Ge Q, Xu L, Pan X, Li W, Zhu Y, Xu J (2013b) Spatial-temporal analysis of zinc homeostasis reveals the response mechanisms to acute zinc deficiency in Sorghum bicolor. New Phytol 200(4):1102–1115

Li S, Zhou X, Li H, Liu Y, Zhu L, Guo J, Liu X, Fan Y, Chen J, Chen R (2015) Overexpression of ZmIRT1 and ZmZIP3 enhances iron and zinc accumulation in transgenic arabidopsis. PLoS One 10(8):e0136647. https://doi.org/10.1371/journal.pone.0136647. PMID: 26317616; PMCID: PMC4552944.

Liao F, Lilay GH, Castro PH, Azevedo H, Assunção AGL (2022) Regulation of the zinc deficiency response in the legume model Medicago truncatula. Front Plant Sci 13:916168

Lilay GH, Castro PH, Guedes JG, Almeida DM, Campilho A, Azevedo H, Aarts MGM, Saibo NJM, Assunção AGL (2020) Rice F-bZIP transcription factors regulate the zinc deficiency response. J Exp Bot 71(12):3664–3677

Lilay GH, Persson DP, Castro PH, Liao F, Alexander RD, Aarts MG, Assunção AG (2021) Arabidopsis bZIP19 and bZIP23 act as zinc sensors to control plant zinc status. Nat Plants 7(2):137–143

Liu XS, Feng SJ, Zhang BQ, Wang MQ, Cao HW, Rono JK, Chen X, Yang ZM (2019) OsZIP1 functions as a metal efflux transporter limiting excess zinc, copper and cadmium accumulation in rice. BMC Plant Biol 19(1):283

Livak KJ, Schmittgen TD (2001) Analysis of relative gene expression data using real-time quantitative PCR and the 2(-Delta Delta C(T)) method. Methods 25(4):402–408

Lloyd A (2003) Vector construction for gene overexpression as a tool to elucidate gene function. In: Grotewold E (ed) Plant functional genomics: methods and protocols, Methods in molecular biology, vol 236. Humana Press, Totowa, pp 329–344

Lonergan PF, Pallotta MA, Lorimer M, Paull JG, Barker SJ, Graham RD (2009) Multiple genetic loci for zinc uptake and distribution in barley (Hordeum Vulgare). New Phytol 184(1):168–179

Lörz H, Baker B, Schell J (1985) Gene transfer to cereal cells mediated by protoplast transformation. Mol Gen Genet 199:178–182

Lu L, Tian S, Zhang J, Yang X, Labavitch JM, Webb SM et al (2013) Efficient xylem transport and phloem remobilization of Z n in the hyperaccumulator plant species S edum alfredii. New Phytol 198(3):721–731

Lu X, Liu S, Zhi S, Chen J, Ye G (2021) Comparative transcriptome profile analysis of rice varieties with different tolerance to zinc deficiency. Plant Biol J 23:375–390

Ludwig Y, Dueñas C Jr, Arcillas E, Macalalad-Cabral RJ, Kohli A, Reinke R, Slamet-Loedin IH (2024) CRISPR-mediated promoter editing of a cis-regulatory element of OsNAS2 increases Zn uptake/translocation and plant yield in rice. Front Genome Ed 5:1308228

Lv Z, Liu F, Zhang P, Li W, Song Q, Xu J (2023) Revealing the molecular mechanisms of zinc accumulation and zinc deficiency responses in quinoa, analyzed by high-throughput gene expression profiling under zinc depletion and resupply. Environ Exp Bot 210:105330

Ma Y, Wen Y, Wang C, Wu Z, Yuan X, Xiong Y et al (2023) ZIP genes are involved in the retransfer of zinc ions during the senescence of zinc-deficient rice leaves. Int J Mol Sci 24(18):13989

Mager S, Schönberger B, Ludewig U (2018) The transcriptome of zinc deficient maize roots and its relationship to DNA methylation loss. BMC Plant Biol 18:372

Mahendrakar MD, Parveda M, Kishor PK, Srivastava RK (2020) Discovery and validation of candidate genes for grain iron and zinc metabolism in pearl millet [Pennisetum glaucum (L.) R. Br.]. Sci Rep 10(1):16562

Malavolta E (2006) Sintomas visuais de deficiência e excesso. In: Malavolta E (ed) Manual de nutrição mineral de plantas. Agronômica Ceres, São Paulo, pp 548–567

Masuda H, Usuda K, Kobayashi T et al (2009) Overexpression of the barley nicotianamine synthase gene HvNAS1 increases iron and zinc concentrations in rice grains. Rice 2:155–166. https://doi.org/10.1007/s12284-009-9031-1

Molla KA, Sretenovic S, Bansal KC, Qi Y (2021) Precise plant genome editing using base editors and prime editors. Nat Plants 7(9):1166–1187

Mondal TK, Ganie SA, Rana MK, Sharma TR (2014) Genome-wide analysis of zinc transporter genes of maize (Zea mays). Plant Mol Biol Report 32:605–616

Mossa AW, Young SD, Crout NM (2020) Zinc uptake and phyto-toxicity: comparing intensity-and capacity-based drivers. Sci Total Environ 699:134314

Mu S, Yamaji N, Sasaki A, Luo L, Du B, Che J et al (2021) A transporter for delivering zinc to the developing tiller bud and panicle in rice. Plant J 105(3):786–799

Nagarathna TK, Shankar AG, Udayakumar M (2010) Assessment of genetic variation in zinc acquisition and transport to seed in diversified germplasm lines of rice (*Oryza sativa* L.). J Agric Technol 6(1):171–178

Nakagawa T, Ishiguro S, Kimura T (2009) Gateway vectors for plant transformation. Plant Biotechnol 26:275–284

Nanda AK, Pujol V, Wissuwa M (2017) Patterns of stress response and tolerance based on transcriptome profiling of rice crown tissue under zinc deficiency. J Exp Bot 68(7):1715–1729

Natasha N, Shahid M, Bibi I, Iqbal J, Khalid S, Murtaza B et al (2022) Zinc in soil-plant-human system: a data-analysis review. Sci Total Environ 808:152024

Neeraj J, Rajani N, Shigaki T (2009) Mining cation (CAX) transporter diversity for nutrition-enhanced crops and phytoremediation. Int J Integr Biol 7(1):22–25

Nishimura A, Aichi I, Matsuoka M (2006) A protocol for agrobacterium-mediated transformation in rice. Nat Protoc 1:2797–2802

Noman A, Aqeel M, Khalid N, Islam W, Sanaullah T, Anwar M et al (2019) Zinc finger protein transcription factors: integrated line of action for plant antimicrobial activity. Microb Pathog 132:141–149

Noulas C, Tziouvalekas M, Karyotis T (2018) Zinc in soils, water and food crops. J Trace Elem Med Biol 49:252–260
Ova EA, Kutman UB, Ozturk L, Cakmak I (2015) High phosphorus supply reduced zinc concentration of wheat in native soil but not in autoclaved soil or nutrient solution. Plant Soil 393:147–162
Pan C, Li G, Malzahn AA, Cheng Y, Leyson B, Sretenovic S, Gurel F, Coleman GD, Qi Y (2022) Boosting plant genome editing with a versatile CRISPR-combo system. Nat Plants 8(5):513–525
Pasricha S, Mathur V, Garg A, Lenka S, Verma K, Agarwal S (2021) Molecular mechanisms underlying heavy metal uptake, translocation and tolerance in hyperaccumulators-an analysis: heavy metal tolerance in hyperaccumulators. Environ Challenges 4:100197
Pausch P, Al-Shayeb B, Bisom-Rapp E, Tsuchida CA, Li Z, Cress BF, Knott GJ, Jacobsen SE, Banfield JF, Doudna JA (2020) CRISPR-CasΦ from huge phages is a hypercompact genome editor. Science 369(6501):333–337
Pegoraro C, Storch TT, Crizel GR, Rombaldi CV, Girardi CL (2016) Identification of biomarkers associated to 'gala' apples ripening and postharvest quality. Comunicata Scientiae 7(4):494–503
Permyakova NV, Deineko EV (2024) Crop improvement: comparison of transgenesis and gene editing. Horticulturae 10:57
Pfaffl MW, Tichopad A, Prgomet C, Neuvians TP (2004) Determination of stable housekeeping genes, differentially regulated target genes and sample integrity: BestKeeper–excel-based tool using pair-wise correlations. Biotechnol Lett 26:509–515
Pinar H, Bulbul C, Simsek D, Shams M, Mutlu N, Ercisli S (2022) Development of molecular markers lınked to QTL/genes controllıng Zn effıcıency. Mol Biol Rep 49(6):5273–5281
Pradhan SK, Pandit E, Pawar S, Naveenkumar R, Barik SR, Mohanty SP, Nayak DK, Ghritlahre SK, Sanjiba Rao D, Reddy JN, Patnaik SSC (2020) Linkage disequilibrium mapping for grain Fe and Zn enhancing QTLs useful for nutrient dense rice breeding. BMC Plant Biol 20:57
Rajeevkumar S, Anunanthini P, Sathishkumar R (2015) Epigenetic silencing in transgenic plants. Front Plant Sci 6:693
Ram H, Soni P, Salvi P, Gandass N, Sharma A, Kaur A, Sharma TR (2019) Insertional mutagenesis approaches and their use in rice for functional genomics. Plan Theory 8:310
Ramesh SA, Choimes S, Schachtman DP (2004) Over-expression of an Arabidopsis zinc transporter in hordeum vulgare increases short-term zinc uptake after zinc deprivation and seed zinc content. Plant Mol Biol 54(3):373–385. https://doi.org/10.1023/B:PLAN.0000036370.70912.34. PMID: 15284493
Rao MJ, Wang L (2021) CRISPR/Cas9 technology for improving agronomic traits and future prospective in agriculture. Planta 254(4):68
Rellán-Álvarez R, Giner-Martínez-Sierra J, Orduna J, Orera I, Rodríguez-Castrillón JÁ, García-Alonso JI et al (2010) Identification of a tri-iron (III), tri-citrate complex in the xylem sap of iron-deficient tomato resupplied with iron: new insights into plant iron long-distance transport. Plant Cell Physiol 51(1):91–102
Rengasamy B, Manna M, Jonwal S, Sathiyabama M, Thajuddin NB, Sinha AK (2024) A simplified and improved protocol of rice transformation to cater wide range of rice cultivars. Protoplasma. https://doi.org/10.1007/s00709-023-01925-8
Rose TJ, Impa SM, Rose MT, Pariasca-Tanaka J, Mori A, Heuer S, Johnson-Beebout SE, Wissuwa M (2013) Enhancing phosphorus and zinc acquisition efficiency in rice: a critical review of root traits and their potential utility in rice breeding. Ann Bot 112(2):331–345
Rössner C, Lotz D, Becker A (2022) VIGS goes viral: how VIGS transforms our understanding of plant science. Annu Rev Plant Biol 73:703–728
Sadeghzadeh B (2013) A review of zinc nutrition and plant breeding. J Soil Sci Plant Nutr 13(4):905–927
Sahoo K, Tripathi AK, Pareek A, Sopory SK, Singla-Pareek SL (2011) An improved protocol for efficient transformation and regeneration of diverse indica rice cultivars. Plant Methods 7:1–11
Saleem MH, Usman K, Rizwan M, Al Jabri H, Alsafran M (2022) Functions and strategies for enhancing zinc availability in plants for sustainable agriculture. Front Plant Sci 13:1033092

Sammons R, Ivashuta S, Liu H, Wang D, Feng P, Kouranov A, Andersen S (2011) Polynucleotide molecules for gene regulation in plants. Patent n°: US20110296556A1
Sasaki A, Yamaji N, Ma JF (2014) Overexpression of OsHMA3 enhances Cd tolerance and expression of Zn transporter genes in rice. J Exp Bot 65(20):6013–6021. https://doi.org/10.1093/jxb/eru340. Epub Aug 23. PMID: 25151617; PMCID: PMC4203134
Sasaki A, Yamaji N, Mitani-Ueno N, Kashino M, Ma JF (2015) A node-localized transporter Os ZIP 3 is responsible for the preferential distribution of Zn to developing tissues in rice. Plant J 84(2):374–384
Senthil-Kumar M, Anand A, Uppalapati SR, Mysore KS (2008) Virus-induced gene silencing and its applications. CAB Rev Perspect Agric Vet Sci Nutr Nat Resour 3(011):1–18
Sharma A, Patni B, Shankhdhar D, Shankhdhar SC (2013) Zinc–an indispensable micronutrient. Physiol Mol Biol Plants 19:11–20
Shi C, Zou W, Liu X, Zhang H, Li X, Fu G, Fei Q, Qian Q, Shang L (2024) Programmable RNA N6-methyladenosine editing with CRISPR/dCas13a in plants. Plant Biotechnol J 22(7):1867–1880
Shirazi Z, Khakdan F, Rafiei F, Balalami MY, Ranjbar M (2023) Genome-wide identification and expression profile analysis of metal tolerance protein gene family in Eucalyptus grandis under metal stresses. BMC Plant Biol 23(1):240
Sidorov V, Duncan D (2009) Agrobacterium-mediated maize transformation: immature embryos versus callus. In: Scott MP (ed) Transgenic maize, Methods in molecular biology, vol 526. Humana Press, Totowa
Sinclair SA, Krämer U (2012) The zinc homeostasis network of land plants. Biochim Biophys Acta Mol Cell Res 1823(9):1553–1567
Sinha D, Tandon PK, Srivastava GP, Srivastava SK, Mukherjee S (2022) Role of heavy metal ATPases in transport of cadmium and zinc in plants. In: Plant metal and metalloid transporters. Singapore, Springer Nature, pp 109–131
Slamet-Loedin IH, Chadha-Mohanty P, Torrizo L (2013) Agrobacterium-mediated transformation: rice transformation. In: Cereal genomics: methods and protocols, methods in molecular biology, vol 1099. Springer, New York. 2014
Song WY, Martinoia E, Lee J, Kim D, Kim DY, Vogt E et al (2004) A novel family of cys-rich membrane proteins mediates cadmium resistance in Arabidopsis. Plant Physiol 135(2):1027–1039
Song WY, Choi KS, Kim DY, Geisler M, Park J, Vincenzetti V et al (2010) Arabidopsis PCR2 is a zinc exporter involved in both zinc extrusion and long-distance zinc transport. Plant Cell 22(7):2237–2252
Song WY, Lee HS, Jin SR, Ko D, Martinoia E, Lee Y et al (2015) Rice PCR1 influences grain weight and Z n accumulation in grains. Plant Cell Environ 38(11):2327–2339
Stachel SE, Timmerman B, Zambryski P (1986) Generation of single-stranded T-DNA molecules during the initial stages of T-DNA transfer from Agrobacterium tumefaciens to plant cells. Nature 322:706–712
Stanton C, Sanders D, Krämer U, Podar D (2022) Zinc in plants: integrating homeostasis and biofortification. Mol Plant 15(1):65–85
Sturikova H, Krystofova O, Huska D, Adam V (2018) Zinc, zinc nanoparticles and plants. J Hazard Mater 349:101–110
Su W, Xu M, Radani Y, Yang L (2023) Technological development and application of plant genetic transformation. Int J Mol Sci 24:10646
Tan S, Han R, Li P, Yang G, Li S, Zhang P, Wang WB, Zhao WZ, Yin LP (2014) Over-expression of the MxIRT1 gene increases iron and zinc content in rice seeds. Transgenic Res 24(1):109–122
Tan L, Qu M, Zhu Y, Peng C, Wang J, Gao D, Chen C (2020) ZINC TRANSPORTER5 and ZINC TRANSPORTER9 function synergistically in zinc/cadmium uptake. Plant Physiol 183(3):1235–1249. https://doi.org/10.1104/pp.19.01569. Epub 2020 Apr 27. PMID: 32341004; PMCID: PMC7333683
Tang L, Dong J, Tan L, Ji Z, Li Y, Sun Y et al (2021) Overexpression of OsLCT2, a low-affinity cation transporter gene, reduces cadmium accumulation in shoots and grains of rice. Rice 14(1):1–15

Tian Y, Shen R, Li Z, Yao Q, Zhang X, Zhong D, Tan X, Song M, Han H, Zhu JK, Lu Y (2022) Efficient C-to-G editing in rice using an optimized base editor. Plant Biotechnol J 20(7):1238–1240
Tiong J, McDonald G, Genc Y, Shirley N, Langridge P, Huang CY (2015) Increased expression of six ZIP family genes by zinc (Zn) deficiency is associated with enhanced uptake and root-to-shoot translocation of Zn in barley (*Hordeum vulgare*). New Phytol 207:1097–1109
Upton RN, Correr FH, Lile J, Reynolds GL, Falaschi K, Cook JP, Lachowiec J (2023) Design, execution, and interpretation of plant RNA-seq analyses. Front Plant Sci 14:1135455
Van Huffel K, Stock M, Ruttink T, De Baets B (2022) Covering the combinatorial design space of multiplex CRISPR/Cas experiments in plants. Front Plant Sci 13:907095
van Kregten M, de Pater S, Romeijn R, van Schendel R, Hooykaas PJ, Tijsterman M (2016) T-DNA integration in plants results from polymerase-θ-mediated DNA repair. Nat Plants 2(11):16164
Vandesompele J, De Preter K, Pattyn F, Poppe B, Van Roy N, De Paepe A, Speleman F (2002) Accurate normalization of real-time quantitative RT-PCR data by geometric averaging of multiple internal control genes. Genome Biol 3:research0034.1
Vega JM, Yu W, Kennon AR, Chen X, Zhang ZJ (2008) Improvement of Agrobacterium-mediated transformation in Hi-II maize (Zea mays) using standard binary vectors. Plant Cell Rep 27:297–305
Veitia RA (2008) Exploring the molecular etiology of dominant-negative mutations. Plant Cell 19:3843–3851
Velu G, Singh RP, Crespo-Herrera L, Juliana P, Dreisigacker S, Valluru R, Stangoulis J, Sohu VS, Mavi GS, Mishra VK, Balasubramaniam A, Chatrath R, Gupta V, Singh GP, Joshi AK (2018) Genetic dissection of grain zinc concentration in spring wheat for mainstreaming biofortification in CIMMYT wheat breeding. Sci Rep 8:13526
Verma PK, Verma S, Chakrabarty D, Pandey N (2021) Biotechnological approaches to enhance zinc uptake and utilization efficiency in cereal crops. J Soil Sci Plant Nutr 21:2412–2424
Viana VE, Pegoraro C, Busanello C, Costa de Oliveira A (2019) Mutagenesis in rice: the basis for breeding a new super plant. Front Plant Sci 10:1326
Viana VE, Maltzahn LE, Costa de Oliveira A, Pegoraro C (2021) Genetic approaches for iron and zinc biofortification and arsenic decrease in Oryza sativa L. grains. Biol Trace Elem Res 200(10):4505–4523
Wang HX, Li M, Lee CM, Chakraborty S, Kim HW, Bao G, Leong KW (2017) CRISPR/Cas9-based genome editing for disease modeling and therapy: challenges and opportunities for non-viral delivery. Chem Rev 117(15):9874e9906
Wang N, Qiu W, Dai J, Guo X, Lu Q, Wang T, Li S, Liu T, Zuo Y (2019) AhNRAMP1 enhances manganese and zinc uptake in plants. Front Plant Sci 10:415
Wang FH, Qiao K, Shen YH, Wang H, Chai TY (2021a) Characterization of the gene family encoding metal tolerance proteins in Triticum urartu: phylogenetic, transcriptional, and functional analyses. Metallomics 13(7):mfab038
Wang Y, Xum X, Hao Y, Zhang Y, Liu Y, Pu Z, Tian Y, Xu D, Xia X, He Z, Zhang Y (2021b) QTL mapping for grain zinc and iron concentrations in bread wheat. Front Nutr 8:680391
Wang J, He Z, Wang G, Zhang R, Duan J, Gao P, Lei X, Qiu H, Zhang C, Zhang Y, Yin H (2022a) Efficient targeted insertion of large DNA fragments without DNA donors. Nat Methods 19(3):331e340
Wang X, Wang L, Huang Y, Deng Z, Lia C, Zhanga J, Zheng M, Yan S (2022b) A plant-specific module for homologous recombination repair. Proc Natl Acad Sci 119(16):e2202970119
Waters BM, Chu HH, DiDonato RJ, Roberts LA, Eisley RB, Lahner B et al (2006) Mutations in Arabidopsis yellow stripe-like1 and yellow stripe-like3 reveal their roles in metal ion homeostasis and loading of metal ions in seeds. Plant Physiol 141(4):1446–1458
Wu S, Kyaw H, Tong Z, Yang Y, Wang Z, Zhang L, Deng L, Zhang Z, Xiao B, Quick WP, Lu T, Xiao G, Qin G, Cui X (2024) A simple and efficient CRISPR/Cas9 system permits ultra-multiplex genome editing in plants. Crop J 12:569–582
Xu Y, Li P, Yang Z, Xu C (2017) Genetic mapping of quantitative trait loci in crops. Crop J 5(2):175–184

Xu J, Wang X, Zhu H, Yu F (2021) Maize genotypes with different zinc efficiency in response to low zinc stress and heterogeneous zinc supply. Front Plant Sci 12:736658

Xu J, Qin X, Ni Z, Chen F, Fu X, Yu F (2022a) Identification of zinc efficiency-associated loci (ZEALs) and candidate genes for Zn deficiency tolerance of two recombination inbred line populations in maize. Int J Mol Sci 23:4852

Xu J, Ni Z, Chen F, Fu X, Yu F (2022b) Integrated linkage mapping and genome-wide association study to dissect the genetic basis of zinc deficiency tolerance in maize at seedling stage. Crop J 10(6):1807–1818

Yamaji N, Xia J, Mitani-Ueno N, Yokosho K, Feng Ma J (2013) Preferential delivery of zinc to developing tissues in rice is mediated by P-type heavy metal ATPase OsHMA2. Plant Physiol 162(2):927–939

Yang M, Li Y, Liu Z, Tian J, Liang L, Qiu Y et al (2020) A high activity zinc transporter OsZIP9 mediates zinc uptake in rice. Plant J 103(5):1695–1709

Yilmaz H, Özer G, Baloch FS, Çiftçi V, Chung YS, Sun HJ (2023) Genome-wide identification and expression analysis of MTP (metal ion transport proteins) genes in the common bean. Plan Theory 12(18):3218

Younas N, Fatima I, Ahmad IA, Ayyaz MK (2023) Alleviation of zinc deficiency in plants and humans through an effective technique; biofortification: a detailed review. Acta Ecol Sin 43(3):419–425

Zeng H, Zhang X, Ding M, Zhang X, Zhu Y (2019a) Transcriptome profiles of soybean leaves and roots in response to zinc deficiency. Physiol Plant 167:330–351

Zeng H, Zhang X, Ding M, Zhu Y (2019b) Integrated analyses of miRNAome and transcriptome reveal zinc deficiency responses in rice seedlings. BMC Plant Biol 19(1):585

Zeng H, Wu H, Yan F, Yi K, Zhu Y (2021) Molecular regulation of zinc deficiency responses in plants. J Plant Physiol 261:153419

Zetsche B, Gootenberg JS, Abudayyeh OO, Slaymaker IM, Makarova KS, Essletzbichler P, Volz SE, Joung J, van der Oost J, Regev A, Koonin EV, Zhang F (2015) Cpf1 is a single RNA-guided endonuclease of a class 2 CRISPR-Cas system. Cell 163:759–771

Zhan J, Zou W, Li S, Tang J, Lu X, Meng L, Ye G (2022) OsNAC15 regulates tolerance to zinc deficiency and cadmium by binding to OsZIP7 and OsZIP10 in rice. Int J Mol Sci 23:11771

Zhang JZ (2003) Overexpression analysis of plant transcription factors. Curr Opin Plant Biol 6:430–440

Zhang Y, Xu YH, Yi HY, Gong JM (2012) Vacuolar membrane transporters *OsVIT1* and *OsVIT2* modulate iron translocation between flag leaves and seeds in rice. Plant J 72(3):400–410

Zhang D, Zhong C, Smith NA, de Feyter R, Greaves IK, Swain SM, Zhang R, Wang M-B (2022) Nucleotide mismatches prevent intrinsic self-silencing of hpRNA transgenes to enhance RNAi stability in plants. Nat Commun 13:3926

Zhao JH, Guo HS (2022) RNA silencing: from discovery and elucidation to application and perspectives. J Integr Plant Biol 64:476–498

Zhu H, Li C, Gao C (2020) Applications of CRISPReCas in agriculture and plant biotechnology. Nat Rev Mol Cell Biol 21(11):661e677

Zia MAB, Yousaf MF, Asim A, Naeem M (2022) An overview of genome-wide association mapping studies in Poaceae species (model crops: wheat and rice). Mol Biol Rep 49:12077–12090

Agronomic Approaches to Zinc Biofortification in Soil and Soilless Crops

Pradip Poudel, Erin L. Connolly, and Francesco Di Gioia

Abstract

Hidden hunger is a pressing global issue, affecting approximately 2 billion people who suffer from malnutrition, including zinc (Zn) deficiency. As an essential micronutrient, Zn plays a vital role in various physiological functions, such as gene expression, cell division, and immune system development. Zinc deficiency can even lead to death, primarily due to Zn deficiency-related diarrhea, pneumonia, and malaria. Zinc deficiency can be attributed to several factors, with inadequate dietary intake and a predominantly cereal-based diet being the main causes. The biofortification of food crops with Zn represents a promising solution to alleviate this persistent problem. Among the various biofortification approaches, agronomic biofortification is considered one of the most practical ways to enhance Zn content in crops and ultimately increase the consumption of bioavailable Zn through the diet. This chapter aims to discuss the major reasons behind Zn deficiency, and the factors affecting Zn plant uptake, and provide an overview of alternative Zn agronomic biofortification techniques applicable to a range of crops grown in soil-based and soilless cultivation systems. Besides increasing Zn content, in most cases, Zn agronomic biofortification has the potential to enhance Zn bioaccessibility and, subsequently, bioavailability, showing the potential of this solution to fulfill the ultimate goal of reducing Zn deficiency in both high- and low-income countries. Additional research is needed to develop new biofortification strategies or combine different approaches and increase Zn content while minimizing undesired effects on the content of other minerals or the crop yield performance. The agronomic biofortification of soil and soilless crops can be an effective short-term strategy to address Zn deficiency, but the successful implementation of this solution depends on the careful

P. Poudel · E. L. Connolly · F. Di Gioia (✉)
Department of Plant Science, Pennsylvania State University, University Park, PA, USA
e-mail: fxd92@psu.edu

S. Srivastava, H. Kaur (eds.), *Zinc in Soil-Plant Continuum*,
https://doi.org/10.1007/978-981-96-4253-3_9

selection and optimization of the agronomic biofortification techniques for each target crop, considering environmental conditions, and resources available. Integrating agronomic biofortification with genetic biofortification approaches may offer the opportunity to increase the efficiency and long-term sustainability of Zn biofortification efforts.

Keywords

Agronomic biofortification · Zn bioavailability · Nutrient solution · Seed treatment · Foliar application · Nano-fertilizer · Biofertilizers

1 Introduction

Zinc is a micronutrient essential for human health as it is involved in different critical physiological functions including gene expression, cell division, and immune system development (Wessels et al. 2021). Zinc deficiency is a common malnutrition issue and, if severe, can even lead to death, determined primarily by diarrhea, pneumonia, and malaria. Zinc deficiency accounts for 14.4% of diarrhea deaths, 10.4% of malaria deaths, and 6.7% of pneumonia deaths among children between 6 months and 5 years of age in Latin America, Africa, and Asia (Walker et al. 2008). Zn deficiency affects nearly 17.1% of the global population (Wessells and Brown 2012). This problem is considerably more prevalent when viewed on a regional level; for instance, 20–40% of the population in Nepal, primarily children (6–59 months) and women, suffer from Zn insufficiency (Ministry of Health and Population, Nepal; New ERA; UNICEF; EU; USAID; and CDC 2018). In India, Zn insufficiency among adolescents aged 10 to 19 years ranged from 4% in Nagaland to 55% in Gujarat (Ministry of Health and Family Welfare (MoHFW), Government of India, UNICEF, and Population Council 2019). This issue is pervasive not only in developing regions but also in developed countries like North America and Europe. About 15% of the adult population in the United States consumes less Zn than the estimated average requirement (Reider et al. 2020). Industrialization and urbanization have contributed to changes in food production and consumption patterns, with a shift towards highly processed foods that are often low in Zn and other essential nutrients (Kruger et al. 2014).

Zinc deficiency is caused by several factors, among which inadequate dietary intake is the most significant factor, especially in low- and middle-income countries. Many people in low- and middle-income countries have limited access to Zn-rich foods like animal products, oysters, and shellfish due to economic, cultural, or religious reasons (Maxfield et al. 2023). Even when Zn intake is adequate, the bioavailability of Zn can be greatly reduced by the presence of antinutrients like phytic acid. Phytic acid forms insoluble complexes with Zn in the human digestive system, limiting its absorption. A diet rich in phytic acid may cause a deficiency of Zn, even if the Zn intake is adequate (Couzy et al. 1998). Conditions like pregnancy, rapid growth in children, and certain chronic diseases can increase the body's Zn requirements, leading to deficiency if dietary intake is not adjusted based on the body requirement. Additionally, conditions that impair intestinal integrity, such as diarrheal diseases, can result in increased endogenous losses of Zn, especially when dietary intake is already marginal (Maywald and Rink 2022).

Various strategies can be employed to mitigate Zn deficiency. These include dietary diversification, inclusion of foods abundant in Zn, food fortification, supplementation, and biofortification of food crops. One of the most sustainable solutions is dietary diversification. A significant number of Zn deficiency cases can be traced back to excessive dependence on cereal-based diets or staple food commodities, particularly in developing nations (De Groote et al. 2021; Gibson and Ferguson 1998). Although cereals are a good source of energy and minerals, including Zn, the Zn bioavailability is relatively low because of the phytic acid present in the grains. Increasing the consumption of nutrient-dense foods, like fruits and vegetables, which have a low content of phytic acid, can contribute to diversify the diet and enhance Zn intake and absorption. Food fortification is an additional strategy that has demonstrated potential. Zinc supplementation into frequently consumed foods such as flour, oil, or condiments can increase Zn intake, particularly among populations with limited dietary diversity. Consumer acceptance and consistent availability of fortified foods are, nevertheless, critical factors in determining the effectiveness of fortification programs (Brown 2007). In the short term, Zn supplementation can be beneficial, especially in addressing acute deficiencies that affect vulnerable populations such as pregnant women and small children (Berhe et al. 2019). Although Zn supplements can increase Zn levels promptly, they fail to target the root causes of deficiency and may not be sustainable in the long term. Biofortification of food crops is another promising strategy to increase Zn dietary intake. Biofortification refers to the systematic approach of increasing the nutritional content of edible crops via genetic engineering, conventional breeding, or agronomic methods. Each method has advantages and disadvantages. Developing a new biofortified crop through a transgenic approach in theory could be a very effective long-term solution and could be accomplished in a relatively short time; however, genetically modified organisms (GMOs) are highly regulated and even banned in many countries. Due to regulations and the long approval process, the development of transgenic crops requires significant investment and time, which may not be feasible in resource-constrained settings. Even in countries like the United States, where the cultivation and consumption of GMOs are legal, many consumers continue to be skeptical about the consumption of genetically engineered crops (Öz et al. 2018). Although a considerable research effort has been made and is still ongoing for the genetic biofortification of cereal crops, extensively grown worldwide, this approach is considered less suitable for most specialty crops like vegetables grown at smaller scale. Even when using a conventional breeding approach, the return on investment in resources and time needed to achieve tangible results could be substantially lower for vegetable crops, given the overall lower acreage investment compared to cereal crops. Developing nutrient-rich cultivars of a given crop through conventional breeding approaches could be a good long-term solution; however, the achievement of substantial improvements requires the availability of genetic resources that have a nutrient-rich profile, or the ability to accumulate higher levels of Zn in the specific case. It might require a long time for research, release, and adaptation of the improved cultivars. Yet, even assuming that an improved crop is capable of taking up and accumulating more Zn, the availability of Zn in the soil may still represent a limiting factor. On the other hand, employing agronomic methods for the biofortification of food crops can be a relatively simple short-term solution to address Zn malnutrition issues in different socioeconomic contexts, using genetic resources already available and accepted by consumers.

This chapter aims to discuss the major factors causing Zn deficiency at the global level, examine the factors affecting Zn plant uptake, and provide an overview of the alternative Zn agronomic biofortification techniques applicable to a range of crops grown in soil-based and soilless cultivation systems while discussing their advantages and disadvantages and future perspective.

We hypothesize that producing food crops enriched with Zn can substantially enhance dietary Zn intake and bioavailability. In areas characterized by a cereal-based diet, this strategy may be especially advantageous, as the increase in Zn content can compensate for the poor Zn bioavailability typical of cereals. This approach has the potential to enhance Zn intake without necessitating dietary modifications or supplementation. Beyond grain staple crops, different categories of biofortified vegetables may provide a new opportunity to address Zn deficiency in different socioeconomic contexts using primarily agronomic biofortification techniques. The overview of the alternative agronomic biofortification methods provided in the following sections should assist in the selection of the optimal biofortification approach for different regions and crops, while also considering consumers' preferences, socioeconomic aspects, efficiency, and long-term sustainability.

2 Soil Zinc Availability and Plant Uptake

The availability of Zn in the soil is a critical factor in determining crop Zn uptake and accumulation and ultimately affects the level of Zn intake of the human population in a given region (Broadley et al. 2007; De Groote et al. 2021). Zinc is present in the soil in various forms, including adsorbed, exchangeable, water-soluble, complexed, and chelated forms (Gupta et al. 2016; Lindsay 1972). The bioavailability of these different Zn pools can vary significantly. Zinc can be adsorbed to soil particles, such as clay minerals and organic matter. Adsorbed Zn is generally less available for plant uptake, as it is tightly bound to the soil matrix. Exchangeable Zn is held on the cation exchange sites of soil particles and is more readily available for plant uptake. However, the availability of exchangeable Zn is influenced by soil pH and by the presence of competing cations. Zinc can be chelated by organic compounds, such as synthetic chelating agents (e.g., EDTA) or natural chelators (e.g., Phytosiderophores), and chelated Zn is generally more available for plant uptake, as the chelation helps to keep Zn in a soluble and plant-accessible form (Lindsay 1972; Northover et al. 2020). Zinc is also complexed by organic compounds such as humic and fulvic acids, or inorganic ligands present in the soil. The availability of complexed Zn for plant uptake depends on the stability and solubility of the Zn complex. The most readily available form of Zn is water-soluble Zn.

At high pH (>7), Zn forms insoluble complexes and is not readily available for absorption, while lower soil pH (5–7) increases the availability of Zn (McBride and Blasiak 1979; Sims 1986). Soil organic matter can adsorb and retain Zn, increasing its availability for plants (Maqueda et al. 2015). Finer-textured soils containing more clay and silt generally have more Zn adsorption sites and may provide more Zn compared to coarser and sandier soils. Soil redox potential is another factor influencing Zn solubility and availability for plants. Prolonged flooding or low redox potential can decrease Zn availability by promoting the formation of insoluble Zn compounds (Wissuwa et al. 2006). Soil moisture content can influence Zn availability affecting its mobility in the soil, with drier soils potentially having lower Zn availability (Labanauskas et al. 1968). The presence of other divalent cations like Fe, Cd, and Mn can limit Zn uptake as these cations compete for the same plant transporter proteins (Connolly et al. 2002; Eide et al. 1996; Grotz et al. 1998). A high level of P in the soil or nutrient solution can also lead to the formation of insoluble compounds like Zn phosphate [$Zn_3(PO_4)_2$] decreasing Zn availability for absorption (Sánchez-Rodríguez et al. 2017). Soil with higher cation exchange capacity (CEC) can retain more cations like Zn^{2+} and slowly release them, increasing the availability for plant absorption in the rhizosphere (Sumner and Miller 1996). Similarly, microbial communities present in the rhizosphere could influence Zn uptake directly and indirectly via various mechanisms including acidification, siderophore production, chelation, phytohormone production, and other mechanisms (Agnolucci et al. 2020; Kaur et al. 2020).

Plant roots typically absorb Zn in the form of divalent cation Zn^{2+} (Cakmak et al. 2023; Epstein and Bloom 2004). Nevertheless, under some circumstances, plant roots can also assimilate organic ligand-Zn complexes (Suzuki et al. 2006). Plants employ both active and passive transport pathways to assist the absorption and distribution of Zn in different organs. Plant roots primarily acquire Zn via carrier proteins (as opposed to channel proteins), which involves specific transporter proteins facilitating the entry of Zn^{2+} ions into the root cells. The transporters mostly involved in the abovementioned process mainly belong to the ZIP (ZRT-IRT-like Protein) family; however, the ZIP transporters also transport other divalent metal cations such as Fe^{+2}, Mn^{2+}, Cu^{2+}, and Cd^{2+} (Connolly et al. 2002; Potocki et al. 2014). Other transporter families like heavy metal ATPase (HMA), metal tolerance protein (MTP), natural resistance-associated macrophage protein (NRAMP), and the vacuolar iron transporter (VIT)-like family can help in the translocation, storage, and overall homeostasis of the divalent cations including Zn within the plant system and could play an important role in Zn biofortification (Gupta et al. 2016; Kaur and Garg 2021; Ricachenevsky et al. 2013). Plants can also take up Zn in the form of organic ligand-Zn complexes (Arnold et al. 2010; Ptashnyk et al. 2011; Suzuki et al. 2006; Widodo et al. 2010). Zinc is carried from the roots to the shoots mostly through the xylem. Zinc can be translocated within plants in the form of free Zn^{2+} ions, and as Zn-chelate complexes, including Zn-nicotianamine (Zn-NA), or Zn-organic acid complexes (Deinlein et al. 2012). Zinc is also delivered to different plant portions through the phloem in the form of Zn-chelate complexes, such as Zn-cysteine (Epstein and Bloom 2004; Lindsay 1972; Stanton et al. 2022).

3 Agronomic Zinc Biofortification as a Potential Solution

Agronomic biofortification is a promising approach aimed at improving the nutritional value of food crops by increasing their content of specific nutrients by employing different agronomic strategies (Agnolucci et al. 2020; Marques et al. 2021). Agronomic biofortification allows to: (i) directly increase Zn content in the edible portion of a variety of crops, (ii) enrich cultivars and species already adopted and preferred by consumers, and (iii) consider the most efficient and cost-effective solution to increase the nutritional profile of different crops in relatively short time. Although agronomic biofortification can target a specific nutrient like Zn, this technique allows the enhancement of the content of multiple micronutrients and bioactive compounds at the same time. Such versatility makes agronomic biofortification a valuable solution that can be employed to produce biofortified crops in a wide range of scenarios, including under emergency situations when immediate action should be taken to prevent the emergence of severe malnutrition issues (Bhardwaj et al. 2022; Di Gioia et al. 2021). The possibility to rapidly and efficiently enhance the nutritional profile of existing crop varieties through the application of fertilizers or nutrient solutions is a key strength of the agronomic biofortification approach. The potential opportunity to address the micronutrient deficiency of a population, without the need for extensive breeding programs or the development of GMO crops, which can face regulatory and acceptance challenges in many countries is a critical advantage of the agronomic biofortification method when it comes to implementation and consumer acceptance (Bartsch 2014; Öz et al. 2018). The main disadvantage of this approach is the need to reapply the biofortification treatment every time.

Zinc agronomic biofortification can be achieved through various methods, such as the application of Zn fertilizers in soil or soilless systems, through seed nutri-priming, foliar Zn fertilizer applications, and employing nano-fertilizers or Zn solubilizing rhizobacteria (Fig. 1). Hereafter the most common agronomic Zn biofortification approaches are described summarizing the results of the published literature.

4 Agronomic Zinc Biofortification Approaches

4.1 Seed Treatment

The most common seed treatment techniques used for Zn biofortification purposes are seed nutrient priming and seed coating. Soaking the seed in a nutrient solution containing macro- or micronutrients is known as seed nutri-priming. Instead, seed coating is implemented by coating the seed surface with a fine layer of fertilizer

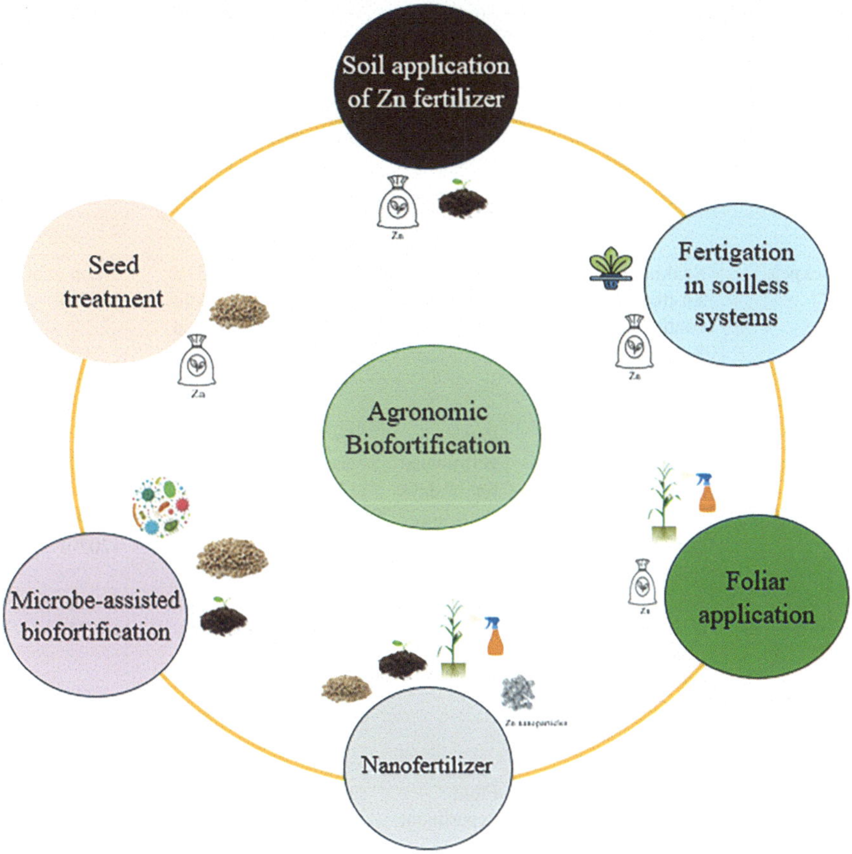

Fig. 1 Agronomic biofortification approaches that can be used for Zn biofortification

with the help of adhesive materials (Veena and Puthur 2022). Seed nutrient priming enhances seed vitality, germination, seedling growth, and the nutrient content of plants (Farooq et al. 2018). Recent studies have proven that very simple and easy-to-use techniques like seed nutri-priming could play a significant role in enriching plants with Zn (Table 1). With the nutrient priming technique, soaking seeds in a Zn-enriched solution facilitates some of the physiological changes happening during the germination process, and being Zn a cofactor of many enzymes involved in RNA expression and cell division processes, the supply of Zn can enhance seedling establishment, seed germination vigor, and tolerance to stress conditions like salinity or drought stress (Wei et al. 2023). Zinc absorbed by seeds soaked in Zn-enriched solutions is stored in different parts of the seed, and post-germination can easily translocate to the seedling and to edible plant portions (Imran et al. 2017). For this reason, seed Zn-priming is more efficient when applied to crops consumed at an early growth stage like sprouts and microgreens. Instead, at a more advanced growth stage, crop Zn concentration may decrease as a consequence of a dilution effect

Table 1 Main findings of representative studies focused on food crop Zn biofortification through the seed nutrient priming technique

Crop	Treatment	Growing condition	Results	References
Broccoli, pea, and sunflower sprouts	Seed soaked in 10–30 μg/mL for 24 h	Germinated in the paper	Increased Zn content in broccoli, sunflower, and peas 25-, 1.2-, and four-fold compared to the control	Bączek-Kwinta et al. (2020)
Chickpea	Soaking in 0.001 M Zn from $ZnSO_4$ and coating seed with 5 mg Zn per kg seed, $ZnSO_4$ in gum Arabic	Field experiment	Increased Zn by 31% through seed soaking and by 64% via seed coating	Ullah et al. (2020)
Maize seedling	Seed soaked in 4 mM Zn for 24 h from $ZnSO_4$	Paper role germinated for 10 days	Increased Zn content in seedlings by 65%	Imran et al. (2017)
Mung bean	Soaked seed in 0.05 M $ZnSO_4$ for 6 h	Soils in pot	Increased Zn content by 60%	Haider et al. (2020)
Pea microgreens	Soaked seed overnight in 25, 50, 100, and 200 mg/L Zn solution from $ZnSO_4$, ZnO nanoparticle, and Zn-EDTA	Peat perlites mix in the greenhouse	200 mg/L of Zn from $ZnSO_4$ increased Zn content by 126%	Poudel et al. (2023)
Rice	Seed soaked in 0.5 M $ZnSO_4$ for 24 h	Field experiment	Increased grain Zn concentration by 15–70% depending on growing site and year	Farooq et al. (2018)
Rice	Seed priming in 0.25 M $ZnSO_4$ for 16 h	Field experiment	Increased grain Zn concentration by 80%	Zulfiqar et al. (2021)
Sunflower microgreens	Soaked seed overnight in 25, 50, 100, and 200 mg/L Zn solution from $ZnSO_4$, ZnO nanoparticle, and Zn-EDTA	Peat perlites mix in the greenhouse	200 mg/L of $ZnSO_4$ increased Zn content by 230%	Poudel et al. (2023)
Wheat	Seed coating 250 mg Zn per kg of seed	Field experiment	Increased grain Zn content by 61% in Faisalabad-2008 but no effect in Lasani-2008, suggesting a variety of effect	Hassan et al. (2019)

associated with the increase in crop biomass. Nevertheless, seed Zn-priming has been found to increase Zn content also in grain crops like rice (Farooq et al. 2018; Zulfiqar et al. 2021), wheat (Hassan et al. 2019), chickpeas (Ullah et al. 2020), and mung bean (Haider et al. 2020). Like Zn nutri-priming, seed coating with Zn fertilizer also increased Zn contained in grain crops like chickpeas. Ullah et al. (2020) reported almost a 64% increase in Zn content in chickpeas when seeds were coated with $ZnSO_4$ at the rate of 5 mg Zn per kg of seeds with the help of gum Arabic as adhesive material. However, the seed coating technique is less suitable for crops like sprouts and microgreens for which the seed integuments can be part of the edible portion.

While being relatively simple to implement, different factors like sources, concentration rate, soaking time, plant genotype, growing environment, and soil physical and chemical properties affect the efficiency of seeds Zn-priming (Farooq et al. 2018; Poudel et al. 2023). Poudel et al. (2023) reported a higher efficiency of Zn accumulation in different microgreen species when $ZnSO_4$ was used as a source of Zn compared to Zn-EDTA and ZnO nanoparticles. Most of the studies reported a positive relationship between Zn concentration and Zn enrichment; however, the rate of enrichment differs by species (Bączek-Kwinta et al. 2020; Poudel et al. 2023). Attention should be given to the level of Zn used to avoid an excess of Zn causing negative effects on plant growth and nutritional quality and to ensure that the level of Zn accumulated is within the safe range for human consumption. In the study by Bączek-Kwinta et al. (2020), Zn enrichment in broccoli sprouts soaked in 30 µg/mL of $ZnSO_4$ determined a 25 times increase in Zn content compared to the untreated control; however, the calculated hazard quotient (estimated daily intake/oral reference dose value) suggested a potential negative effect for human health due to the risk of Zn intake beyond recommended daily levels. A side effect of increasing crop Zn content may be the reduced content of other micronutrients like Fe (Poudel et al. 2023). Minerals characterized by similar charge and atomic size compete with each other for plant uptake as they share transporter proteins in the root regions and for translocation within plants (Connolly et al. 2002; Eide et al. 1996). Nevertheless, the accumulation of Zn and the effect on other minerals are modulated by the plant genotype. For example, Poudel et al. (2023) observed a higher accumulation of Zn in sunflower microgreens compared to pea shoots. Moreover, the accumulation of Zn in sunflower microgreens had a very low effect on yield parameters and Fe content, while in peas the accumulation of Zn reduced yield and Fe content.

4.2 Zinc Fertilizer Application in Soil Crops

A common approach for the biofortification of soil-grown crops is the application of fertilizer containing essential minerals like Zn. Supplementary Zn fertilization is particularly important in regions with soil characterized by relatively low Zn levels, which have been associated with Zn deficiency in the population (De Groote et al. 2021; Morton et al. 2023). This approach has been proposed primarily for grain crops like cereals and legumes, considered staple crops in several developing countries with populations affected by Zn deficiency. Besides increasing Zn content in the grain crops, Zn biofortification can reduce phytic acid content and the phytic acid/Zn molar ratio, thus enhancing Zn bioavailability in biofortified crops (Saha et al. 2017; Silva et al. 2021). Many studies have proven the efficacy of Zn fertilization in increasing Zn content in a range of soil-grown crops (Table 2). However, the efficiency of soil Zn fertilization is influenced by multiple agronomic and environmental factors, and in the long term, an excessive application of Zn fertilizer can lead to soil and water pollution with consequent risks to the environment and human health (Marques et al. 2021).

The soil's physical and chemical properties play an important role in determining the efficacy of Zn fertilizer applications for biofortification purposes (de Almeida et al. 2020; Almendros et al. 2015; Farooq et al. 2018). Variations of Zn accumulation observed across different regions and growing areas are associated primarily to soil properties like soil texture, soil pH and calcium carbonate content, soil organic matter content, soil Zn level, and the content of other minerals (Alloway 2009; Farooq et al. 2018; Zou et al. 2012). Soil pH is a critical factor, considering that Zn is more soluble and bioavailable for plant uptake at low (acidic) pH levels, and its availability is substantially decreased with increasing soil pH above 7. Soil characterized by fine soil texture containing clay and silt increases Zn adsorption and can reduce Zn availability compared to coarser and sandier soils. Soil organic matter can enhance Zn solubility and availability for plants and has been positively correlated to plant Zn accumulation (Cakmak 2008). Soil characterized by high calcium carbonates ($CaCO_3$), high phosphorous (P), high Fe and aluminum (Al) oxides, or low soil moisture generally decreases the availability of Zn for plants and can reduce the efficacy of agronomic biofortification efforts (Cakmak 2008; Moreno-Lora and Delgado 2020). If soil conditions limit the efficacy of Zn fertilization and are not easy to address, alternative approaches like foliar Zn applications should be considered.

Besides the soil physicochemical properties, other agronomic factors like the form of Zn fertilizer and application rate and the crop genotype play a critical role in determining the success of Zn biofortification in soil crops. The source and rate of Zn fertilizer applied are primary factors affecting plant Zn uptake (Almendros et al. 2015; Du et al. 2019). Du et al. (2019) reported higher Zn content in wheat grain when using ZnO nanoparticles compared to the $ZnSO_4$ as a Zn source. The optimal Zn source and application rate may be influenced by soil properties. For instance, Zn-amino lignosulfonate (Zn-AML) was the best source of fertilizer in acidic soil, while Zn-DTPA-HEDTA-EDTA was the best source of Zn for the biofortification of onion in calcareous soil (Almendros et al. 2015). When it comes to Zn application rate, a major concern about soil application of Zn fertilizer is the possible accumulation of Zn over time which could pose risks for the environment and human health. Liu et al. (2020) assessed the potential health risk of repeated Zn application over 8 years on wheat and found that applications of Zn over 11.4 kg/ha resulted in grain Zn concentration > 50 mg/kg which is higher than the US-EPA standard, suggesting that the long-term application of Zn could pose a risk for human health. However, the same authors reported that a Zn application rate below 5.7 kg/ha can fulfill the purpose of Zn biofortification without increasing Zn and other heavy metals and thus not posing a health risk for consumers even if applying Zn fertilizer for 8 years.

Table 2 Main findings of representative studies focused on food crop Zn biofortification through soil application of Zn fertilizer

Crop	Treatment	Results	References
Cabbage	22.7 kg/ha of Zn, $ZnSO_4$, sprayed in soil-making solution and later mixed in topsoil before transplanting	Increased Zn content by 205%	Mao et al. (2014)
Canola	22.7 kg/ha of Zn, $ZnSO_4$, sprayed in soil-making solution and later mixed in topsoil before transplanting	Increased Zn content by 25%	Mao et al. (2014)
Cowpea	25 mg/dm^3, $ZnSO_4$, after 45 days after emergence	Increased Zn content 0–100% in grain based on genotype, Zn increment reduced phytic acid and phytic acid/Zn molar ratio	Silva et al. (2021)
Lettuce	5–30 mg Zn/kg soil, $ZnSO_4$, before planting in three lettuce genotypes and two soil type	Genotype, soil, and concentration interactively influenced Zn accumulation, increased Zn accumulation in shoot up to 600%	Almeida et al. (2020)
Onion	0, 5 and 10 mg/kg of soil of eight alternative Zn sources applied in acidic and calcareous soils: Zn-amino lignosulfonate (Zn-AML), Zn-EDDHSA (Zn-ethylenediamine-di-(2-hydroxy-5sulfophenylacetate), Zn-EDTA (Zn-ethylenediaminetetraacetate), Zn-polyhydroxyphenylcarboxylate (Zn-PHP), Zn-HEDTA (Zn-N-2hydroxyethyl-ethylenediaminetriacetate), Zn-EDTA-HEDTA, ZnDTPA-HEDTA-EDTA (Zn-DTPA, Zn-diethylenetriaminepentaacetate) and Zn-EDDS (Zn-ethylenediamine disuccinate)	Total Zn and soluble Zn in onion leaves were affected by soil and fertilizer interactively; Zn content in leaves was the highest with 10 mg/kg Zn soil application	Almendros et al. (2015)
Rice	10 kg/ha of Zn applied at the bed preparation stage with $ZnSO_4$	Grain Zn concentration increased by 40–70% depending on location and growing year	Farooq et al. (2018)

(continued)

Table 2 (continued)

Crop	Treatment	Results	References
Rice	Six cultivars, soil, and foliar application; 5 kg Zn/ha in field preparation phase	Grain Zn increased by 11% to 44% depending on the cultivar; phytic acid and Fe contained in grain decreased with increasing Zn content	Saha et al. (2017)
Rice	10 kg/ha of Zn applied at the bed preparation stage with $ZnSO_4$	Grain Zn concentration increased by 60%	Zulfiqar et al. (2021)
Wheat	0, 10, 20, 50, 100, 200, 1000 mg/kg Zn from ZnO-nanoparticle and $ZnSO_4$	Zn content increased with increasing Zn application rate, and ZnO nanoparticle fertilizer was more effective	Du et al. (2019)
Wheat	50 kg/ha of $ZnSO_4$	Increased Zn content by 12% across many regions, with no effect on yield except for Pakistan	Zou et al. (2012)
Tomato	0, 2, 3, 4, 5, and 6 kg/ha; 100% seedling priming, 50% soil þ 50% foliar spray at seedling stage, 50% seedling priming þ 50% foliar spray, 100% foliar spray at seedling stage, 100% foliar spray at flowering stage, and 100% foliar spray at the fruiting stage	5 kg Zn/ha as 50% soil application +50% foliar spray at seedling stage resulted in higher Zn content in fruits	Rabbi et al. (2024)

While influenced by soil, agronomic, and environmental factors, the Zn accumulation potential is modulated by the plant genotype and varies among plant species and cultivars of the same species leading to different responses to Zn supplementation as observed by Saha et al. (2017) and Silva et al. (2021) in rice and cowpeas, respectively. These results suggest the importance of genotype selection for the successful Zn biofortification of soil-grown crops.

4.3 Zinc Fertilizer Application in Soilless Crops

An alternative to the biofortification of soil crops is the use of Zn-enriched nutrient solution for the biofortification of soilless crops. Compared to soil, soilless growing systems allow precise control of crop nutrition, resulting in consistent and predictable uptake, translocation, and accumulation of nutrients in plants (Marques et al. 2021). Excluding the complex interaction effect of all the soil physicochemical properties, the agronomic biofortification of soilless crops is potentially more efficient, allows for standardization of the agronomic biofortification process, and can minimize the risks of nutrient loss and environmental pollution (Di Gioia et al. 2019; Marques et al. 2021).

The easy manipulation of the nutrient solution allows to increase or lower the content of specific minerals to obtain crops with a tailored mineral profile (Kyriacou et al. 2016), and several studies have demonstrated the efficacy of the biofortification with Zn of a variety of soilless crops through the simple manipulation of the nutrient solution (Table 3). While the biofortification target in soil growing systems is primarily grain staple crops grown on large acreage, soilless growing systems are typically used to grow specialty crops like vegetables, which have different physiology.

Besides the Zn concentration, the source of Zn can affect plant Zn uptake, and for example, comparing Zn-EDTA, ZnO nanoparticles, and ZnSO4 applied via fertigation as a Zn source, the latter resulted in a higher accumulation of Zn in pea, radish, and sunflower microgreens (Poudel et al. 2024). However, working on stevia, Velázquez Gamboa et al. (2021) observed a higher efficacy of ZnO nanoparticles over ZnSO4 as a Zn source using plant-derived nanoparticles. Like for other agronomic

Table 3 Main findings of representative studies focused on Zn biofortification through Zn-enriched nutrient solutions in soilless food crops

Crop	Treatment	Growing condition	Results	References
Basil	12.5, 25, 37.5, and 50 μM Zn	DWC	Zn content increased with Zn application rate (up to 1.4 times), but yield decreased in all treatments	Ciriello et al. (2022)
Brassica microgreens	5, 10, and 15 mg/L of Zn	NFT growing channel	Zn content increased up to 23 times, with an adverse effect on yield parameters at 15 mg/L Zn	Di Gioia et al. (2019)
Cabbage	10, 20, 80, 40, 80, and 100 mg/L Zn	DWC	Zn concentration increased with Zn applied (up to 8 times), but yield decreased above 40 mg/L	Barrameda-Medina et al. (2017a, b)
Lettuce	10, 20, 80, 40, 80, and 100 μM Zn	DWC	Zn content increased with Zn applied (up to 3 times) without affecting yield	Barrameda-Medina et al. (2017a, b)
Peas, radish, and sunflower microgreens	5, 10, and 15 mg/L of Zn from $ZnSO_4$, Zn-EDTA and ZnO nanoparticles	NFT growing channel	Zn content increased up to 5–13 times depending on the species and Zn application rate. $ZnSO_4$ was the most efficient source followed by ZnO and Zn-EDTA. Radish accumulated more Zn followed by sunflower and peas	Poudel et al. (2024)
Rocket and purslane	0.13 and 5.2 mg/L of Zn	DWC	Zn content increased 2–3 times depending on the species	D'Imperio et al. (2022)
Stevia	50, 75, and 100 mg/L Zn	Polyethylene bags filled with moss/agrolite	Zn content increased up to 14 times with Zn applied without affecting yield	Velázquez-Gamboa et al. (2021)

biofortification approaches, also in soilless growing systems, the crop genotype plays a critical role in determining the efficiency of the Zn biofortification process as reported by D'imperio et al. (2022) and Poudel et al. (2024). For instance, Poudel et al. (2024) observed that while exposed to the same level and source of Zn, radish microgreens accumulated higher levels of Zn compared to sunflower and pea microgreens. Similarly, comparing three Brassicaceae species, Di Gioia et al. (2019) observed a different capacity to accumulate Zn between red cabbage, red mustard, and arugula microgreens. Considering the variability of the mineral profile across species of different botanical families, and even within the same botanical family, the selection of the target biofortification crop is critical and may allow the use of hyperaccumulator species as well as the selection of species with a more balanced mineral profile (Di Gioia et al. 2023). While considering the overall Zn accumulation capacity, the presence of anti-nutritional factors and the Zn bioaccessibility should be accounted for in the selection of the target biofortification crop. As observed for other Zn biofortified crops, D'imperio et al. (2022) reported an increased Zn bioaccessibility in biofortified arugula and purslane, suggesting that the Zn biofortification process per se can enhance the Zn bioaccessibility also in soilless leafy vegetables.

Although the Zn biofortification of soilless crops can be very efficient, a challenge of this approach is the effect that the accumulation of Zn can have on the uptake of other essential minerals like Fe (Di Gioia et al. 2019; Poudel et al. 2024). Supplementing the nutrient solution with 20 mg/L of Zn, Di Gioia et al. (2019) observed a 32.7% decrease in Fe content in brassica microgreens. Such outcome is a consequence of the antagonism between Zn and Fe, which have similar ionic charge and atomic size compete for plant uptake at the root level and for the same transporter proteins (Kaur and Garg 2021; Rietra et al. 2017). Depending on the specific biofortification objective, the source and level of Zn applied through the nutrient solution may be defined to maximize Zn accumulation or to increase Zn content without reducing the content of other essential minerals.

4.4 Foliar Zinc Fertilizer Application

Although roots are the primary organ devoted to absorbing minerals in plants, nutrient uptake also occurs in leaves through the cuticle and stomata openings (Eichert 2013). Crop nutrition through foliar fertilizer application can be used as a biofortification strategy, especially in the presence of factors that limit nutrient uptake through the roots (Fernández et al. 2021). For instance, in the presence of soil characterized by high pH, high carbonate, and low organic matter content limiting the solubility and availability of microminerals like Zn, foliar Zn applications may be an alternative approach to increasing crop Zn content (Alshaal and El-Ramady 2017; Lyons and Cakmak 2012), and in some cases, foliar and soil Zn fertilizer applications can be combined for effective crop Zn enrichment (Nayak et al. 2023). The application of nutrients via foliar spraying is better suited for micro minerals, which are required in relatively low concentrations and whose availability is limited more by soil physicochemical properties (Alshaal and El-Ramady 2017; Eichert 2013).

The potential of enriching crops with Zn through foliar fertilizer applications has been demonstrated for a variety of crops including staple crops as well as vegetables

Table 4 Main findings of representative studies focused on food crop Zn biofortification through foliar Zn fertilizer applications

Crop	Treatment	Growing condition	Results	References
Arugula	0.5, 1, and 1.5 kg/ha Zn foliar sprayed 5 times after 15 days of emergence at 5 days interval	Field experiment	Zn concentration increased with application rate and was the highest when spraying at 25 days after emergence	Rugeles-Reyes et al. (2019)
Cauliflower	0, 25, 50, 75, and 100 μM ZnO applied after curd formation	Field experiment	Zn content increased with Zn application rate	Raza et al. (2022)
Lettuce	0, 25, 50, 75, and 100 μM Zn, every 15 days after transplanting	NFT system	Zn content increased with Zn applied, but nutritional quality and yield were best at 75 μM Zn	Preciado-Rangel et al. (2021)
Maize	Foliar application of ZnO nanoparticle, $ZnSO_4$, and Zn-CNP mix with urea; 3 times every 7–10 days from the tasseling stage	Field experiment	Grain Zn content increased up to 36–134% when using urea+ZnONPs, urea+ZnCNPs, urea+$ZnSO_4$ in different varieties	Xue et al. (2023)
Rice	0.5% Zn solution sprayed at the peak tillering stage	Field experiment	Grain Zn concentration increased by 20–70% depending on location and growing year	Farooq et al. (2018)
Rice	0.5% Zn solution sprayed at the booting stage with $ZnSO_4$	Field experiment	Increased grain Zn content by 52%	Zulfiqar et al. (2021)
Rice	0.5% $ZnSO_4$ twice at maximum tillering and flowering stage	Field experiment	Grain Zn increased by 27.4–92.6%	Saha et al. (2017)
Wheat	0.5% (w/v) of aqueous solution of $ZnSO_4{\cdot}7H_2O$ with generally 600–800 L per ha, 2 times heading and milk stage	Field experiment	Grain Zn concentration increased by 17–260% based on location, and growing year	Zou et al. (2012)
Wheat	0.1–0.5% Zn from $ZnSO_4$, 2 times, when seedlings had 4–6 weeks and after 14 days	Pot experiment with field soil	Grain Zn content almost doubled while reducing cd content by 60%.	Qian et al. (2023)

(Table 4). The key factors affecting foliar nutrient absorption are the source and concentration of Zn solution used and the permeability of the leaf surface, which is influenced by the plant genotype, leaf morphologic traits, and growth stage as well as by environmental conditions such as the relative humidity, evapotranspiration demand, and temperature (Eichert 2013; Fernández et al. 2021; Oosterhuis 2009).

Most of the studies reported an increase in Zn accumulation with increasing the Zn application rate; however, also in this case, defining the Zn application rate is critical to achieve an adequate Zn accumulation level without causing phytotoxicity damage on the leaves or negatively impacting the plant growth and yield. The source of Zn is another factor influencing the absorption and accumulation of Zn in cereal crops through foliar application. Comparing foliar applications of $ZnSO_4$ and Zn-EDTA in wheat leaves, Doolette et al. (2018) observed that while both forms of Zn were absorbed in the leaves, Zn-EDTA was less mobile than $ZnSO_4$. However, the effects of alternative Zn sources may be different for different species and environmental conditions. Shivay et al. (2015) reported higher Zn concentrations in chickpea (*Cicer arietinum* L.) when using Zn-EDTA for foliar applications compared to the use of $ZnSO_4$. Moreover, Shivay et al. (2015) reported an increase in Zn content of the chickpea with increasing the Zn foliar application frequency, considering that three applications resulted in higher Zn content compared to a single or double Zn application. The timing of the Zn application is a critical factor considering that the leaf permeability and the Zn accumulation capacity may change over different phenological stages. Increasing the Zn content or Zn pool in the vegetative tissue during the grain-filling period can lead to higher Zn content in the grains, particularly in the endosperm (Cakmak et al. 2010). This is an important consideration, as the endosperm is the predominant grain fraction consumed in many countries for grain crops like wheat and rice, which could also avoid the loss of Zn associated with grain processing. In fact, during the grain milling process, the aleurone layer and embryo are often removed, leading to a significant loss of Zn content in the final refined milled products (Singh et al. 2023). Increasing Zn content in the endosperm can therefore increase Zn absorption in populations with a cereal-based diet. The foliar application of Zn can also decrease the Zn: phytic acid molar ratio increasing the overall Zn bioaccessibility (Nayak et al. 2023; Xue et al. 2023). The foliar application of Zn also has the potential to limit the uptake and accumulation of heavy metals. For instance, the foliar application of Zn at 0.5% Zn two times (when seedlings had 4–6 leaves and after 14 days) reduced the grain Cd content in wheat grown in Cd-contaminated soil (Qian et al. 2023).

Although Zn foliar application is most commonly proposed for the biofortification of grain crops, it has been successfully applied also for the biofortification of vegetable crops like arugula, cauliflower, and lettuce (Table 4). A major concern associated with the foliar application of Zn to vegetables is the potential fertilizer residue left on the surface of the leaves which may require washing the product before it is marketed. Moreover, foliar fertilizer applications may require the use of adhesive materials to improve the efficiency of the treatment, and the use of food-grade adhesive or surfactant materials is recommended to minimize any food safety concerns.

Although the efficacy of agronomic biofortification through foliar Zn applications has been proven, there are several limitations and disadvantages associated with this approach. Environmental conditions, such as relative humidity and wind speed, have been shown to significantly impact nutrient absorption through leaves (Eichert 2013; Fageria et al. 2009; Fernández et al. 2021). Relative humidity plays a crucial role in determining the concentration gradient and permeability of the leaf membrane, influencing nutrient uptake (Eichert 2013; Fernández et al. 2021). Low relative humidity can lead to increased evaporation, which may result in the crystallization of applied nutrients on the leaf surface (Fernández et al. 2021). To avoid this issue, it is recommended to carry out foliar applications early in the morning or late in the evening when the air relative humidity is higher (Fernández et al. 2021). The efficiency of foliar fertilizer applications is also dependent on the leaf surface expansion and leaf area index (Fageria et al. 2009). A short contact time between the nutrient solution and the leaf surface is another concern, and using adjuvants or surfactants may help increase the contact time, thereby enhancing nutrient absorption (Oosterhuis 2009). On the other hand, it is critical to avoid applying solutions with nutrient concentrations that are too high and that can be phytotoxic for the crop, which may cause leaf burning (Fageria et al. 2009; Oosterhuis 2009; Niu et al. 2021). Fertilizer residues remaining on the surface of the leaves can be another challenge especially for edible leafy vegetables often consumed raw.

4.5 Use of Zinc Nano-Fertilizers

The rapid advancement of nanoscience and nanotechnologies capable of manipulating structures and molecules at the scale of nanometers ranging from 1 to 100 nm is leading to innovations and new applications in every field, including agriculture. Among the different agricultural applications of nanotechnologies, the use of nano-fertilizers characterized by nanometer-size particles has the potential to improve the delivery of nutrients to plants and can play a key role in improving the sustainability of current crop production systems by increasing nutrient use efficiency and enhancing crop growth and yield along with quality (Pande and Arora 2019; Rana et al. 2021; Shweta et al. 2021). The small particle size and high surface-to-volume ratio of nano-fertilizers allow nutrients to quickly permeate into plant roots or leaf cells (F. Zulfiqar et al. 2019). Given their enhanced properties, nano-fertilizers have the potential to be used also for agronomic biofortification purposes, and several studies have proven their efficacy in increasing plant Zn content more efficiently than standard soluble fertilizers (Table 5). For example, Umar et al. (2021) reported higher Zn content in maize grain with the foliar application of ZnO nanoparticles compared to soluble ZnO and $ZnSO_4$.

Nano-fertilizer can be applied directly in the soil or in soilless systems for root absorption, or can be applied through foliar applications. Sharifan et al. (2019) and Zoufan et al. (2020) reported increases in Zn content almost 5–29 times higher with the application of ZnO nanoparticles via nutrient solution in lettuce and Australian spinach (*Chenopodiastrum murale*), respectively. Besides increasing Zn content, Sharifan et al. (2019) also observed an increase in Fe content and a reduction of Cd in lettuce supplemented with ZnO nanoparticles. Such results are particularly interesting because normally Zn and Fe are competing for each other, and supplementing Zn, Fe absorption is expected to decrease, but both increased in this case. ZnO nanoparticles can also be applied to soilless peat-based media as reported by

Table 5 Main findings of representative studies focused on food crop Zn biofortification employing Zn nano-fertilizers

Crop	Treatments	Growing condition	Results	References
Chickpea	ZnO nanoparticle 0.5% at the pre-flowering stage	Field experiment	Increased Zn content by 16% compared to control	Dhaliwal et al. (2021)
Lettuce	ZnO nanoparticles were applied along with heavy metals; 100 mg/L of Zn	Hydroponics; in tube with treatment solution	Increased Zn content in shoot 7–29 times; reduced the Cd uptake by 30% while increasing Pb uptake by 44% in the shoot, Fe content also increased	Sharifan et al. (2019)
Maize	ZnO nanoparticles: Foliar applied 50, 100, 200, 400, 600, 800, 1000, 1500, and 2000 mg/L; 2 times tasselling and milking stage	Field experiment	Increased Zn content by 13–126%	Subbaiah et al. (2016)
Maize	ZnO nanoparticle, ZnO salts, $ZnSO_4$; soil applied at 8 kg Zn/ha, foliar applied at 2% at 6–9 leaf stage	Field experiment	ZnO nanoparticle foliar spray increased Zn content by 87% compared to the control and was higher than other salts and soil applications	Umar et al. (2021)
Maize	40, 80, 120, and 160 mg/kg Zn soil applied: 10, 20, 30, and 40 mg/L of Zn foliar applied	Field experiment	80 mg/kg Zn soil application and 30 mg/L of Zn foliar application resulted in higher yield, photosynthetic pigments, and nutritional quality	Azam et al. (2022)
Nettle-leaved goosefoot	ZnO nanoparticles; 0, 10, 50, and 250 mg/L	Hydroponics in 2.5 L vessels with aeration, similar to DWC	Shoot Zn content increased about 5–15 times	Zoufan et al. (2020)
Potato	ZnO nanoparticle foliar sprayed at the tuber initiation stage, 0, 25, 50, 100, and 250 mg/L	Field experiment	Tuber Zn content increased 15–27% with increasing Zn application rate	Nazir et al. (2024)

(continued)

Table 5 (continued)

Crop	Treatments	Growing condition	Results	References
Rice	ZnO nanoparticle foliar sprayed; 4.5, 9.0, 18.0, 36.0, and 72.0 g/L of Zn at the gestation stage	Field experiment	Increased Zn content in higher proportion in edible parts (brown rice and polished rice, 45.1–79.4%) than inedible parts like glume and rice bran (7.1–41.7%) and improved rice aroma and taste	Wang et al. (2023)
Rice	ZnO nanoparticles vs $ZnSO_4$; foliar applied 5 times every week after anthesis; 25, 50, 75, and 100 mg/L Zn	Field experiment	ZnO NPs increased Zn content in rice grain by 55% compared to $ZnSO_4$ (35%)	Parashar et al. (2023)
Rice	ZnO nanoparticle vs $ZnSO_4$; application time to soil basal, tillering, and panicle stage; 25 and 50 mg Zn/kg of soil	A pot experiment with field soil	Zn content in brown rice was higher with ZnO while $ZnSO_4$ resulted in higher Zn in glume; Zn application at the panicle stage increased Zn content higher	Yang et al. (2021)
Stevia	Phyto ZnO nanoparticles; 0, 50, 75 and 100 mg/L Zn	Fertigation in growing media of peat moss/aerolite	Increased Zn content about 14 times compared to control	Velázquez-Gamboa et al. (2021)

Velázquez-Gamboa et al. (2021). Plant-sourced ZnO nanoparticles increased the Zn content in stevia up to 14 times when applied at the concentration of 100 mg/L in the peat moss/aerolite media without affecting crop yield.

Umar et al. (2021) reported slightly higher Zn content in maize grain when 2% ZnO nanoparticles were applied through foliar spray compared to the soil application of 8 kg/ha of Zn before transplanting. However, the outcome could differ depending on Zn application rate and frequency of the application. Yang et al. (2021) reported that compared to conventional fertilization, foliar applications of ZnO nanoparticles increased the concentration of Zn in brown rice by 13.5–39.4% and had no negative impact on human health. The foliar application of Zn nanoparticles has been tested primarily on grain crops like rice, maize, and chickpeas; however, recently the use of nano Zn fertilizer has been tested also on other crop categories, including potatoes and vegetables (Table 5). Foliar application of Zn nanoparticles (25–250 Zn mg/L) at the tuber initiation stage increased potato tuber Zn content by 15–27% (Nazir et al. 2024).

Zinc nano-fertilizers are manufactured primarily as ZnO and are sold as powders with different particle size formulations and different levels of purity and in some cases marketed as liquid fertilizer. The formulation and particularly the size of the

ZnO nanoparticles are key factors in determining the application method and the efficacy of absorption by plants. Smaller particle sizes are generally considered easier to absorb, but are also more expensive; therefore, a good compromise between efficiency and cost is needed in most cases. Often, ZnO nanoparticles may be not soluble in water, and suspensions are formed, or additional treatments may be needed to obtain a true solution. Besides the formulation of the nano-fertilizer, the major factors influencing the efficacy of Zn nanoparticle applications are the rate and timing of the application. Yang et al. (2021) reported that the efficacy of ZnO nanoparticles on the Zn biofortification of brown rice was stage-dependent, and the application of ZnO nanoparticles at the panicle stage had a higher efficacy in improving Zn concentration than applications at basal and tillering stage. Similarly, Wang et al. (2023) observed an increase in Zn content in all parts of the grain but more notably in the edible portion of the rice when ZnO nanoparticles were applied through foliar spray at the gestation stage. Overall, the foliar application of ZnO nanoparticle was found to be more effective when applied at later stages of rice crops like panicle initiation and flowering stage compared to the basal and tillering stages (Parashar et al. 2023; Yang et al. 2021).

The use of Zn nanoparticles for crop biofortification purposes has been proposed also via seed nutrient priming, although limited research has been conducted so far on this particular application method. Poudel et al. (2023) explored the potential of using ZnO nanoparticles for the biofortification of pea and sunflower microgreens via seed nutri-priming by soaking the seeds of peas and sunflower with ZnO nanoparticle solutions prepared at different concentrations (25–200 mg/L Zn) overnight. Compared to the untreated control, ZnO nanoparticles increased Zn content in pea and sunflower microgreens up to 80% and 120%, respectively, performing much better than Zn-EDTA; however, $ZnSO_4$ was the best source of Zn. Similar results were obtained by Poudel et al. (2024) comparing ZnO nanoparticles, $ZnSO_4$, and Zn-EDTA applied via fertigation on pea, radish, and sunflower microgreens. Instead, contrasting results were reported by Khanm et al. (2017) who working on tomato plants observed that ZnO nanoparticles used to soak seeds and via foliar applications increased Zn content and improved tomato yield compared to $ZnSO_4$.

Phytotoxicity and food safety are critical concerns associated with the use of nanoparticles in agriculture. Over-application of nanoparticles may pollute the environment, and high accumulation levels in plant tissue can cause stress and phytotoxicity but also raise concerns for human health (Zulfiqar et al. 2019). Using pure sources of Zn nano-fertilizer and adequately defining the optimal application rate is therefore critical from a food safety and environmental perspective. While demonstrating potential for the biofortification of several food crops, the use of Zn nanoparticles provided different results depending on the targeted crop, application methods, formulation, and composition of the nano-fertilizer used (Pande and Arora 2019; Rana et al. 2021; Zulfiqar et al. 2019). These results suggest that more systematic research is needed to assess the efficacy and economic viability of nano-fertilizers for the biofortification of different food crops and to further assess their potential impact on human health and the environment.

4.6 Use of Microorganisms and Biofertilizers

Zinc biofortification mediated by microorganisms is considered a viable strategy for the biofortification of soil-grown food crops. The living component of the soil, particularly in the rhizosphere, is constituted by a rich, diverse, complex, and dynamic population of microorganisms that respond to edaphic factors and exert a series of functions and play a critical role in determining soil health and fertility. The rhizosphere is where plant-microbe interactions occur, and microbes either exist independently or establish a relationship, often mutualistic or symbiotic with plants through the roots (Kaur et al. 2020). Microbes belonging to the genera *Arthrobacter*, *Azotobacter*, *Azospirillum*, *Bacillus*, *Pseudomonas*, *Rhizobium*, and arbuscular mycorrhizal fungi can solubilize and mobilize nutrients including Zn in the soil, making it more accessible to plant roots (Agnolucci et al. 2020). These microbes are considered biostimulants and can generate plant growth phytoregulators, improving the overall plant health, enhancing plant response to stress conditions, and improving nutrient uptake (Asif et al. 2023). Some of these microbes produce chelating compounds that form complexes with Zn, increasing its bioavailability in the rhizosphere. Some microbes even influence plant gene expression related to nutrient uptake. Yadav et al. (2022) reported that *Bacillus tequilensis* CRS-38 and *Paenarthrobacter nicotinovorans* CRS-30 inoculation improved the expression of eight TaZIP transporter genes in the roots after 30 days of wheat growth and increased grain Zn content and grain yield in wheat. Further, endophytic microbes that colonize plant tissues can directly transport and accumulate Zn in the edible parts of the crop (Ku et al. 2019). Recent studies have shown that most endophytic bacteria promote growth and yield while only a few of the bacteria strains promote Zn bioavailability in the soil as well as Zn accumulation in plants. For example, Liao et al. (2024) tested 30 strains of Zn solubilizing endophytic bacteria and reported that only four strains SaBA1, SaPS2, SaEN1, and SaPA1 increased soil bioavailable Zn and Zn content in the plant along with promoting plant growth, while most of the other strains only promoted tomato seedling growth. Therefore, finding a suitable strain of bacteria is important to promote Zn biofortification by inoculating the soil with endophytic bacteria successfully. The most common ways to use beneficial microorganisms for Zn biofortification include the direct inoculation of specific strains of bacteria in the soil or growing media, seed inoculation, or dipping of seedling roots in beneficial microorganism solutions before planting. Some of the recent studies of using microorganisms for Zn biofortification are summarized in Table 6.

Microbe-assisted Zn biofortification is an eco-friendly and sustainable approach compared to the use of synthetic fertilizer; however, its efficiency could be lower compared to the exclusive use of synthetic Zn fertilizer. Nevertheless, beneficial microbes can also be used in synergy with synthetic fertilizers to increase their efficiency. The use of beneficial microbes and biofertilizers could be an affordable solution for resource-poor farmers in developing countries that have limited access to synthetic fertilizers. Moreover, microbe-mediated Zn biofortification can be combined with other strategies for synergistic effects. For example, Jalal et al.

Table 6 Main findings of representative studies focused on food crop Zn biofortification employing microorganisms and biofertilizers

Crop	Treatment	Growing condition	Results	References
Chickpea	Zn-solubilizing bacteria (ZSB) *Bacillus altitudinis*: Tested 24 strains of ZSB for the Zn solubilizing properties from 3 sources of Zn; ZnO, $Zn_3(PO_4)_2$, and $ZnCO_3$ and seeds were inoculated with BT3 and CT8	Pot experiment	Only two strains (BT3 and CT8) were able to solubilize all three Zn sources and increased soil Zn availability by 57–60% and plant Zn uptake by 3.9–6.0%	Kushwaha et al. (2021)
Common bean	*Rhizobium tropici*, *R. tropici* + *Azospirillum brasilense*, *R. tropici* + *Bacillus subtilis*, *R. tropici* + *Pseudomonas fluorescens*, *R. tropici* + *A. brasilense* + *B. subtilis*, and *R. tropici* + *A. brasilense* + *P. fluorescens* inoculated in seeds, and ZnO nanoparticle foliar spayed; 1.5 and 3 kg/ha split into two sprays at R5 and R8 stage	Field experiment	*R. tropici* + *B. subtilis* inoculation with 1.5 kg/ha Zn application resulted in higher grain and shoot Zn content	Jalal et al. (2023)
Lettuce	Fungus: *Talaromyces flavus* var. *flavus* CBS 310.38 inculcation in lettuce seed, $ZnSO_4$, $ZnCO_3$ and mix of $ZnSO_4$ and $ZnCO_3$ *Talaromyces flavus:* Zn: 25 kg/ha	A pot experiment with soil with 35% manure	A mix of *Talaromyces flavus* seed inoculation and Zn treatment resulted in higher Zn accumulation and better yield	Gupta et al. (2024)
Rice	7 Zn solubilizing bacteria strains isolated from rice field: Genus *Burkholderia* spp. (EIKU12, EIKU13, EIKU14, EIKU15, EIKU16, and EIKU17); seed primed in bacterial solubilized Zn	Hydroponic culture	Increased Zn content in seedlings dry biomass by 8–9 times	Bhakat et al. (2024)
Tomato	30 Zn solubilizing bacteria isolates from vermicompost were screened; the six bacterial strains were examined for Zn solubilization; VZB4 and VZB17 were examined on the growth of tomato plants dipping seedling root (14 days old) in a bacterial solution	Pot experiment	*Pseudomonas* sp. and *P. stutzeri* VBZ17 increased Zn content in fruits about 5 and 4 times, respectively	Karnwal (2021)
Tomato seedlings	Endophytic bacteria: *Sedum alfredii;* 30 strains were tested mixing with irrigated water at root regions	Pot experiment	Only strains SaBA1, SaPS2, SaEN1, and SaPA1 increased soil bioavailable Zn and plant Zn content and plant growth	Liao et al. (2024)
Wheat	Two varieties of seeds inoculated with ten indigenous bacterial strains and grown with and without Zn treatment in irrigated and rainfed conditions	Field condition	Beneficial microbes increased Zn in grain by 32% while increasing the efficiency of Zn fertilizer almost by 28–42%	Ali et al. (2024)

Wheat	Screened 15 Zn-solubilizing Rhizobacteria isolates, tested four bacterial isolates: *Bacillus glycinifermentans* CRS-9, *microbacterium oxydans* CRS-17, *Paenarthrobacter nicotinovorans* CRS-30, and *B. tequilensis* CRS-38	Pot and field experiment	Eight Zn transporter (TaZIP) genes expressed in roots, B. *glycinifermentans* CRS-9, *M. oxydans* CRS-17, and *P. nicotinovorans* CRS-30 showed Zn solubilizing traits; increased grain Zn content by 35–56% along with improving yield	Yadav et al. (2022)
Wheat	Zn-solubilizing *bacillus* spp. *Bacillus altitudinis* AJW-3, *B. subtilis* ABW-30, *B. megaterium* CHW-22, *B. licheniformis* MJW-38, *Brevibacillus borstelensis* CHW-2, and *B. xiamenensis* BLW-7	Pot and field experiment	*B. megaterium* CHW-22 followed by *B. licheniformis* MJW-38 increased Zn content in plants and grains increasing overall nutrient use efficiency	Yadav et al. (2023)

(2023) observed a higher Zn content in common beans when the foliar application of ZnO nanoparticles was combined with the inoculation of the soil with *Rhizobium tropici* + *Bacillus subtilis*. Utilizing naturally occurring microbes can be more cost-effective compared to conventional methods of Zn supplementation, such as soil fertilization, or can contribute to reducing the use of synthetic fertilizer (Ali et al. 2024; Jalal et al. 2023). Ali et al. (2024) reported around 28–42% increase in Zn fertilizer use efficiency with the use of 10 indigenous bacterial strains when they compared Zn fertilizer with or without the microbe inoculated seed in rainfed and irrigated field conditions, demonstrating the ability to biofortify crops while reducing the use of synthetic Zn fertilizer with the help of microorganisms.

Microbe-assisted Zn biofortification has the potential to play a huge role in the sustainable biofortification with Zn of various crops; however, the use of microbial products is not free of challenges. Identifying the most effective microbes and their optimal application methods for different crop-soil systems is crucial for the successful biofortification of food crops (Bhakat et al. 2024; Karnwal 2021; Liao et al. 2024; Yadav et al. 2022). Ensuring consistent performance and stability of microbial inoculants under field conditions is fundamental, considering that results obtained under controlled environments might not always translate under field conditions (Ali et al. 2024). Scaling up the production and delivery of microbial inoculants to reach a large population of farmers, and defining sustainable approaches to foster the growth of beneficial microbes, especially in areas characterized by low levels of bioavailable Zn in the soils could help address Zn deficiency issues at the global level(De Groote et al. 2021; Morton et al. 2023).

5 Zinc Bioavailability and Selection of Alternative Biofortification Target Crops

Zinc is generally stored in plant tissues and grains or seeds in different forms including primarily Zn-phytate, Zn-protein, chelated Zn complexes, and in small quantities as free ionic Zn (Zn^{2+}), which is the most bioavailable form of Zn for the plant metabolism (Stanton et al. 2022). The concentration of free Zn^{2+} is tightly regulated by specialized membrane transporters to maintain Zn homeostasis in plants. Zn-phytate complex is more common in seeds and grains and this complex can limit the bioavailability of Zn for human nutrition. Other forms of Zn like the Zn-protein complexes and chelated-Zn complexes are the most common forms present in plants. Zn-containing proteins make up around 8–9% of the proteome in plants and play crucial roles in plant metabolism, signaling, and stress response (Clemens 2010). Similarly, Zn is stored and transported within plants in chelated complexes, often formed with organic ligands such as nicotinamide and histidine. These chelated forms of Zn facilitate its uptake, distribution, and sequestration within plant tissues and cells. Depending on the prevalent form of Zn present in different plant portions, the bioaccessibility and bioavailability of Zn for human nutrition can vary significantly. For example, seeds and grain contain substantial amounts of Zn, but because of the presence of phytic acid in the seeds, Zn-phytate complexes are formed (Masum Akond et al. 2011), which precipitate in the human digestive system making Zn less bioavailable for absorption. Therefore, the Zn present in seeds and grains is less bioavailable than Zn accumulated in other plant portions like stem, leaves, and roots characterized by relatively low phytic acid content and a higher proportion of other forms of Zn-like Zn-protein, chelated Zn, and free Zn^{2+} ions, which are more available for absorption. This is one of the reasons behind the higher bioaccessible and bioavailable fractions of Zn in vegetable crops and leafy greens compared to seeds and grains (Gupta et al. 2015; Poudel et al. 2023). The most common indicator of the bioaccessible fraction of Zn present in grain and other plant edible portions is the phytic acid/Zn molar ratio; the higher the ratio, the lower the Zn bioavailability (Morris and Ellis 1989).

The goal of Zn biofortification is not just to increase the Zn level in the edible plant parts but also to increase the bioavailable fraction of Zn present in the edible portion of plants. Therefore, the selection of the target crop for Zn biofortification is critical in fulfilling the ultimate goal of increasing dietary intake of bioavailable Zn. The primary options to consider in the selection of target crops are (i) the capacity to accumulate Zn and (ii) the content of antinutritional factors like phytic acid which may limit Zn bioavailability. Considering the phytic acid content in different food crops, having relatively low phytic acid content, leafy vegetables could be a better target crop for Zn biofortification compared to grain crops. However, the selection of the target crop should be also done considering socioeconomic aspects and consumer preferences. Vegetables could be a good target crop for Zn biofortification in regions characterized by relatively high consumption of vegetables, while may not be the most suitable option for communities that follow primarily a cereal-based diet and that do not commonly consume vegetables. In such a scenario, the easier focus could be decreasing the phytic acid: Zn molar ratio either by reducing the phytic acid content in the grain or increasing Zn content via Zn biofortification (Poudel et al. 2023; Pramitha et al. 2021). The selection of genetic lines with low phytic acid seed content or the genetic engineering of grain crops characterized by a low phytic acid content could provide a lot of benefits from a nutritional standpoint (Raboy 2020). However, phytic acid itself plays a significant role in plant growth and development, stress response, and cellular signaling and protects seeds from oxidative stress (Kumar et al. 2021; Pramitha et al. 2021), and there are some arguments against breeding for the low-phytate trait, which could reduce seed germination and vigor, affecting crop field performance and ultimately yield (Raboy 2020). Nevertheless, combining the selection of low phytic acid crop lines with agronomic Zn biofortification could be a potential solution to increase the fraction of bioavailable Zn and its dietary intake. An alternative way to approach the same issue is the selection of Zn hyperaccumulator species or genotypes which could increase the efficiency of agronomic biofortification and reduce the phytic acid: Zn molar ratio primarily by increasing the accumulation of Zn (Clemens 2017). Several studies have reported that simple agronomic biofortification can decrease the phytic acid: Zn molar ratio in biofortified plants compared to non-biofortified ones (Saha et al. 2017; Silva et al. 2021). Moreover, in the case of grain crops, some processing techniques like milling, soaking, dehulling, fermentation, germination, and thermal processing can be used to increase the fraction of bioavailable Zn (Rousseau et al. 2020).

6 Summary and Future Perspectives

Agronomic biofortification is a promising approach including a series of alternative methods to enrich with Zn a range of food crops grown in soil or soilless systems under a variety of environmental conditions. The opportunity to choose different ways of supplementing Zn fertilizer through the soil or the nutrient solution, treating seeds or spraying leaves, and employing fertilizer technologies ranging from simple water-soluble salts to more advanced chelated products, nano-fertilizers, and even biofertilizers offers the flexibility needed to address the complex issues that are at the origin of Zn deficiency. While relatively simple compared to other approaches, the range of agronomic methods currently available has a real potential to contribute to addressing Zn deficiency under a variety of circumstances and limitations affecting malnourished populations in different regions of the world.

Agronomic biofortification should not be considered as a stand-alone solution but rather as a component of a larger long-term strategy deployed to address and defeat Zn deficiency. The successful implementation of agronomic Zn biofortification strategies requires an attentive analysis of all the factors (environmental, dietary, socioeconomic) that contribute to the cause of the Zn deficiency issue which should be considered in selecting the biofortification approach or combination of methods considered the most suitable and effective for a given situation. Taking into account environmental, social, and economic limitations and/or requirements, the first step is to carefully select the biofortification target crop or group of crops and then select the most efficient and sustainable agronomic biofortification method applicable in the specific case by examining the advantages and disadvantages of the available options. Each of the agronomic biofortification methods examined has advantages and disadvantages; additional work is generally required to optimize the solution being implemented defining all the details (target crop, biofortification method, type of fertilizer, application rate, application timing, etc.) to maximize efficacy and evaluate the long-term sustainability.

The ongoing research effort and the new technologies and solutions being developed will continue to improve the efficacy and sustainability of the Zn agronomic biofortification strategies developed so far; however, additional efforts are needed for integrating agronomic biofortification with genetic biofortification approaches. An integrated multi-strategy biofortification approach may offer the opportunity to increase the efficiency and long-term sustainability of Zn biofortification efforts and may allow to prevent some of the disadvantages of each approach.FundingThis work was supported by Open Philanthropy through the grant "Food Resilience in the Face of Catastrophic Global Events" and by the Pennsylvania Department of Agriculture through the project "Employing Agronomic Biofortification Techniques to Produce High-Value Functional Vegetable Crops" Grant Agreement #C940001096. The contribution of PP was supported by the Strategic Networks and Initiatives

Program (SNIP) Soil, Plants, and Consumers, funded by the Penn State College of Agricultural Sciences and the Penn State Department of Plant Science. FD contribution was supported by the USDA National Institute of Food and Agriculture and Hatch Appropriations under Project #PEN04723 and Accession #1020664.

References

Agnolucci M, Avio L, Palla M, Sbrana C, Turrini A, Giovannetti M (2020) Health-promoting properties of plant products: the role of Mycorrhizal fungi and associated bacteria. Agronomy 10(12). https://doi.org/10.3390/agronomy10121864

Ali M, Sharif M, Ahmed I (2024) Enhancement of wheat productivity and zinc accumulation through integrated use of zinc and beneficial microbes under irrigated and Rainfed conditions. Commun Soil Sci Plant Anal 55(4):517–528. https://doi.org/10.1080/00103624.2023.2274034

Alloway BJ (2009) Soil factors associated with zinc deficiency in crops and humans. Environ Geochem Health 31(5):537–548. https://doi.org/10.1007/s10653-009-9255-4

Almendros P, Obrador A, Gonzalez D, Alvarez JM (2015) Biofortification of zinc in onions (*Allium cepa* L.) and soil Zn status by the application of different organic Zn complexes. Sci Hortic 186:254–265. https://doi.org/10.1016/j.scienta.2015.02.023

Alshaal T, El-Ramady H (2017) Foliar application: from plant nutrition to biofortification. Environ Biodivers Soil Secur 1:71–83. https://doi.org/10.21608/jenvbs.2017.1089.1006

Arnold T, Kirk GJD, Wissuwa M, Frei M, Zhao F-J, Mason TFD, Weiss DJ (2010) Evidence for the mechanisms of zinc uptake by rice using isotope fractionation. Plant Cell Environ 33(3):370–381. https://doi.org/10.1111/j.1365-3040.2009.02085.x

Asif A, Ali M, Qadir M, Karthikeyan R, Singh Z, Khangura R, Di Gioia F, Ahmed ZFR (2023) Enhancing crop resilience by harnessing the synergistic effects of biostimulants against abiotic stress. Front Plant Sci 14. https://doi.org/10.3389/fpls.2023.1276117

Azam M, Bhatti HN, Khan A, Zafar L, Iqbal M (2022) Zinc oxide nano-fertilizer application (foliar and soil) effect on the growth, photosynthetic pigments and antioxidant system of maize cultivar. Biocatal Agric Biotechnol 42:102343. https://doi.org/10.1016/j.bcab.2022.102343

Bączek-Kwinta R, Baran A, Simlat M, Lang J, Bieniek M, Florek B (2020) Enrichment of different plant seeds with zinc and assessment of health risk of Zn-fortified sprouts consumption. Agronomy 10(7). https://doi.org/10.3390/agronomy10070937

Barrameda-Medina Y, Blasco B, Lentini M, Esposito S, Baenas N, Moreno DA, Ruiz JM (2017a) Zinc biofortification improves phytochemicals and amino-acidic profile in Brassica oleracea cv Bronco. Plant Sci 258:45–51. https://doi.org/10.1016/j.plantsci.2017.02.004

Barrameda-Medina Y, Lentini M, Esposito S, Ruiz JM, Blasco B (2017b) Zn-biofortification enhanced nitrogen metabolism and photorespiration process in green leafy vegetable Lactuca sativa L. J Sci Food Agric 97(6):1828–1836. https://doi.org/10.1002/jsfa.7983

Bartsch D (2014) GMO regulatory challenges and science: a European perspective. J Verbr Lebensm 9(1):51–58. https://doi.org/10.1007/s00003-014-0885-9

Berhe K, Gebrearegay F, Gebremariam H (2019) Prevalence and associated factors of zinc deficiency among pregnant women and children in Ethiopia: a systematic review and meta-analysis. BMC Public Health 19(1):1663. https://doi.org/10.1186/s12889-019-7979-3

Bhakat K, Islam E, Mukherjee SK (2024) Biosolubilized zinc from ZnO in bacterial culture supernatant enhanced seed germination, early growth, and zinc accumulation in rice (Oryza sativa L.) under hydroponic system through seed priming method. J Plant Nutr 47(1):65–81. https://doi.org/10.1080/01904167.2023.2262501

Bhardwaj AK, Chejara S, Malik K, Kumar R, Kumar A, Yadav RK (2022) Agronomic biofortification of food crops: an emerging opportunity for global food and nutritional security. Front Plant Sci 13. https://doi.org/10.3389/fpls.2022.1055278

Broadley MR, White PJ, Hammond JP, Zelko I, Lux A (2007) Zinc in plants. New Phytol 173(4):677–702
Brown WH (2007) Zinc bioavailability from zinc-fortified foods. Int J Vitam Nutr Res 77(3):174–181. https://doi.org/10.1024/0300-9831.77.3.174
Cakmak I (2008) Enrichment of cereal grains with zinc: agronomic or genetic biofortification? Plant Soil 302(1):1–17. https://doi.org/10.1007/s11104-007-9466-3
Cakmak I, Kalayci M, Kaya Y, Torun AA, Aydin N, Wang Y, Arisoy Z, Erdem H, Yazici A, Gokmen O, Ozturk L, Horst WJ (2010) Biofortification and localization of zinc in wheat grain. J Agric Food Chem 58(16):9092–9102. https://doi.org/10.1021/jf101197h
Cakmak I, Brown P, Colmenero-Flores JM, Husted S, Kutman BY, Nikolic M, Rengel Z, Schmidt SB, Zhao F-J (2023) Micronutrients. In: Rengel Z, Cakmak I, White PJ (eds) Marschner's mineral nutrition of plants, 4th edn. Academic Press, pp 283–385. https://doi.org/10.1016/B978-0-12-819773-8.00017-4
Ciriello M, Formisano L, Kyriacou M, Soteriou GA, Graziani G, De Pascale S, Rouphael Y (2022) Zinc biofortification of hydroponically grown basil: stress physiological responses and impact on antioxidant secondary metabolites of genotypic variants. Front Plant Sci 13:1–17. https://doi.org/10.3389/fpls.2022.1049004
Clemens S (2010) Zn – a versatile player in plant cell biology. In: Hell R, Mendel R-R (eds) Cell biology of metals and nutrients. Springer, pp 281–298. https://doi.org/10.1007/978-3-642-10613-2_12
Clemens S (2017) How metal hyperaccumulating plants can advance Zn biofortification. Plant Soil 411(1):111–120. https://doi.org/10.1007/s11104-016-2920-3
Connolly EL, Fett JP, Guerinot ML (2002) Expression of the IRT1 metal transporter is controlled by metals at the levels of transcript and protein accumulation. Plant Cell 14(6):1347–1357. https://doi.org/10.1105/tpc.001263
Couzy F, Mansourian R, Labate A, Guinchard S, Montagne DH, Dirren H (1998) Effect of dietary phytic acid on zinc absorption in the healthy elderly, as assessed by serum concentration curve tests. Br J Nutr 80(2):177–182. https://doi.org/10.1017/S0007114598001081
D'imperio M, Montesano FF, Serio F, Santovito E, Parente A (2022) Mineral composition and bioaccessibility in rocket and purslane after Zn biofortification process. Food Secur 11(3). https://doi.org/10.3390/foods11030484
de Almeida HJ, Vergara Carmona VM, Ferreira Inocêncio M, Furtini Neto AE, Cecílio Filho AB, Mauad M (2020) Soil type and zinc doses in agronomic biofortification of lettuce genotypes. Agronomy 10(1). https://doi.org/10.3390/agronomy10010124
De Groote H, Tessema M, Gameda S, Gunaratna NS (2021) Soil zinc, serum zinc, and the potential for agronomic biofortification to reduce human zinc deficiency in Ethiopia. Sci Rep 11(1):8770. https://doi.org/10.1038/s41598-021-88304-6
Deinlein U, Weber M, Schmidt H, Rensch S, Trampczynska A, Hansen TH, Husted S, Schjoerring JK, Talke IN, Krämer U, Clemens S (2012) Elevated Nicotianamine levels in Arabidopsis halleri roots play a key role in zinc Hyperaccumulation. Plant Cell 24(2):708–723. https://doi.org/10.1105/tpc.111.095000
Dhaliwal SS, Sharma V, Shukla AK, Verma V, Behera SK, Singh P, Alotaibi SS, Gaber A, Hossain A (2021) Comparative efficiency of mineral, chelated and nano forms of zinc and iron for improvement of zinc and iron in chickpea (Cicer arietinum L.) through biofortification. Agronomy 11(12). https://doi.org/10.3390/agronomy11122436
Di Gioia F, Renna M, Santamaria P (2017) Sprouts, microgreens and "baby leaf" vegetables. In: Yildiz F, Wiley RC (eds) Minimally processed refrigerated fruits and vegetables. Springer US, pp 403–432. https://doi.org/10.1007/978-1-4939-7018-6_11
Di Gioia F, Petropoulos SA, Ozores-Hampton M, Morgan K, Rosskopf EN (2019) Zinc and iron agronomic biofortification of Brassicaceae microgreens. Agronomy 9(11). https://doi.org/10.3390/agronomy9110677
Di Gioia F, Petropoulos SA, Ferreira ICFR, Rosskopf EN (2021) Microgreens: from trendy vegetables to functional food and potential nutrition security resource. Acta Hortic 1321:235–242. https://doi.org/10.17660/ActaHortic.2021.1321.31

Di Gioia F, Hong JC, Pisani C, Petropoulos SA, Bai J, Rosskopf EN (2023) Yield performance, mineral profile, and nitrate content in a selection of seventeen microgreen species. Front Plant Sci 14:1220691. https://doi.org/10.3389/fpls.2023.1220691

Doolette CL, Read TL, Li C, Scheckel KG, Donner E, Kopittke PM, Schjoerring JK, Lombi E (2018) Foliar application of zinc sulphate and zinc EDTA to wheat leaves: differences in mobility, distribution, and speciation. J Exp Bot 69(18):4469–4481. https://doi.org/10.1093/jxb/ery236

Du W, Yang J, Peng Q, Liang X, Mao H (2019) Comparison study of zinc nanoparticles and zinc sulphate on wheat growth: from toxicity and zinc biofortification. Chemosphere 227:109–116. https://doi.org/10.1016/j.chemosphere.2019.03.168

Eichert T (2013) Foliar nutrient uptake – of myths and legends. Acta Hortic 984:69–75. https://doi.org/10.17660/ActaHortic.2013.984.5

Eide D, Broderius M, Fett J, Guerinot ML (1996) A novel iron-regulated metal transporter from plants identified by functional expression in yeast. Proc Natl Acad Sci 93(11):5624–5628. https://doi.org/10.1073/pnas.93.11.5624

Epstein E, Bloom AJ (2004) Mineral nutrition of plants: principles and perspectives, 2nd edn. Sinauer Associates, Inc

Fageria NK, Filho MPB, Moreira A, Guimarães CM (2009) Foliar fertilization of crop plants. J Plant Nut 32(6):1044–1064. https://doi.org/10.1080/01904160902872826

Farooq M, Ullah A, Rehman A, Nawaz A, Nadeem A, Wakeel A, Nadeem F, Siddique KHM (2018) Application of zinc improves the productivity and biofortification of fine grain aromatic rice grown in dry seeded and puddled transplanted production systems. Field Crop Res 216:53–62. https://doi.org/10.1016/j.fcr.2017.11.004

Fernández V, Gil-Pelegrín E, Eichert T (2021) Foliar water and solute absorption: an update. Plant J 105(4):870–883. https://doi.org/10.1111/tpj.15090

Gibson RS, Ferguson EL (1998) Nutrition intervention strategies to combat zinc deficiency in developing countries. Nutr Res Rev 11(1):115–131. https://doi.org/10.1079/NRR19980008

Grotz N, Fox T, Connolly E, Park W, Guerinot ML, Eide D (1998) Identification of a family of zinc transporter genes from *Arabidopsis* that respond to zinc deficiency. Proc Natl Acad Sci 95(12):7220–7224. https://doi.org/10.1073/pnas.95.12.7220

Gupta RK, Gangoliya SS, Singh NK (2015) Reduction of phytic acid and enhancement of bioavailable micronutrients in food grains. J Food Sci Technol 52(2):676–684. https://doi.org/10.1007/s13197-013-0978-y

Gupta N, Ram H, Kumar B (2016) Mechanism of zinc absorption in plants: uptake, transport, translocation and accumulation. Rev Environ Sci Biotechnol 15(1):89–109. https://doi.org/10.1007/s11157-016-9390-1

Gupta S, Srivastava PK, Singh RP (2024) Growth promotion and zinc biofortification in lettuce (*Lactuca sativa* L.) by the application of *Talaromyces* strain as a biostimulant. Sci Hortic 323:112534. https://doi.org/10.1016/j.scienta.2023.112534

Haider MU, Hussain M, Farooq M, Nawaz A (2020) Optimizing zinc seed priming for improving the growth, yield, and grain biofortification of mungbean. J Plant Nutr 43(10):1438–1446. https://doi.org/10.1080/01904167.2020.1730895

Hassan N, Irshad S, Saddiq MS, Bashir S, Khan S, Wahid MA, Khan RR, Yousra M (2019) Potential of zinc seed treatment in improving stand establishment, phenology, yield and grain biofortification of wheat. J Plant Nutr 42(14):1676–1692. https://doi.org/10.1080/01904167.2019.1630429

Imran M, Garbe-Schönberg D, Neumann G, Boelt B, Mühling KH (2017) Zinc distribution and localization in primed maize seeds and its translocation during early seedling development. Environ Exp Bot 143:91–98. https://doi.org/10.1016/j.envexpbot.2017.09.001

Jalal A, Mortinho ES, da Silva Oliveira CE, Fernandes GC, Junior EF, de Lima BH, Moreira A, Nogueira TAR, Galindo FS, Filho MCMT (2023) Nano-zinc and plant growth-promoting bacteria is a sustainable alternative for improving productivity and agronomic biofortification of common bean. Chem Biol Technol Agric 10(1):77. https://doi.org/10.1186/s40538-023-00440-5

Karnwal A (2021) Pseudomonas spp., a zinc-solubilizing vermicompost bacteria with plant growth-promoting activity moderates zinc biofortification in tomato. Int J Veg Sci 27(4):398–412. https://doi.org/10.1080/19315260.2020.1812143

Kaur H, Garg N (2021) Zinc toxicity in plants: a review. Planta 253(6):129. https://doi.org/10.1007/s00425-021-03642-z

Kaur T, Rana KL, Kour D, Sheikh I, Yadav N, Kumar V, Yadav AN, Dhaliwal HS, Saxena AK (2020) Microbe-mediated biofortification for micronutrients: present status and future challenges. In: Rastegari AA, Yadav AN, Yadav N (eds) New and future developments in microbial biotechnology and bioengineering. Elsevier, pp 1–17. https://doi.org/10.1016/B978-0-12-820528-0.00002-8

Khanm H, Vaishnavi BA, Namratha MR, Shankar AG (2017) Nano zinc oxide boosting growth and yield in tomato: the rise of "Nano fertilizer era.". Int J Agric Sci Res 7(3):197–206. https://doi.org/10.24247/ijasrjun201724

Kruger DJ, Greenberg E, Murphy JB, DiFazio LA, Youra KR (2014) Local concentration of fast-food outlets is associated with poor nutrition and obesity. Am J Health Promot 28(5):340–343

Ku Y-S, Rehman HM, Lam H-M (2019) Possible roles of Rhizospheric and Endophytic microbes to provide a safe and affordable means of crop biofortification. Agronomy 9(11). https://doi.org/10.3390/agronomy9110764

Kumar A, Singh B, Raigond P, Sahu C, Mishra UN, Sharma S, Lal MK (2021) Phytic acid: blessing in disguise, a prime compound required for both plant and human nutrition. Food Res Int 142:110193. https://doi.org/10.1016/j.foodres.2021.110193

Kushwaha P, Srivastava R, Pandiyan K, Singh A, Chakdar H, Kashyap PL, Bhardwaj AK, Murugan K, Karthikeyan N, Bagul SY, Srivastava AK, Saxena AK (2021) Enhancement in plant growth and zinc biofortification of chickpea (Cicer arietinum L.) by Bacillus altitudinis. J Soil Sci Plant Nutr 21(2):922–935. https://doi.org/10.1007/s42729-021-00411-5

Kyriacou MC, Rouphael Y, Di Gioia F, Kyratzis A, Serio F, Renna M, De Pascale S, Santamaria P (2016) Micro-scale vegetable production and the rise of microgreens. Trends Food Sci Technol 57:103–115. https://doi.org/10.1016/j.tifs.2016.09.005

Labanauskas CK, Stolzy LH, Zentmyer GA, Szuszkiewicz TE (1968) Influence of soil oxygen and soil water on the accumulation of nutrients in avocado seedlings (persea Americana mill). Plant Soil 29(3):391–406

Liao J, Li Z, Huang L, Xu S, Liu Q, Pan A, Liu Y, Yu C, Feng Y (2024) Enhancing the growth and zinc efficiency of cherry tomatoes by inoculating endophytic bacteria from zinc hyperaccumulator, *Sedum alfredi* Hance. Sci Hortic 332:113227. https://doi.org/10.1016/j.scienta.2024.113227

Lindsay WL (1972) Zinc in soils and plant nutrition. In: Brady NC (ed) Advances in agronomy, vol 24. Academic Press, pp 147–186. https://doi.org/10.1016/S0065-2113(08)60635-5

Liu Y-M, Liu D-Y, Zhang W, Chen X-X, Zhao Q-Y, Chen X-P, Zou C-Q (2020) Health risk assessment of heavy metals (Zn, Cu, Cd, Pb, as and Cr) in wheat grain receiving repeated Zn fertilizers. Environ Pollut 257:113581. https://doi.org/10.1016/j.envpol.2019.113581

Lyons G, Cakmak I (2012) Agronomic biofortification of food crops with micronutrients. In: Bruulsema TW, Heffer P, Welch RM, Cakmak I, Moran K (eds) Fertilizing crops to improve human health: a scientific review. International Plant Nutrition Institute, Paris, pp 97–122

Mao H, Wang J, Wang Z, Zan Y, Lyons G, Zou C (2014) Using agronomic biofortification to boost zinc, selenium, and iodine concentrations of food crops grown on the loess plateau in China. J Soil Sci Plant Nutr 14(2):459–470. https://doi.org/10.4067/S0718-95162014005000036

Maqueda C, Morillo E, Lopez R, Undabeytia T, Cabrera F (2015) Influence of organic amendments on Fe, Cu, Mn, and Zn availability and clay minerals of different soils. Arch Agron Soil Sci 61(5):599–613. https://doi.org/10.1080/03650340.2014.946409

Marques E, Darby HM, Kraft J (2021) Benefits and limitations of non-transgenic micronutrient biofortification approaches. Agronomy 11(3):464

Masum Akond ASMG, Crawford H, Berthold J, Talukder ZI, Hossain K (2011) Minerals (Zn, Fe, Ca and Mg) and Antinutrient (Phytic acid) constituents in common bean. Am J Food Technol 6(3):235–243. https://doi.org/10.3923/ajft.2011.235.243

Maxfield, L., Shukla, S., & Crane, J. S. (2023, June 28). Zinc deficiency. StatPearls. http://www.ncbi.nlm.nih.gov/books/NBK493231/

Maywald M, Rink L (2022) Zinc in human health and infectious diseases. Biomol Ther 12(12):1748. https://doi.org/10.3390/biom12121748

McBride MB, Blasiak JJ (1979) Zinc and copper solubility as a function of pH in an acid soil. Soil Sci Soc Am J 43(5):866–870. https://doi.org/10.2136/sssaj1979.03615995004300050009x

Ministry of Health and Family Welfare (MoHFW), Government of India, UNICEF and Population Council (2019) Comprehensive National Nutrition Survey: 2016–2018

Ministry of Health and Population, Nepal; New ERA; UNICEF; EU; USAID; and CDC (2018). Nepal National Micronutrient Status Survey, 2016. https://www.unicef.org/nepal/media/1206/file/Nepal%20National%20Micronutrient%20Status%20Survey%20Report%202016.pdf

Moreno-Lora A, Delgado A (2020) Factors determining Zn availability and uptake by plants in soils developed under Mediterranean climate. Geoderma 376:114509. https://doi.org/10.1016/j.geoderma.2020.114509

Morris ER, Ellis R (1989) Usefulness of the dietary phytic acid/zinc molar ratio as an index of zinc bioavailability to rats and humans. Biol Trace Elem Res 19(1):107–117. https://doi.org/10.1007/BF02925452

Morton CM, Pullabhotla H, Bevis L, Lobell DB (2023) Soil micronutrients linked to human health in India. Sci Rep 13(1):13591. https://doi.org/10.1038/s41598-023-39084-8

Nayak S, Shivay YS, Prasanna R, Mandi S, Parveen S, Baral K, Reddy KS (2023) Soil and foliar application of Zn enhances its biofortification, bioavailability and productivity in both biofortified and non-biofortified wheat varieties. J Food Compos Anal 124:105691. https://doi.org/10.1016/j.jfca.2023.105691

Nazir T, Anwar S, Mahmood A, Shafiq F (2024) Nano-zinc oxide foliar application at the tuber initiation stage enhances the starch synthesis and zinc bioaccumulation in potato tubers. Int J Appl Exp Biol 3(1):51. https://doi.org/10.56612/ijaaeb.v1i1.74

Niu J, Liu C, Huang M, Liu K, Yan D (2021) Effects of foliar fertilization: a review of current status and future perspectives. J Soil Sci Plant Nut 21(1):104–118. https://doi.org/10.1007/s42729-020-00346-3

Northover GHR, Garcia-España E, Weiss DJ (2020) Unravelling the modus operandi of phytosiderophores during zinc uptake in rice: the importance of geochemical gradients and accurate stability constants. J Exp Bot 72(5):1517–1526. https://doi.org/10.1093/jxb/eraa580

Oosterhuis D (2009) Foliar fertilization: mechanisms and magnitude of nutrient uptake. Proc Fluid Forum 15–17

Öz B, Unsal F, Movassaghi H (2018) Consumer attitudes toward genetically modified food in the United States: are millennials different? J Transnatl Manag 23(1):3–21. https://doi.org/10.1080/15475778.2017.1373316

Pande A, Arora S (2019) Nanotechnological interventions for improving plant health and productivity. In: Khurana SMP, Gaur RK (eds) Plant biotechnology: progress in genomic era. Springer, pp 375–395. https://doi.org/10.1007/978-981-13-8499-8_17

Parashar R, Afzal S, Mishra M, Singh NK (2023) Improving biofortification success rates and productivity through zinc nanocomposites in rice (Oryza sativa L.). Environ Sci Pollut Res 30(15):44223–44233. https://doi.org/10.1007/s11356-023-25293-1

Potocki S, Valensin D, Kozlowski H (2014) The specificity of interaction of Zn $^{2+}$, Ni $^{2+}$ and Cu $^{2+}$ ions with the histidine-rich domain of the TjZNT1 ZIP family transporter. Dalton Trans 43(26):10215–10223. https://doi.org/10.1039/C4DT00903G

Poudel P, Di Gioia F, Lambert JD, Connolly EL (2023) Zinc biofortification through seed nutri-priming using alternative zinc sources and concentration levels in pea and sunflower microgreens. Front Plant Sci 14

Poudel P, Connolly EL, Kwasniewski M, Lambert JD, Di Gioia F (2024) Zinc biofortification via fertigation using alternative zinc sources and concentration levels in pea, radish, and sunflower microgreens. Sci Hortic 331:113098. https://doi.org/10.1016/j.scienta.2024.113098

Pramitha JL, Rana S, Aggarwal PR, Ravikesavan R, Joel AJ, Muthamilarasan M (2021) Chapter three – diverse role of phytic acid in plants and approaches to develop low-phytate grains to

enhance bioavailability of micronutrients. In: Kumar D (ed) Advances in genetics, vol 107. Academic Press, pp 89–120. https://doi.org/10.1016/bs.adgen.2020.11.003
Preciado-Rangel P, Campos-Ortiz A, Chávez ES, Reyes-Gonzalez A, Ruiz-Espinoza F, Ojeda-Barrios D, Hernandez-Montiel L (2021) Zinc biofortification improves yield, nutraceutical quality and antioxidant capacity in lettuce. Trop Subtrop Agroecosystems 24(3)
Ptashnyk M, Roose T, Jones DL, Kirk GJD (2011) Enhanced zinc uptake by rice through phytosiderophore secretion: a modelling study. Plant Cell Environ 34(12):2038–2046. https://doi.org/10.1111/j.1365-3040.2011.02401.x
Qian L, Dawar K, Ullah I, Irfan M, Zhang Z, Mian IA, Khan B, Gul N, Fahad S, Jalal A, Danish S, Iqbal RK, Alarfaj AA (2023) Zinc foliar application mitigates cadmium-induced growth inhibition and enhances wheat growth, chlorophyll contents, and yield. ACS Omega 8(36):32372–32381. https://doi.org/10.1021/acsomega.3c01511
Rabbi RHM, Chowdhury MAH, Uddin MK, Saha BK (2024) Agronomic biofortification of zinc in tomato. J Plant Nutr:1–15. https://doi.org/10.1080/01904167.2024.2320215
Raboy V (2020) Low phytic acid crops: observations based on four decades of research. Plan Theory 9(2). https://doi.org/10.3390/plants9020140
Rana RA, Siddiqui MN, Skalicky M, Brestic M, Hossain A, Kayesh E, Popov M, Hejnak V, Gupta DR, Mahmud NU, Islam T (2021) Prospects of nanotechnology in improving the productivity and quality of horticultural crops. Horticulturae 7(10). https://doi.org/10.3390/horticulturae7100332
Raza SH, Shahzadi A, Iqbal M, Shafiq F, Mahmood A, Anwar S, Ashraf M (2022) Foliar application of nano-zinc oxide crystals improved zinc biofortification in cauliflower (Brassica oleracea L. var. Botrytis). Appl Nanosci 12(6):1803–1813. https://doi.org/10.1007/s13204-022-02455-0
Reider CA, Chung RY, Devarshi PP, Grant RW, Hazels Mitmesser S (2020) Inadequacy of immune health nutrients: intakes in US adults, the 2005–2016 NHANES. Nutrients 12(6):1735. https://doi.org/10.3390/nu12061735
Ricachenevsky FK, Menguer PK, Sperotto RA, Williams LE, Fett JP (2013) Roles of plant metal tolerance proteins (MTP) in metal storage and potential use in biofortification strategies. Front Plant Sci 4. https://doi.org/10.3389/fpls.2013.00144
Rietra RPJJ, Heinen M, Dimkpa CO, Bindraban PS (2017) Effects of nutrient antagonism and synergism on yield and fertilizer use efficiency. Commun Soil Sci Plant Anal 48(16):1895–1920. https://doi.org/10.1080/00103624.2017.1407429
Rousseau S, Kyomugasho C, Celus M, Hendrickx MEG, Grauwet T (2020) Barriers impairing mineral bioaccessibility and bioavailability in plant-based foods and the perspectives for food processing. Crit Rev Food Sci Nutr 60(5):826–843. https://doi.org/10.1080/10408398.2018.1552243
Rugeles-Reyes SM, Cecílio AB, López Aguilar MA, Silva PHS (2019) Foliar application of zinc in the agronomic biofortification of arugula. Food Sci Technol 39:1011–1017. https://doi.org/10.1590/fst.12318
Saha S, Chakraborty M, Padhan D, Saha B, Murmu S, Batabyal K, Seth A, Hazra GC, Mandal B, Bell RW (2017) Agronomic biofortification of zinc in rice: influence of cultivars and zinc application methods on grain yield and zinc bioavailability. Field Crop Res 210:52–60. https://doi.org/10.1016/j.fcr.2017.05.023
Sánchez-Rodríguez AR, del Campillo MC, Torrent J (2017) Phosphorus reduces the zinc concentration in cereals pot-grown on calcareous Vertisols from southern Spain. J Sci Food Agric 97(10):3427–3432. https://doi.org/10.1002/jsfa.8195
Sharifan H, Ma X, Moore JM, Habib MR, Evans C (2019) Zinc oxide nanoparticles alleviated the bioavailability of cadmium and Lead and changed the uptake of iron in hydroponically grown lettuce (Lactuca sativa L. var. Longifolia). ACS Sustain Chem Eng 7(19):16401–16409. https://doi.org/10.1021/acssuschemeng.9b03531
Shivay YS, Prasad R, Pal M (2015) Effects of source and method of zinc application on yield, zinc biofortification of grain, and Zn uptake and use efficiency in chickpea (Cicer arietinum L.). Commun Soil Sci Plant Anal 46:2191

Shweta SS, Sharma A, Chadha S, Guleria V (2021) Nanotechnology: a cutting-edge technology in vegetable production. J Hortic Sci Biotechnol 96(6):682–695. https://doi.org/10.1080/14620316.2021.1902864

Silva VM et al (2021) Agronomic biofortification of cowpea with zinc: variation in primary metabolism responses and grain nutritional quality among 29 diverse genotypes. Plant Physiol Biochem 162:378–387. https://doi.org/10.1016/j.plaphy.2021.02.020

Sims JT (1986) Soil pH effects on the distribution and plant availability of manganese, copper, and zinc. Soil Sci Soc Am J 50(2):367–373. https://doi.org/10.2136/sssaj1986.03615995005000020023x

Singh S, Kaur J, Ram H, Singh J, Kaur S (2023) Agronomic bio-fortification of wheat (Triticum aestivum L.) to alleviate zinc deficiency in human being. Rev Environ Sci Biotechnol 22(2):505–526. https://doi.org/10.1007/s11157-023-09653-4

Stanton C, Sanders D, Krämer U, Podar D (2022) Zinc in plants: integrating homeostasis and biofortification. Mol Plant 15(1):65–85. https://doi.org/10.1016/j.molp.2021.12.008

Subbaiah LV, Prasad TNVKV, Krishna TG, Sudhakar P, Reddy BR, Pradeep T (2016) Novel effects of Nanoparticulate delivery of zinc on growth, productivity, and zinc biofortification in maize (Zea mays L.). J Agric Food Chem 64(19):3778–3788. https://doi.org/10.1021/acs.jafc.6b00838

Sumner ME, Miller WP (1996) Cation exchange capacity and exchange coefficients. In: Methods of soil analysis. Wiley, pp 1201–1229. https://doi.org/10.2136/sssabookser5.3.c40

Suzuki M, Takahashi M, Tsukamoto T, Watanabe S, Matsuhashi S, Yazaki J, Kishimoto N, Kikuchi S, Nakanishi H, Mori S, Nishizawa NK (2006) Biosynthesis and secretion of mugineic acid family phytosiderophores in zinc-deficient barley. Plant J 48(1):85–97. https://doi.org/10.1111/j.1365-313X.2006.02853.x

Ullah A, Farooq M, Nadeem F, Rehman A, Nawaz A, Naveed M, Wakeel A, Hussain M (2020) Zinc seed treatments improve productivity, quality and grain biofortification of desi and kabuli chickpea (Cicer arietinum). Crop Pasture Sci 71(7):668–678. https://doi.org/10.1071/CP19266

Umar W, Hameed MK, Aziz T, Maqsood MA, Bilal HM, Rasheed N (2021) Synthesis, characterization and application of ZnO nanoparticles for improved growth and Zn biofortification in maize. Arch Agron Soil Sci 67(9):1164–1176. https://doi.org/10.1080/03650340.2020.1782893

Veena M, Puthur JT (2022) Seed nutripriming with zinc is an apt tool to alleviate malnutrition. Environ Geochem Health 44(8):2355–2373. https://doi.org/10.1007/s10653-021-01054-2

Velázquez-Gamboa MC, Rodríguez-Hernández L, Abud-Archila M, Gutiérrez-Miceli FA, González-Mendoza D, Valdez-Salas B, González-Terreros E, Luján-Hidalgo MC (2021) Agronomic biofortification of Stevia rebaudiana with zinc oxide (ZnO) Phytonanoparticles and antioxidant compounds. Sugar Tech 23(2):453–460. https://doi.org/10.1007/s12355-020-00897-w

Walker C, Ezzati M, Black R (2008) Global and regional child mortality and burden of disease attributable to zinc deficiency. Eur J Clin Nutr 63:591–597. https://doi.org/10.1038/ejcn.2008.9

Wang R, Mi K, Yuan X, Chen J, Pu J, Shi X, Yang Y, Zhang H, Zhang H (2023) Zinc oxide nanoparticles foliar application effectively enhanced zinc and aroma content in Rice (Oryza sativa L.) grains. Rice 16(1):36. https://doi.org/10.1186/s12284-023-00653-0

Wei W, Lu L, Bian X-H, Li Q-T, Han J-Q, Tao J-J, Yin C-C, Lai Y-C, Li W, Bi Y-D, Man W-Q, Chen S-Y, Zhang J-S, Zhang W-K (2023) Zinc-finger protein GmZF351 improves both salt and drought stress tolerance in soybean. J Integr Plant Biol 65(7):1636–1650. https://doi.org/10.1111/jipb.13474

Wessells KR, Brown KH (2012) Estimating the global prevalence of zinc deficiency: results based on zinc availability in National Food Supplies and the prevalence of stunting. PLoS One 7(11):e50568. https://doi.org/10.1371/journal.pone.0050568

Wessels I, Fischer HJ, Rink L (2021) Dietary and physiological effects of zinc on the immune system. Annu Rev Nutr 41(1):133–175. https://doi.org/10.1146/annurev-nutr-122019-120635

Widodo B et al (2010) Response to zinc deficiency of two rice lines with contrasting tolerance is determined by root growth maintenance and organic acid exudation rates, and not by zinc-transporter activity. New Phytol 186(2):400–414. https://doi.org/10.1111/j.1469-8137.2009.03177.x

Wissuwa M, Ismail AM, Yanagihara S (2006) Effects of zinc deficiency on rice growth and genetic factors contributing to tolerance. Plant Physiol 142(2):731–741. https://doi.org/10.1104/pp.106.085225

Xue Y-F, Li X-J, Yan W, Miao Q, Zhang C-Y, Huang M, Sun J-B, Qi S-J, Ding Z-H, Cui Z-L (2023) Biofortification of different maize cultivars with zinc, iron and selenium by foliar fertilizer applications. Front Plant Sci 14

Yadav RC, Sharma SK, Varma A, Rajawat MVS, Khan MS, Sharma PK, Malviya D, Singh UB, Rai JP, Saxena AK (2022) Modulation in biofertilization and biofortification of wheat crop by inoculation of zinc-solubilizing Rhizobacteria. Front Plant Sci 13. https://doi.org/10.3389/fpls.2022.777771

Yadav RC, Sharma SK, Varma A, Singh UB, Kumar A, Bhupenchandra I, Rai JP, Sharma PK, Singh HV (2023) Zinc-solubilizing bacillus spp. in conjunction with chemical fertilizers enhance growth, yield, nutrient content, and zinc biofortification in wheat crop. Front Microbiol 14. https://doi.org/10.3389/fmicb.2023.1210938

Yang G, Yuan H, Ji H, Liu H, Zhang Y, Wang G, Chen L, Guo Z (2021) Effect of ZnO nanoparticles on the productivity, Zn biofortification, and nutritional quality of rice in a life cycle study. Plant Physiol Biochem 163:87–94. https://doi.org/10.1016/j.plaphy.2021.03.053

Zou CQ, Zhang YQ, Rashid A, Ram H, Savasli E, Arisoy RZ, Ortiz-Monasterio I, Simunji S, Wang ZH, Sohu V, Hassan M, Kaya Y, Onder O, Lungu O, Mujahid MY, Joshi AK, Zelenskiy Y, Zhang FS, Cakmak I (2012) Biofortification of wheat with zinc through zinc fertilization in seven countries. Plant Soil 361(1):119–130. https://doi.org/10.1007/s11104-012-1369-2

Zoufan P, Baroonian M, Zargar B (2020) ZnO nanoparticles-induced oxidative stress in Chenopodium murale L, Zn uptake, and accumulation under hydroponic culture. Environ Sci Pollut Res 27(10):11066–11078. https://doi.org/10.1007/s11356-020-07735-2

Zulfiqar F, Navarro M, Ashraf M, Akram NA, Munné-Bosch S (2019) Nanofertilizer use for sustainable agriculture: advantages and limitations. Plant Sci 289:110270. https://doi.org/10.1016/j.plantsci.2019.110270

Zulfiqar U, Hussain S, Maqsood M, Ishfaq M, Ali N (2021) Zinc nutrition to enhance rice productivity, zinc use efficiency, and grain biofortification under different production systems. Crop Sci 61(1):739–749. https://doi.org/10.1002/csc2.20381

Understanding Hyperaccumulation of Zinc in Plants

Fizanaj Sheikh, Swati Sachdev, and Kuldeep Bauddh

Abstract

Metal pollution as a result of escalated industrial activities has become a global issue. Accumulation of metals in the environment can induce hazardous impact on human health, reduce plant productivity, and alter microbial community dynamic in soil, owing to nondegradable and toxic nature. Existing metals on earth are either essential or nonessential based on whether needed or not by living being to perform their functions. Although essential metals such as Zn are required, their presence above threshold limit may cause toxicity. In plants, Zn orchestrates various crucial processes like photosynthesis and cell division; however, excess quantity may trigger oxidative stress resulting in cellular damage, disrupt respiration and photosynthesis, abridge nutrient uptake, and alter the structure and functions of enzymes and pigments. The content of Zn in soil, which is the primary site for metal uptake by plants, influences its bioavailability. Employing phytoremediation technique like phytoextraction that involves hyperaccumulator plants could help in alleviating toxicity. Hyperaccumulators recruit various mechanisms to uptake and accumulate excess Zn in aboveground parts, without exhibiting phytotoxicity. Exploring the potential of such plants can sustainably facilitate remediation of contaminated soil and restore their fertility. Furthermore, encompassing research in improving capacity of such plants through integration of new techniques like genetic engineering could pave way for obtaining better outcomes.

F. Sheikh · K. Bauddh (✉)
Institute of Environment and Sustainable Development, RGSC, Banaras Hindu University, Mirzapur, Uttar Pradesh, India
e-mail: kuldeep.bauddh@bhu.ac.in

S. Sachdev
Department of Liberal Education, Era University, Lucknow, Uttar Pradesh, India

S. Srivastava, H. Kaur (eds.), *Zinc in Soil-Plant Continuum*,
https://doi.org/10.1007/978-981-96-4253-3_10

Keywords

Food security · Genetic engineering · Hyperaccumulators · Phytoremediation · Uptake and accumulation · Zinc

1 Introduction

Heavy metals (HMs) are elements with higher atomic numbers (above 20) and greater density (5 g/cm^3) such as cadmium (Cd), lead (Pb), nickel (Ni), chromium (Cr), copper (Cu), and zinc (Zn) (Leonard et al. 2004). Although they are naturally present in the earth's crust and released in the environment by natural processes like erosion, weathering, and volcano eruption, human activities such as industrial processes, mining, use of agro-chemicals, and inappropriate waste disposal have increased the emission and concentration in the environment, causing pollution (Fig. 1.) (Barakat 2011; Dixit et al. 2015; Mani and Bharagava 2016; Mishra et al. 2019). These metals are classified into non-essential and essential metals. Non-essential metals including Cd, Cr, arsenic (As), mercury (Hg), and Pb are not required by living beings as they have no structural or functional role, whereas essential metals like molybdenum (Mo), manganese (Mn), Cu, Ni, Fe, and Zn are crucially needed to perform certain roles such as serving as nutrients, enzyme cofactors, and participating in enzymatic and metabolic processes, but in minimal quantities (Mildvan 1970). Therefore, occurrence of nonessential metals and even essential metals above threshold limit can mediate harmful effects on all living organisms (Mishra et al. 2019), owing to inherent toxic nature, high reactivity, and ability to inhibit biodegradation. Heavy metal toxicity in plants, animals, humans, and

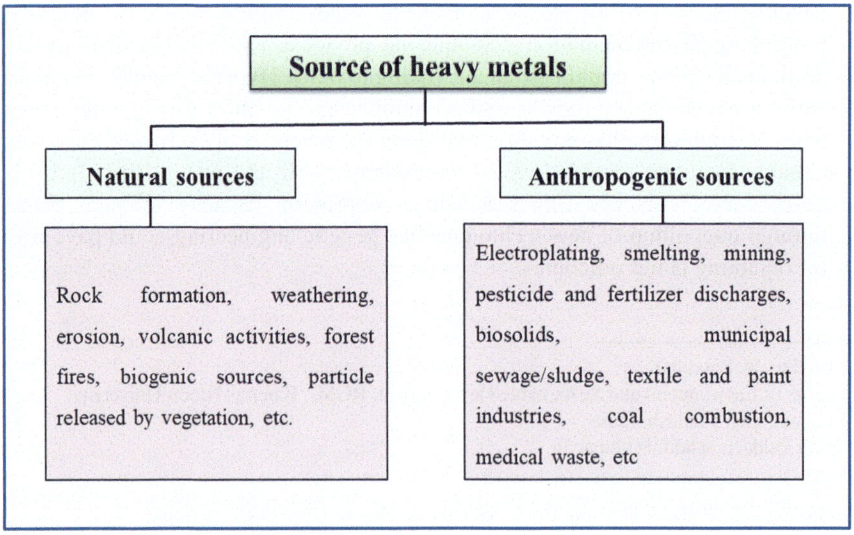

Fig. 1 Environmental sources of HMs

microorganisms can result in cellular damage and oxidative stress, disrupt photosynthetic and respiratory activities, hinder nutrient uptake, and alter structures as well as functions of enzymes, pigments, proteins, etc., which eventually reduces plant growth and yield and increases mortality and morbidity (Henry 2000; Nagajyoti et al. 2010; Gupta et al. 2014; Mishra et al. 2019; Rajkumar et al. 2020). Moreover, in plants, the toxic impact of heavy metals varies depending on plant species influenced, particular metal involved, concentration and chemical form of the metal, and soil composition and pH levels.

The excess presence of both types of HMs, i.e., essential and nonessential, causes reduction in growth and productivity of the crops and severe disease in human beings. Therefore, the remediation of such elements from the contaminated sites is needed. Among several HM removal technologies, phytoremediation has gained excellent attraction due to its environment and eco-friendly nature (Bauddh and Singh 2015a, b; Bauddh et al. 2016; Pandey and Bauddh 2018; Bauddh et al. 2020; Bauddh and Ma 2022). Several plants have efficiency to accumulate the HMs including Zn in excess amount and also transport them into their aerial parts, and these plants are known as metal hyperaccumulator plants. In the present book chapter, the Zn hyperaccumulator plants and their mechanisms have been explored.

2 Sources of Zn

Zinc is a transition metal that is essentially and abundantly required by plants, after iron (Fe) to mediate growth, reproduction, and signaling processes owing to structural, catalytic, and activating properties (Marschner 2011; Lehmann et al. 2014). The sources of Zn in the environment include both natural and anthropogenic processes. The natural sources include leaching from parent rocks through pedogenetic processes, volcanic eruptions, and soil erosion, primarily through soil particles (Fig. 2) (Mishra et al. 2019). There is an estimate that about 5.9 million tons of Zn

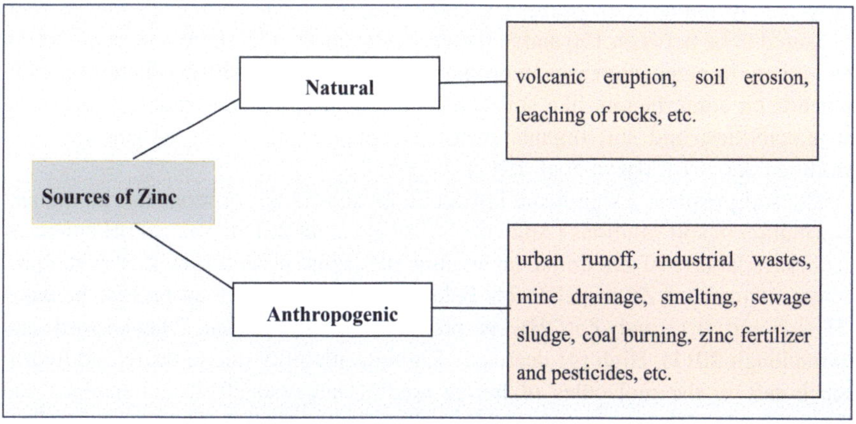

Fig. 2 Environmental sources of Zn

is emitted from natural sources annually (Prasad 2008). Surface and groundwater as well as soil contain Zn, which enter these ecosystems through anthropogenic sources like urban runoff, industrial wastes, mine drainage, smelting, sewage sludge, coal-burning power plants, Zn-related manufacturing processes, transportation, atmospheric fallout, and continuous use of Zn fertilizer (Wuana and Okieimen 2011; Dong et al. 2012; Mateos-Naranjo et al. 2014; Noulas et al. 2018). These activities have resulted in concentration build-up of Zn in agricultural soil to the level that can affect soil properties like pH and microbial diversity (Gautam et al. 2016). The use of contaminated water for irrigation also allows Zn to leach into soils. Due to the contamination of soil and water, crops growing in such contaminated sites result in incorporation of Zn in food chain, affecting humans and other organisms.

3 Factors Influencing Zn Bioavailability in Soil

Soil plays an important role in the terrestrial environment as it influences element cycling, and heavy metals can harm soil organisms and plants (Wyszkowski and Modrzewska 2016). High concentrations of heavy metals in the soil can negatively impact productivity and fertility and have the potential to enter into the food chain, increasing the human health hazards risk (Mishra et al. 2019). Zinc is one of the essential metal that occurs abundantly in nature, with an average concentration of 70 mg/kg in the earth's crust (Ikediobi et al. 2004). It is present in soils in various forms, including free ions (Zn^{2+} and $ZnOH^{+}$); complexes with organic matter; colloidal fractions, for example, neutral sulphate ($ZnSO_4$) and ($ZnSO_4$) phosphate; and secondary minerals (Chahal et al. 2005; Hawkesford and Barraclough 2011). The concentrations of Zn in soils and rocks usually range from 10 to 300 ppm, whereas those in rivers are less than 0.01–0.2 ppm (Noulas et al. 2018). The normal range of Zn concentrations needed for most crops is roughly 15–20 mg/kg dry weight (Cakmak et al. 1996). Above these ranges, Zn can be hazardous to humans, animals, and plants (Cambier et al. 2009). The concentration of Zn in polluted soils has been measured to be between 150 and 300 mg/kg (Davies et al. 2002; Warne et al. 2008). Zn toxicity in plants increases in proportion to its bioavailability, which ultimately depends on some factors like soil pH, root exudates, soil microbial ecology and their activities, and soil organic matter (Duplay et al. 2014; Wyszkowski and Modrzewska 2016; Kwon et al. 2017).

The soil pH plays a significant impact on Zn solubility, sorption, speciation, and distribution in soil solutions (Adamczyk-Szabela et al. 2015). An acid pH enhances the bioavailability of Zn, unlike an alkaline pH (Pinto et al. 2014). Zn^{2+} is the predominant form of Zn at pH levels below 7.7, while $ZnOH^{+}$ is present between pH 7.7 and 9.1, and $Zn(OH)_2$ is predominant above that (Hawkesford and Barraclough 2011). High pH decreases Zn bioavailability due to decreased hydrogen bonds in the molecules of humic acids (DalCorso 2012; Leitenmaier and Küpper 2013). The concentration of Zn in *Valeriana officinalis* was found to reduce with increase in soil pH up to 10, due to reduced bioavailability (Adamczyk-Szabela et al. 2015). Increased pH values increase soil negative charge, promote co-precipitation of Zn in ferric oxide, mediate chemisorption on calcite, and facilitate conversion into hydrolyzed form, which increases its adsorption on soil particles, thereby reducing bioavailability (Rutkowska et al. 2015; Abedi et al. 2022). Soil organic matter greatly affects Zn availability by gradually releasing it and modifying the soil, which enhances root absorption through physicochemical changes (Cakmak 2009). Soil microbes can influence Zn solubilization and availability. Some microbes generate organic acids that solubilize zinc, enhancing its accessibility to plants, whereas others can sequester it, diminishing its availability (Singh et al. 2024). Soils with a high clay content bind Zn more strongly, reducing its availability to plants due to the high cation exchange capacity of clay particles (Alloway 2009). High concentrations of elements like phosphorus, calcium, magnesium, and bicarbonates can compete with Zn for absorption by plants, forming insoluble complexes and reducing its availability (Wan et al. 2024).

4 Adverse Effect of Excessive Zn on Soil Ecosystem and Plants

Zinc is a micronutrient that is essential for the growth of the plants, which partake in photosynthesis and energy processes, maintain bio-membrane integrity, and promote seed and generative organ development (Sturikova et al. 2018). Zinc is considered as a key element that plays an important role in the production of auxin, nucleic acid, and proteins, as well as the nitrogen metabolism regulations, cell division, photosynthesis, and nitrogen and phosphorous use during seed development (Rout and Das 2009; Shier 1994). It functions as a plant nutrient, but at higher concentrations, it becomes toxic (Welch and Shuman 1995; Liu et al. 2016; Meng et al. 2017; Zhang et al. 2017; Zhong et al. 2022) (Table 1). Unhealthy soil, stunted plant development, and impaired ecosystem functioning can result from high Zn concentrations in soil ecosystems.

Excessive Zn can alter soil structure by affecting soil colloids nutrient holding capacity, leading to reduced macronutrient availability (Abedi et al. 2022), decreased soil fertility, and impaired plant growth (Kaur and Garg 2021; Meng et al. 2023). Excess Zn in plants can result in abnormalities related to morphology, metabolism, and physiology (Balafrej et al. 2020). Zinc toxicity in vegetation inhibited root and shoot development (Kaur and Garg 2021; Meng et al. 2023). It can cause chlorosis in young leaves and spread to older ones after long-term exposure to high soil Zn levels (Ebbs and Kochian 1997). Variations in concentration, exposure duration, genotype, and developmental stage can impact chlorosis, cell death, and growth in plants due to Zn toxicity (Sturikova et al. 2018). Understanding Zn's impact on seed germination is critical for understanding its effects on plant growth and development, which vary according to plant species and Zn concentration (Balafrej et al. 2020). Elevated level of Zn is also well reported to produce adverse changes in root morphology which results in decreased growth and development of plants

Table 1 Adverse effect of Zn on plants

Sl. No.	Plant	Conc. of Zn	Adverse effect	References
(i)	*Solanum lycopersicum*	150–250 mg/kg	Decreased in growth and yield parameters such as length of the root and shoot, area of leaves	Vijayarengan and Mahalakshmi (2013)
(ii)	*Arabidopsis thaliana*	> 0.1 mM	Growth inhibition	Jain et al. (2013)
(iii)	Red amaranth *(Amaranthus* sp.*)*	200 ppm	Length of roots and shoots led to a decrease	Malik et al. (1970)
(iv)	*Triticum aestivum L*	8 mg/L	Resulting in adverse effects on germination, mitosis, and grain yield	Reis et al. (2018)
(v)	*Saccharum* spp.	65 and 130 mg/L	Stunted growth, with leaves appearing a deep shade of green, reduction in length of roots, and decrease in mitotic activity	Jain et al. (2010)
(vi)	*Ipomoea batatas*	40 μM	Caused an imbalance in mineral concentrations resulted in the reduction in the growth	Meng et al. (2023)
(vii)	*Vigna unguiculata*	>50 mg/L	Decreases morphophysiology parameters	Basha and Selvaraju (2015)
(viii)	*Soybean glycine*	750–1250 mg/kg	Exhibited a negative effect on growth factors and germination	Gupta et al. (2016)

Table 2 Adverse effect of Zn on soil microbes

Microbe	Conc. of Zn in soil (mg/kg)	Adverse effect	References
Bacillus cereus	762.6	Inhibition of growth and metabolic activity	Hrynkiewicz et al. (2012)
Organotrophic bacteria	1250	Decline in colony development index	Strachel et al. (2017)
Oligotrophic bacteria	1250	Changes in the diversity of the soil microbiome	Strachel et al. (2017)
Actinomycetes	1250	Reduction in the count of microorganisms	Strachel et al. (2017)

(Bochicchio et al. 2015; Emamverdian et al. 2015; Küpper and Andresen 2016; Mustafa and Komatsu 2016).

High Zn concentration can disrupt soil microbial communities, inhibits beneficial bacteria and fungi growth, reduces nutrient cycling, and decreases soil fertility (Table 2) (Sobolev and Begonia 2008). A high level of Zn has a detrimental effect on soil enzymes, including phosphatase, urease, and dehydrogenase. These enzymes are important for the organic matter breakdown and the nutrients availability in the soil. A decrease in enzyme activity can reduce soil fertility (Nannipieri et al. 2003). Moreover, Zn-polluted soil reduces the activity of earthworms and other microorganisms by interfering with the breakdown of organic matter (Llobet et al. 1988; Lev et al. 2010; Mahmood and Mohammed 2023).

5 Phytoremediation of Zn

Phytoremediation refers to plant-based bioremediation technologies that use green plants and their microbiota to treat contaminated soil and groundwater (Chakravarty et al. 2017; Tiwari et al. 2018; Mishra et al. 2019; Bauddh and Ma 2022). Plants with a high tolerance to toxicity can withstand metal stresses through reduced uptake or increased internal sequestration, so phytoremediation is an affordable solution for sites contaminated by heavy metals (Fig. 1) (Yang et al. 2005). Plant species that accumulate metals demonstrate ecological tolerance and evolution, which may facilitate the creation of new environmental monitoring and remediation technologies. Recent phytoextraction strategies utilize three types of heavy metal-extracting plants: hyperaccumulator plants, high-growth and biomass-producing plants, and genetically engineered plants for enhanced efficiency (Suman et al. 2018; Yan et al. 2020) (Fig. 3).

Different plant species are extensively used for the phytoremedation of Zn from the contaminated sites (Table 3). The plant species from *Brassicaceae* family have been well reported plants for the significant accumulation of Zn. The plant species *Brassica tournefortii*, *Phyla nodiflora*, *Pteris fauriei*, *Scariola orientalis*, and *Stipa barbata* are good examples of Zn accumulator plants.

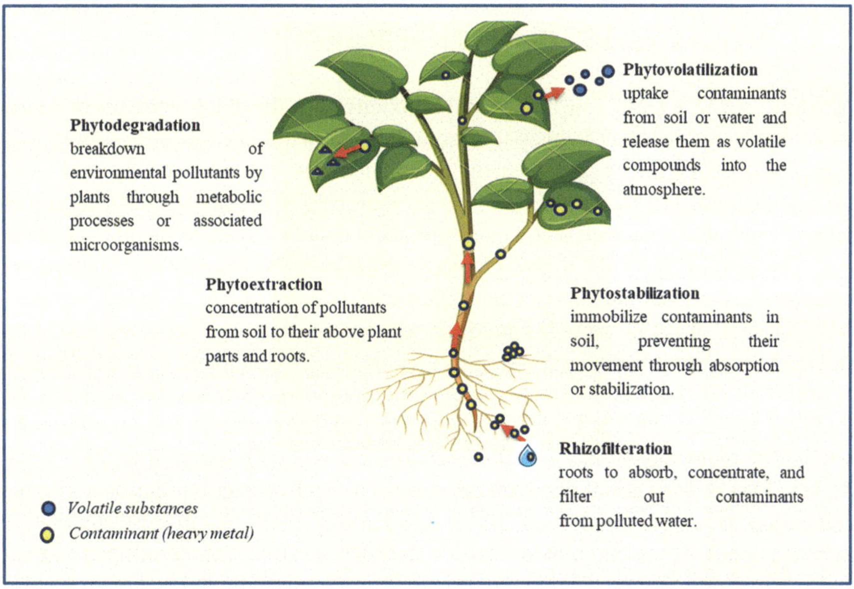

Fig. 3 Phytoremediation and its types. (Bauddh and Singh 2015a, b; Bauddh et al. 2016, 2017)

Table 3 Different plant species used for phytoremediation of Zn

Sl. No.	Plant name	Conc. of Zn in soil/water	Accumulation in (root/shoot)	References
1.	*Brassica tournefortii*	–	262.8 mg/kg in shoot	El-Keblawy et al. (2024)
2.	*Phyla nodiflora*	906 mg/kg	598 mg/kg in root and 110 mg/kg in shoot	Yoon et al. (2006)
3.	*Pteris fauriei*	2380 mg/kg	938 mg/kg in root and 690 mg/kg in shoot	Liu et al. (2008)
4.	*Scariola orientalis*	–	1208.3 mg/kg in root	Nouri et al. (2009)
5.	*Stipa barbata*	–	329.3 mg/kg in shoot	Nouri et al. (2009)
6.	*Alternanthera sessilis*	–	2366–3531 mg/g	Mazumdar and Das (2021)
7.	*Ipomoea aquatica*	–	2366–3531 mg/g	Mazumdar and Das (2021)
8.	*Celosia argentea* L.	300 mg/kg	171.4 mg/kg in leaves (shoot)	Hussain et al. (2024)
9.	*Helianthus annuus*	599 mg/kg	434 mg/kg in root 343 mg/kg in shoot	Paulo et al. (2023)
10.	*Reseda lutea* L.	12963.4 mg/kg	63.8 mg/kg to 12,003 mg/kg in roots	Lorestani et al. (2012)
11.	*Euphorbia macroclada*	12963.4 mg/kg	51 mg/kg to 8771.6 mg/kg in shoot	Lorestani et al. (2012)
12.	*Eleocharis acicularis*	2.0 and 4.0 mg/L	213 mg/kg in shoot/DW	Ha et al. (2011)

6 Hyperaccumulation of Zn in Plants

Although Zn in excess can induce phytotoxicity, certain plants have been found which have significant efficiency to accumulate Zn, without exhibiting sign of toxicity. Such plants are regarded as hyperaccumulators. "Hyper" means excessive, over, or beyond, and "accumulation" means acquiring or gradually gathering something. Plants that accumulate Zn mainly in their shoots and keep metal concentrations low in roots are known as Zn hyperaccumulators (Krämer 2010). Such plants possess array of mechanisms to deal with excess amount of Zn, including accumulation/storage of Zn in vacuoles and in large epidermis storage cells, facilitated by elevated transport protein expression in various tissues (Gupta et al. 2016; Vaculík et al. 2025). Hyperaccumulators can accumulate Cd more than 100 mg kg^{-1}, Ni 1000 mg kg^{-1}, and Pb, Cr, Cu, Co, Zn, and Mn >10,000 mg kg^{-1} (Baker and Brooks 1989). The plant species having efficiency to accumulate Zn higher than 10,000 mg Zn kg^{-1} DM in their aerial parts are considered as Zn hyperaccumulators during its cultivation in natural conditions (Baker and Brooks 1989). As hyperaccumulators exhibit potential to accumulate metals in their shoots, they can be utilized to treat and remediate soil contaminated with excessive amount of Zn. Several plant species have been reported to bear an efficiency to grow in the soil having high Zn

Table 4 Different plant species used for hyperaccumulation of Zn

Sl. No.	Plant name	Conc. of Zn in soil	Accumulation in (root/shoot)	References
(i)	*Thlaspi caerulescens*	50 to 25,000 mg/kg	21,000 to 30,000 mg/kg in shoot	Escarré et al. (2000)
(ii)	*Pisum sativum*	1656 mg/kg	18,700 mg/kg in shoot	Wioleta et al. (2015)
(iii)	*Arabidopsis halleri*	1000 μM	32,000 mg/kg in shoot	Zhao et al. (2000)
(iv)	*Dichapetalum gelonioides Dichapetalum* subsp. *sumatranum*	20 mg/kg	10,730 mg/kg in shoot	Nkrumah et al. (2018)
(v)	*Justicia procumbens*	2733–57,012 mg/kg	>10,000 mg/kg dry mass in shoot	Phaenark et al. (2009)

concentrations and also potential to bioaccumulate high level of Zn in roots along with transfer of Zn into their shoots, higher than 3000 ppm (Table 4) (Prasad 2008; Moreira et al. 2018; Tiecher et al. 2016). Metal accumulation is typically higher in shoots than roots (Baker et al. 1994). Zinc hyperaccumulation is common in various plant species, accounting for 1% of dry weight (Rascio 1977). Hyperaccumulator plants, which perform normal metabolic activities and physiological functions without visible stress symptoms, are classified as plants that grow under excess metal concentrations (Viehweger 2014; Sytar et al. 2021).

Shoot and root ratio greater than 1 categorize plants into Zn hyperaccumulators, while ratio less than 1 designates plants as non-accumulator (Gupta et al. 2016). Most Zn-hyperaccumulating plants are from the family Brassicaceae; however, several plants having excellent efficiency to accumulate Zn belong to Dichapetalaceae and Caryophyllaceae families (Clemens 2017; Reeves et al. 2018).

7 Mechanisms of Hyperaccumulation of Zn in Plants

There are around 500 plant species which are reported to absorb heavy metals/metalloids from the contaminated soils. Among these, more than 10–20 plant species are considered as Zn hyperaccumulators. The plant species which can accumulate Zn in higher amounts belongs to family Brassicaceae (Krämer 2010; White and Brown 2010). There are several factors responsible to make a plant suitable for the hyperaccumulation of Zn along with its transportation from roots to the shoots (Fig. 4). Genetic make-up of the plant species, defense mechanisms, plant roots systems, etc. are some common factors. Through comparative physiological, genomic, and proteomics studies, researchers have gained insight into the molecular mechanisms underlying Zn hyperaccumulation in the plants. Hyperaccumulator plants efficiently absorb Zn, transport them to shoots, and sequester them in ways that do not harm essential biochemical processes like photosynthesis. These plants develop a specialized mechanism to combat Zn toxicity, such as modifying

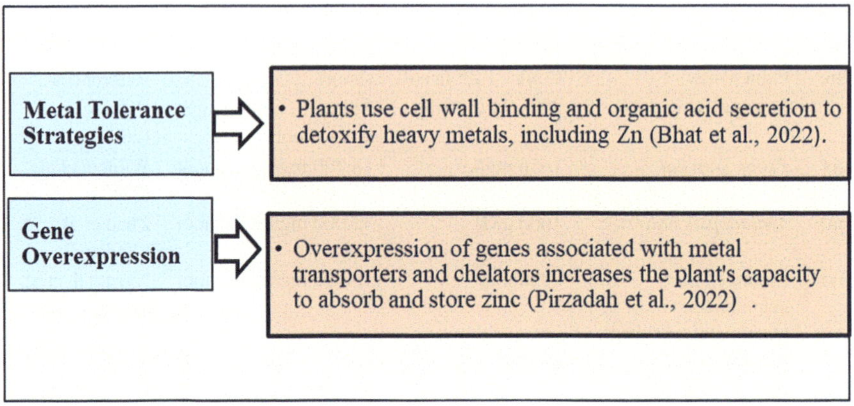

Fig. 4 Strategies to make a Zn hyperaccumulator plant

rhizosphere interactions, using Zn transporters, chelation processes, detoxification and tolerance, and strategies for reducing reactive oxygen species (ROS) caused by Zn (Gupta et al. 2016; Vaculík et al. 2025).

The rhizosphere is the first point of contact for all kind of metals, with Zn found in soil and absorbed in variety of forms, including changeable forms such as Zn^{2+}, $ZnOH^{+}$, and $ZnCl^{+}$ on clay surfaces and organic matter and crystalline forms such as iron-magnesium minerals and sulfide (ZnS), except at high pH, where Zn is being absorbed as ZnOH (Moreira et al. 2018; Balafrej et al. 2020). Plant roots absorb Zn through two main pathways:

(i) Symplastic route which involves the movement of Zn through root cell cytoplasm for energy-efficient transport to xylem.
(ii) Apoplastic route where Zn travels through cell walls and intercellular spaces, blocked by endodermis' Casparian strip (Gupta et al. 2016).

Several Zn transporters from different families play important roles in Zn uptake and distribution across the plant parts: Zn-regulated transporter and iron-regulated transporter-like proteins (ZIP). For example, in the plant *Sedum alfredii* SaZIP4 gene is a major Zn transporter found in the plasma membrane, allowing Zn to be get accumulated by both roots and shoots (Yang et al. 2018). ZIP6 in *Arabidopsis halleri* also promotes Zn uptake in shoots (Becher et al. 2004; Gupta et al. 2016). Several types of metal tolerance proteins play an important role in detoxification and sequestration of Zn.

Genetic manipulation in Zn-hyperaccumulator plants has been found to enhance the Zn accumulation during the cultivation in Zn-contaminated sites. These plants are designed in a way to hold significant potential for phytoremediation without exhibiting the adverse effects of the metals at excess concentrations. Genetic engineering, which employs gene editing, transgenic techniques, metabolic engineering, and other methods to increase a plant's ability to absorb and store Zn, is an effective tool for increasing Zn hyperaccumulation in plants. Subcellular compartmentation of Zn in the aerial parts of the plants is a key Zn detoxification mechanism. Several genes (e.g., *At*HMA3, *At*MTP1, *At*MTP3, *OsHMA3*, etc.) have been found for the vacuolar transport of Zn in different plant species (Desbrosses-Fonrouge et al. 2005; Arrivault et al. 2006; Morel et al. 2009; Miyadate et al. 2011). Subcellular compartmentation of Zn in the aerial parts of the plants is a key Zn detoxification mechanism. Several genes (e.g., *At*HMA3, *At*MTP1, *At*MTP3, *OsHMA3,* etc.) have been found for the vacuolar transport of Zn in different plant species (Desbrosses-Fonrouge et al. 2005; Arrivault et al. 2006; Morel et al. 2009; Miyadate et al. 2011).

Here are some notable examples of transgenic plants engineered for Zn accumulation.

(i.) *Thlaspi caerulescens* (*Alpine Pennycress*)

The plant, a natural Zn hyperaccumulator, has been extensively researched for phytoremediation and biofortification, with transgenic versions developed to further enhance Zn accumulation (Clemens 2017).

(ii.) *Arabidopsis thaliana*

The model organism has been genetically modified to overexpress Zn transporter genes, resulting in increased Zn accumulation in its tissues, surpassing non-modified plants (Lin et al. 2016).

(iii.) *Sedum alfredii*

Sedum alfredii, a plant known for its high Zn tolerance, has been developed into transgenic versions for study of Zn uptake, transport, phytoremediation, and biofortification (Zheng et al. 2024).

8 Conclusions

Excess amount of Zn in soil could have detrimental effect on soil microbes, plant growth, and productivity. Reducing level of Zn from soil is a warrant to improve food security. Several physical and chemical approaches can be used to alleviate the Zn contamination, but eventually raise environmental and economic cost. Employing green approach, however, generates desired result without hampering environmental and social integrity. Green approach like phytoremediation, which involves plants that can withstand toxic levels of metals by employing diverse mechanisms, can facilitate remediation of site. The plants employ phytoextraction, phytostabilization, phytodegradation, and phytovolatilization techniques. Phytoextraction method that involves hyperaccumulator plants that can accumulate excess amount of metal in aboveground parts could effectively facilitate Zn removal. Hyperaccumulators uptake and translocate Zn in aerial parts with the help of transporters and ultimately sequester them in vacuole, thereby rendering Zn-induced toxicity. This approach is sustainable but, however, has low efficiency due to slow growth and/or low biomass, and therefore, techniques like genetic engineering can be utilized to enhance the potential of hyperaccumulator plants. Genetic engineering technique can improve growth rate and biomass production of hyperaccumulator plants. Thus, focusing on research that promotes developing engineered phytoremediators can facilitate restoration of contaminated soil and fortify global food security.

References

Abedi T, Gavanji S, Mojiri A (2022) Lead and zinc uptake and toxicity in maize and their management. Plan Theory 11(15):1922

Adamczyk-Szabela D, Markiewicz J, Wolf WM (2015) Heavy metal uptake by herbs. IV. Influence of soil pH on the content of heavy metals in Valeriana officinalis L. Water Air Soil Pollut 226(4):106. https://doi.org/10.1007/s11270-015-2360-3

Alloway BJ (2009) Soil factors associated with zinc deficiency in crops and humans. Environ Geochem Health 31(5):537–548. https://doi.org/10.1007/s10653-009-9255-4

Arrivault S, Senger T, Krämer U (2006) The *Arabidopsis* metal tolerance protein AtMTP3 maintains metal homeostasis by mediating Zn exclusion from the shoot under Fe deficiency and Zn oversupply. Plant J 46:861–879

Baker AJ, Brooks R (1989) Terrestrial higher plants which hyperaccumulate metallic elements. A review of their distribution, ecology and phytochemistry. Biorecovery 1(2):81–126

Baker AJM, McGrath SP, Sidoli CMD, Reeves RD (1994) The possibility of in situ heavy metal decontamination of polluted soils using crops of metal-accumulating plants. Resour Conserv Recycl 11(1–4):41–49. https://doi.org/10.1016/0921-3449(94)90077-9

Balafrej H, Bogusz D, Triqui Z-EA, Guedira A, Bendaou N, Smouni A, Fahr M (2020) Zinc hyperaccumulation in plants: a review. Plan Theory 9(5):562. https://doi.org/10.3390/plants9050562

Barakat MA (2011) New trends in removing heavy metals from industrial wastewater. Arab J Chem 4(4):361–377. https://doi.org/10.1016/j.arabjc.2010.07.019

Basha SA, Selvaraju M (2015) Toxic effect of zinc on growth and nutrient accumulation of cow pea (*Vigna unguiculata* L.). Int Lett Nat Sci 43:48–53. https://doi.org/10.56431/p-6tf03z

Bauddh K, Ma Y (eds) (2022) Advances in microbe-assisted phytoremediation of polluted sites. Elsevier. https://doi.org/10.1016/C2020-0-00332-4

Bauddh K, Singh RP (2015a) Assessment of metal uptake capacity of castor bean and mustard for phytoremediation of nickel from contaminated soil. Biorem J 19(2):124–138. https://doi.org/10.1080/10889868.2014.979277

Bauddh K, Singh RP (2015b) Effects of organic and inorganic amendments on bio-accumulation and partitioning of cd in Brassica juncea and Ricinus communis. Ecol Eng 74:93–100. https://doi.org/10.1016/j.ecoleng.2014.10.022

Bauddh K, Singh K, Singh RP (2016) Ricinus communis L. a value added crop for remediation of cadmium contaminated soil. Bull Environ Contam Toxicol 96:265–269. https://doi.org/10.1007/s00128-015-1669-3

Bauddh K, Singh B, Korstad J (eds) (2017) Phytoremediation potential of bioenergy plants. Springer Singapore, Singapore. https://doi.org/10.1007/978-981-10-3084-0

Bauddh K, Korstad J, Sharma P (eds) (2020) Phytorestoration of abandoned mining and oil drilling sites. Elsevier. https://doi.org/10.1016/C2019-0-02457-0

Becher M, Talke IN, Krall L, Krämer U (2004) Cross-species microarray transcript profiling reveals high constitutive expression of metal homeostasis genes in shoots of the zinc hyperaccumulator *Arabidopsis halleri*. Plant J 37(2):251–268. https://doi.org/10.1046/j.1365-313X.2003.01959.x

Bochicchio R, Sofo A, Terzano R, Gattullo CE, Amato M, Scopa A (2015) Root architecture and morphometric analysis of Arabidopsis thaliana grown in Cd/Cu/Zn-gradient agar dishes: a new screening technique for studying plant response to metals. Plant Physiol Biochem 91:20–27. https://doi.org/10.1016/j.plaphy.2015.03.010

Cakmak I (2009) Enrichment of fertilizers with zinc: an excellent investment for humanity and crop production in India. J Trace Elem Med Biol 23(4):281–289. https://doi.org/10.1016/j.jtemb.2009.05.002

Cakmak I, Öztürk L, Karanlik S, Marschner H, Ekiz H (1996) Zinc-efficient wild grasses enhance release of phytosiderophores under zinc deficiency. J Plant Nutr 19(3–4):551–563. https://doi.org/10.1080/01904169609365142

Cambier P, Schvartz C, Van Oort F (2009) Contaminations métalliques des agrosystèmes et écosystèmes péri-industriels. Editions Quae

Chahal DS, Sharma BD, Singh PK (2005) Distribution of forms of zinc and their association with soil properties and uptake in different soil orders in semi-arid soils of Punjab, India. Commun Soil Sci Plant Anal 36(19–20):2857–2874. https://doi.org/10.1080/00103620500306031

Chakravarty P, Bauddh K, Kumar M (2017) Phytoremediation: a multidimensional and ecologically viable practice for the cleanup of environmental contaminants. In: Phytoremediation potential of bioenergy plants, pp 1–46. https://doi.org/10.1007/978-981-10-3084-0_1

Clemens S (2017) How metal hyperaccumulating plants can advance Zn biofortification. Plant Soil 411(1):111–120. https://doi.org/10.1007/s11104-016-2920-3

DalCorso G (2012) Heavy metal toxicity in plants. In: Furini A (ed) Plants and heavy metals. Springer, Netherlands, pp 1–25. https://doi.org/10.1007/978-94-007-4441-7_1

Davies FT, Puryear JD, Newton RJ, Egilla JN, Saraiva Grossi JA (2002) Mycorrhizal fungi increase chromium uptake by sunflower plants: influence on tissue mineral concentration, growth, and gas exchange. J Plant Nutr 25(11):2389–2407. https://doi.org/10.1081/PLN-120014702

Desbrosses-Fonrouge AG, Voigt K, Schröder A, Arrivault S, Thomine S, Krämer U (2005) Arabidopsis thaliana MTP1 is a Zn transporter in the vacuolar membrane which mediates Zn detoxification and drives leaf Zn accumulation. FEBS Lett 579(19):4165–4174. https://doi.org/10.1016/j.febslet.2005.06.046

Dixit R, Wasiullah X, Malaviya D, Pandiyan K, Singh UB, Sahu A et al (2015) Bioremediation of heavy metals from soil and aquatic environment: an overview of principles and criteria of fundamental processes. Sustain For 7(2):2189–2212. https://doi.org/10.3390/su7022189

Dong CD, Chen CF, Chen CW (2012) Contamination of zinc in sediments at river mouths and channel in northern Kaohsiung Harbor, Taiwan. Int J Environ Sci Dev 3(6):517. https://doi.org/10.7763/IJESD.2012.V3.278

Duplay J, Semhi K, Errais E, Imfeld G, Babcsanyi I, Perrone T (2014) Copper, zinc, lead and cadmium bioavailability and retention in vineyard soils (Rouffach, France): the impact of cultural practices. Geoderma 230–231:318–328. https://doi.org/10.1016/j.geoderma.2014.04.022

Ebbs SD, Kochian LV (1997) Toxicity of zinc and copper to brassica species: implications for phytoremediation. J Environ Qual 26(3):776–781. https://doi.org/10.2134/jeq1997.00472425002600030026x

El-Keblawy A, Almehdi AM, Elsheikh EAE, Abouleish MY, Sheteiwy MS, Galal TM (2024) Native desert plants have the potential for phytoremediation of phytotoxic metals in urban cities: implications for cities sustainability in arid environments. Sci Rep 14(1):13761. https://doi.org/10.1038/s41598-024-62622-x

Emamverdian A, Ding Y, Mokhberdoran F, Xie Y (2015) Heavy metal stress and some mechanisms of plant defense response. Sci World J 2015(1):756120. https://doi.org/10.1155/2015/756120

Escarré J, Lefèbvre C, Gruber W, Leblanc M, Lepart J, Rivière Y, Delay B (2000) Zinc and cadmium hyperaccumulation by Thlaspi caerulescens from metalliferous and nonmetalliferous sites in the Mediterranean area: implications for phytoremediation. New Phytol 145(3):429–437. https://doi.org/10.1046/j.1469-8137.2000.00599.x

Gautam PK, Gautam RK, Banerjee S, Chattopadhyaya MC, Pandey JD (2016) Heavy metals in the environment: fate, transport, toxicity and remediation technologies. Nova Sci Publishers 60:101–130

Gupta M, Kumari A, Gauri S (2014) Bioremediation of heavy metals polluted environment by using resistance bacteria. Int J Environ Res Dev 8:883–889

Gupta N, Ram H, Kumar B (2016) Mechanism of zinc absorption in plants: uptake, transport, translocation and accumulation. Rev Environ Sci Biotechnol 15(1):89–109. https://doi.org/10.1007/s11157-016-9390-1

Ha NTH, Sakakibara M, Sano S (2011) Accumulation of indium and other heavy metals by Eleocharis acicularis: an option for phytoremediation and phytomining. Bioresour Technol 102(3):2228–2234. https://doi.org/10.1016/j.biortech.2010.10.014

Hawkesford MJ, Barraclough P (2011) The molecular and physiological basis of nutrient use efficiency in crops. Wiley

Henry JR (2000) An overview of the phytoremediation of lead and mercury. National Network of Environmental management Studies (NNEMS) Fellow

Hrynkiewicz K, Dabrowska G, Baum C, Niedojadlo K, Leinweber P (2012) Interactive and single effects of Ectomycorrhiza formation and Bacillus cereus on Metallothionein MT1 expression and Phytoextraction of Cd and Zn by willows. Water Air Soil Pollut 223(3):957–968. https://doi.org/10.1007/s11270-011-0915-5

Hussain U, Afza R, Gul I, Sajad MA, Shah GM, Muhammad Z, Khan SM (2024) Phytoremediation of heavy metals spiked soil by Celosia argentea L.: effect on plant growth and metal stabilization. Environ Sci Pollut Res 31(10):15339–15347. https://doi.org/10.1007/s11356-024-32176-6

Ikediobi CO, Badisa VL, Ayuk-Takem LT, Latinwo LM, West J (2004) Response of antioxidant enzymes and redox metabolites to cadmium-induced oxidative stress in CRL-1439 normal rat liver cells. Int J Mol Med 14(1):87–92. https://doi.org/10.3892/ijmm.14.1.87

Jain R, Srivastava S, Solomon S, Shrivastava AK, Chandra A (2010) Impact of excess zinc on growth parameters, cell division, nutrient accumulation, photosynthetic pigments and oxidative stress of sugarcane (Saccharum spp.). Acta Physiol Plant 32(5):979–986. https://doi.org/10.1007/s11738-010-0487-9

Jain A, Sinilal B, Dhandapani G, Meagher RB, Sahi SV (2013) Effects of deficiency and excess of zinc on morphophysiological traits and spatiotemporal regulation of zinc-responsive genes reveal incidence of cross talk between micro- and macronutrients. Environ Sci Technol 47(10):5327–5335. https://doi.org/10.1021/es400113y

Kaur H, Garg N (2021) Zinc toxicity in plants: a review. Planta 253(6):129. https://doi.org/10.1007/s00425-021-03642-z

Krämer U (2010) Metal hyperaccumulation in plants. Annu Rev Plant Biol 61(1):517–534. https://doi.org/10.1146/annurev-arplant-042809-112156

Küpper H, Andresen E (2016) Mechanisms of metal toxicity in plants. Metallomics 8(3):269–285. https://doi.org/10.1039/C5MT00244C

Kwon MJ, Boyanov MI, Yang JS, Lee S, Hwang YH, Lee JY et al (2017) Transformation of zinc-concentrate in surface and subsurface environments: implications for assessing zinc mobility/toxicity and choosing an optimal remediation strategy. Environ Pollut 226:346–355. https://doi.org/10.1016/j.envpol.2017.01.066

Lehmann A, Veresoglou SD, Leifheit EF, Rillig MC (2014) Arbuscular mycorrhizal influence on zinc nutrition in crop plants–a meta-analysis. Soil Biol Biochem 69:123–131. https://doi.org/10.1016/j.soilbio.2013.11.001

Leitenmaier B, Küpper H (2013) Compartmentation and complexation of metals in hyperaccumulator plants. Front Plant Sci 4:374. https://doi.org/10.3389/fpls.2013.00374

Leonard SS, Harris GK, Shi X (2004) Metal-induced oxidative stress and signal transduction. Free Radic Biol Med 37(12):1921–1942. https://doi.org/10.1016/j.freeradbiomed.2004.09.010

Lev SM, Matthies N, Snodgrass JW, Casey RE, Ownby DR (2010) Effects of zinc exposure on earthworms, Lumbricus terrestris, in an artificial soil. Bull Environ Contam Toxicol 84:687–691. https://doi.org/10.1007/s00128-010-0002-4

Lin Y-F, Hassan Z, Talukdar S, Schat H, Aarts MGM (2016) Expression of the ZNT1 zinc transporter from the metal Hyperaccumulator Noccaea caerulescens confers enhanced zinc and cadmium tolerance and accumulation to Arabidopsis thaliana. PLoS One 11(3):e0149750. https://doi.org/10.1371/journal.pone.0149750

Liu X, Gao Y, Khan S, Duan G, Chen A, Ling L, Zhao L, Liu Z, Wu X (2008) Accumulation of Pb, Cu, and Zn in native plants growing on contaminated sites and their potential accumu-

lation capacity in Heqing, Yunnan. J Environ Sci 20(12):1469–1474. https://doi.org/10.1016/S1001-0742(08)62551-6
Liu H, Soomro A, Zhu Y, Qiu X, Chen K, Zheng T et al (2016) QTL underlying iron and zinc toxicity tolerances at seedling stage revealed by two sets of reciprocal introgression populations of rice (Oryza sativa L.). Crop J 4(4):280–289
Llobet JM, Domingo JL, Colomina MT, Mayayo E, Corbella J (1988) Subchronic oral toxicity of zinc in rats. Bull Environ Contam Toxicol 41(1):36–43. https://doi.org/10.1007/BF01689056
Lorestani B, Cheraghi M, Yousefi N (2012) The potential of phytoremediation using hyperaccumulator plants: a case study at a Lead-zinc mine site. Int J Phytoremediation 14(8):786–795. https://doi.org/10.1080/15226514.2011.619594
Mahmood SM, Mohammed AM (2023) Effects of zinc accumulation on earthworm Aporrectodea caliginosa (Haplotaxida: Lumbricidae). Jurnal Riset Biologi dan Aplikasinya 5(1):8–15. https://doi.org/10.26740/jrba.v5n1.p8-15
Malik N, Chamon A, Mondol M, Elahi S, Faiz S (1970) Effects of different levels of zinc on growth and yield of red amaranth (*Amaranthus sp.*) and rice (*Oryza sativa*, variety-BR49). J Bangladesh Assoc Young Res 1(1):79–91. https://doi.org/10.3329/jbayr.v1i1.6836
Mani S, Bharagava RN (2016) Exposure to crystal violet, its toxic, genotoxic and carcinogenic effects on environment and its degradation and detoxification for environmental safety. Rev Environ Contam Toxicol 237:71–104. https://doi.org/10.1007/978-3-319-23573-8_4
Marschner H (ed) (2011) Marschner's mineral nutrition of higher plants. Academic Press
Mateos-Naranjo E, Castellanos EM, Perez-Martin A (2014) Zinc tolerance and accumulation in the halophytic species Juncus acutus. Environ Exp Bot 100:114–121. https://doi.org/10.1016/j.envexpbot.2013.12.023
Mazumdar K, Das S (2021) Phytoremediation of soil treated with metalliferous leachate from an abandoned industrial site by Alternanthera sessilis and Ipomoea aquatica: metal extraction and biochemical responses. Ecol Eng 170:106349. https://doi.org/10.1016/j.ecoleng.2021.106349
Meng LJ et al (2017) Association mapping of ferrous, zinc, and aluminum tolerance at the seedling stage in *indica* rice using MAGIC populations. Front Plant Sci 8:1822
Meng Y, Xiang C, Huo J, Shen S, Tang Y, Wu L (2023) Toxicity effects of zinc supply on growth revealed by physiological and transcriptomic evidences in sweet potato (Ipomoea batatas (L.) Lam). Sci Rep 13(1):19203. https://doi.org/10.1038/s41598-023-46504-2
Mildvan AS (1970) 9 metals in enzyme Catalsis. In: The enzymes, vol 2. Elsevier, pp 445–536. https://doi.org/10.1016/S1874-6047(08)60188-2
Mishra S, Bharagava RN, More N, Yadav A, Zainith S, Mani S, Chowdhary P (2019) Heavy metal contamination: an alarming threat to environment and human health. In: Environmental biotechnology: for sustainable future, pp 103–125. https://doi.org/10.1007/978-981-10-7284-0_5
Miyadate H, Adachi S, Hiraizumi A, Tezuka K, Nakazawa N, Kawamoto T et al (2011) OsHMA3, a P1B-type of ATPase affects root-to-shoot cadmium translocation in rice by mediating efflux into vacuoles. New Phytol 189(1):190–199. https://doi.org/10.1111/j.1469-8137.2010.03459.x
Moreira A, Moraes LA, dos Reis AR (2018) The molecular genetics of zinc uptake and utilization efficiency in crop plants. In: Plant micronutrient use efficiency, pp 87–108. https://doi.org/10.1016/B978-0-12-812104-7.00006-X
Morel M, Crouzet J, Gravot A, Auroy P, Leonhardt N, Vavasseur A, Richaud P (2009) AtHMA3, a P1B-ATPase allowing Cd/Zn/co/Pb vacuolar storage in Arabidopsis. Plant Physiol 149(2):894–904. https://doi.org/10.1104/pp.108.130294
Mustafa G, Komatsu S (2016) Toxicity of heavy metals and metal-containing nanoparticles on plants. Biochim Biophys Acta BBA Proteins Proteom 1864:932–944. https://doi.org/10.1016/j.bbapap.2016.02.020
Nagajyoti PC, Lee KD, Sreekanth TVM (2010) Heavy metals, occurrence and toxicity for plants: a review. Environ Chem Lett 8(3):199–216. https://doi.org/10.1007/s10311-010-0297-8
Nannipieri P, Ascher J, Ceccherini MT, Landi L, Pietramellara G, Renella G (2003) Microbial diversity and soil functions. Eur J Soil Sci 54(4):655–670. https://doi.org/10.1046/j.1351-0754.2003.0556.x

Nkrumah PN, Echevarria G, Erskine PD, van der Ent A (2018) Contrasting nickel and zinc hyperaccumulation in subspecies of Dichapetalum gelonioides from Southeast Asia. Sci Rep 8(1):9659. https://doi.org/10.1038/s41598-018-26859-7

Noulas C, Tziouvalekas M, Karyotis T (2018) Zinc in soils, water and food crops. J Trace Elem Med Biol 49:252–260. https://doi.org/10.1016/j.jtemb.2018.02.009

Nouri J, Khorasani N, Lorestani B, Karami M, Hassani AH, Yousefi N (2009) Accumulation of heavy metals in soil and uptake by plant species with phytoremediation potential. Environ Earth Sci 59(2):315–323. https://doi.org/10.1007/s12665-009-0028-2

Pandey VC, Bauddh K (eds) (2018) Phytomanagement of polluted sites: market opportunities in sustainable phytoremediation. Elsevier

Paulo AMS, Caetano NS, Castro PML, Marques APGC (2023) Phytomanagement of Zn- and cd-contaminated soil: Helianthus annuus biomass production and metal remediation abilities with plant-growth-promoting microbiota assistance. Soil Syst 7(3):69. https://doi.org/10.3390/soilsystems7030069

Phaenark C, Pokethitiyook P, Kruatrachue M, Ngernsansaruay C (2009) Cd and Zn accumulation in plants from the padaeng zinc mine area. Int J Phytoremediation 11(5):479–495. https://doi.org/10.1080/15226510802656243

Pinto E, Aguiar AARM, Ferreira IMPLVO (2014) Influence of soil chemistry and plant physiology in the phytoremediation of Cu, Mn, and Zn. Crit Rev Plant Sci 33(5):351–373. https://doi.org/10.1080/07352689.2014.885729

Prasad MNV (2008) Trace elements as contaminants and nutrients: consequences in ecosystems and human health. Wiley

Rajkumar V, Lee VR, Gupta V (2020) Heavy metal toxicity. StatPearls Publishing, Treasure Island (FL). Available from: https://www.ncbi.nlm.nih.gov/books/NBK560920/

Rascio N (1977) Metal accumulation by some plants growing on zinc-mine deposits. Oikos 29(2):250. https://doi.org/10.2307/3543610

Reeves RD, Baker AJM, Jaffré T, Erskine PD, Echevarria G, van der Ent A (2018) A global database for plants that hyperaccumulate metal and metalloid trace elements. New Phytol 218(2):407–411

Reis S, Pavia I, Carvalho A, Moutinho-Pereira J, Correia C, Lima-Brito J (2018) Seed priming with iron and zinc in bread wheat: effects in germination, mitosis and grain yield. Protoplasma 255(4):1179–1194. https://doi.org/10.1007/s00709-018-1222-4

Rout GR, Das P (2009) Effect of metal toxicity on plant growth and metabolism: I. Zinc. In: Lichtfouse E, Navarrete M, Debaeke P, Véronique S, Alberola C (eds) Sustainable agriculture. Springer, Netherlands, pp 873–884. https://doi.org/10.1007/978-90-481-2666-8_53

Rutkowska B, Szulc W, Bomze K, Gozdowski D, Spychaj-Fabisiak E (2015) Soil factors affecting solubility and mobility of zinc in contaminated soils. Int J Environ Sci Technol 12:1687–1694

Shier WT (1994) Metals as toxins in plants. J Toxicol Toxin Rev 13(2):205–216. https://doi.org/10.3109/15569549409089960

Singh S, Chhabra R, Sharma A, Bisht A (2024) Harnessing the power of zinc-solubilizing bacteria: a catalyst for a sustainable Agrosystem. Bacteria 3(1):15–29. https://doi.org/10.3390/bacteria3010002

Sobolev D, Begonia M (2008) Effects of heavy metal contamination upon soil microbes: lead-induced changes in general and denitrifying microbial communities as evidenced by molecular markers. Int J Environ Res Public Health 5(5):450–456. https://doi.org/10.3390/ijerph5050450

Strachel R, Wyszkowska J, Baćmaga M (2017) Microbiome of soil contaminated with zinc. J Elem 22(3):907–919. https://doi.org/10.5601/jelem.2016.21.4.1322

Sturikova H, Krystofova O, Huska D, Adam V (2018) Zinc, zinc nanoparticles and plants. J Hazard Mater 349:101–110. https://doi.org/10.1016/j.jhazmat.2018.01.040

Suman J, Uhlik O, Viktorova J, Macek T (2018) Phytoextraction of heavy metals: a promising tool for clean-up of polluted environment? Front Plant Sci 9:1476. https://doi.org/10.3389/fpls.2018.01476

Sytar O, Ghosh S, Malinska H, Zivcak M, Brestic M (2021) Physiological and molecular mechanisms of metal accumulation in hyperaccumulator plants. Physiol Plant 173(1):148–166. https://doi.org/10.1111/ppl.13285

Tiecher TL, Ceretta CA, Tiecher T, Ferreira PAA, Nicoloso FT, Soriani HH, Rossato LV, Mimmo T, Cesco S, Lourenzi CR, Giachini AJ, Brunetto G (2016) Effects of zinc addition to a copper-contaminated vineyard soil on sorption of Zn by soil and plant physiological responses. Ecotoxicol Environ Saf 129:109–119. https://doi.org/10.1016/j.ecoenv.2016.03.016

Tiwari J, Ankit S, Kumar S, Korstad J, Bauddh K (2018) Ecorestoration of polluted aquatic ecosystems through rhizofiltration. In: Pandey VC, Bauddh K (eds) Phytomanagement of polluted sites. Elsevier, pp 179–201. https://doi.org/10.1016/B978-0-12-813912-7.00005-3

Vaculík M, Mišljenović T, Lukačová Z, Jakovljević K, Podar D, Kováč J (2025) Zinc hyperaccumulation in plants: mechanisms and principles. In: Zinc in plants. Academic Press, pp 1–39. https://doi.org/10.1016/B978-0-323-91314-0.00017-X

Viehweger K (2014) How plants cope with heavy metals. Bot Stud 55(1):35. https://doi.org/10.1186/1999-3110-55-35

Vijayarengan P, Mahalakshmi G (2013) Zinc toxicity in tomato plants. World Appl Sci J 24(5):649–653. https://doi.org/10.5829/idosi.wasj.2013.24.05.2249

Wan Y, Liu J, Zhuang Z, Wang Q, Li H (2024) Heavy metals in agricultural soils: sources, influencing factors, and remediation strategies. Toxics 12(1):63. https://doi.org/10.3390/toxics12010063

Warne MSJ, Heemsbergen D, Stevens D, McLaughlin M, Cozens G, Whatmuff M, Broos K, Barry G, Bell M, Nash D, Pritchard D (2008) Modeling the toxicity of copper and zinc salts to wheat in 14 soils. Environ Toxicol Chem 27(4):786–792. https://doi.org/10.1897/07-294.1

Welch RM, Shuman L (1995) Micronutrient nutrition of plants. Crit Rev Plant Sci 14(1):49–82. https://doi.org/10.1080/07352689509701922

White PJ, Brown P (2010) Plant nutrition for sustainable development and global health. Ann Bot 105(7):1073–1080. https://doi.org/10.1093/aob/mcq085

Wioleta W, Anna D, Ilona B, Kamila K, Elżbieta R (2015) Lead induced changes in phosphorylation of PSII proteins in low light grown pea plants. Biometals 28(1):151–162. https://doi.org/10.1007/s10534-014-9811-y

Wuana RA, Okieimen FE (2011) Heavy metals in contaminated soils: a review of sources, chemistry, risks and best available strategies for remediation. Int Schol Res Notices 2011(1):402647. https://doi.org/10.5402/2011/402647

Wyszkowski M, Modrzewska B (2016) Acidity and sorption properties of zinc-contaminated soil following the application of neutralising substances. J Ecol Eng 17(1):63–68. https://doi.org/10.12911/22998993/61191

Yan A, Wang Y, Tan SN, Mohd Yusof ML, Ghosh S, Chen Z (2020) Phytoremediation: a promising approach for revegetation of heavy metal-polluted land. Front Plant Sci 11:359

Yang X-E, Jin X-F, Feng Y, Islam E (2005) Molecular mechanisms and genetic basis of heavy metal tolerance/hyperaccumulation in plants. J Integr Plant Biol 47(9):1025–1035. https://doi.org/10.1111/j.1744-7909.2005.00144.x

Yang Q, Ma X, Luo S, Gao J, Yang X, Feng Y (2018) SaZIP4, an uptake transporter of Zn/Cd hyperaccumulator sedum alfredii Hance. Environ Exp Bot 155:107–117. https://doi.org/10.1016/j.envexpbot.2018.06.021

Yoon J, Cao X, Zhou Q, Ma LQ (2006) Accumulation of Pb, Cu, and Zn in native plants growing on a contaminated Florida site. Sci Total Environ 368(2–3):456–464. https://doi.org/10.1016/j.scitotenv.2006.01.016

Zhang J, Chen K, Pang Y, Naveed SA, Zhao X, Wang X et al (2017) QTL mapping and candidate gene analysis of ferrous iron and zinc toxicity tolerance at seedling stage in rice by genome-wide association study. BMC Genomics 18:1–15. https://doi.org/10.1186/s12864-017-4221-5

Zhao FJ, Lombi E, Breedon TMSP (2000) Zinc hyperaccumulation and cellular distribution in *Arabidopsis halleri*. Plant Cell Environ 23(5):507–514. https://doi.org/10.1046/j.1365-3040.2000.00569.x

Zheng J, Yue Y, Zhu Y, Wang Y, Zheng W, Hu L, Hou D, Wang F, Yang L, Zhang H (2024) Zinc enhances cadmium accumulation in shoots of Hyperaccumulator Solanum nigrum by improving ATP-dependent transport and alleviating toxicity. Plan Theory 13(17):2528. https://doi.org/10.3390/plants13172528

Zhong K, Xie L, Hu S, Shao G, Sheng Z, Jiao G et al (2022) Genome-wide association study of zinc toxicity tolerance within a rice core collection (Oryza sativa L.). Plan Theory 11(22):3138. https://doi.org/10.3390/plants11223138

Microbes as Tools for the Amelioration of Zinc Phytotoxicity

Dhanyashree Rai, Neeksha, J. R. Shubha, K. Shama Prakash, and M. Divyashree

Abstract

Heavy metal contamination significantly threatens agricultural production and environmental sustainability, particularly zinc (Zn) pollution. Excessive zinc accumulation in soil causes phytotoxicity, which affects plant development and soil health. It is the most common microelement phytotoxicity, exceeding Co, Ni, Cd, Cu, and other metals. Anthropogenic contamination from various sources, including pesticides, manures, fertilizers, smelters, incinerators, sewage sludge, galvanized goods, and mines, has led to increased distribution of Zn and reached phytotoxic levels in many soils. Traditional remediation methods fail to address these challenges, prompting the exploration of alternative, eco-friendly strategies. Using beneficial microorganisms to remove heavy metals is a possible strategy for addressing these adverse effects. Microbes play pivotal roles in enhancing plant tolerance to heavy metal stress through metal solubilization, immobilization, and detoxification. Specific groups of microbes, including plant growth-promoting rhizobacteria (PGPR), mycorrhizal fungi, and metal-resistant bacteria, are discussed for their beneficial effects in remediation processes. By leveraging microbial diversity and metabolic versatility, innovative solutions can be developed to address zinc phytotoxicity sustainably, promoting agricultural productivity and environmental health. Harnessing microbial potential for sustainable

D. Rai · Neeksha · J. R. Shubha · M. Divyashree (✉)
Nitte University Center for Science Education and Research, Nitte (Deemed to be University), Derelakatte, Mangaluru, Karnataka, India
e-mail: dhanyashree.21phdbs107@student.nitte.edu.in; neeksha.21phdbs110@student.nitte.edu.in; shubha.jr@nitte.edu.in; divyashree.m@nitte.edu.in

K. Shama Prakash
K. S Hegde Medical Academy, Nitte (Deemed to be University), Derelakatte, Mangaluru, India
e-mail: ksprakash@nitte.edu.in

S. Srivastava, H. Kaur (eds.), *Zinc in Soil-Plant Continuum*,
https://doi.org/10.1007/978-981-96-4253-3_11

remediation of zinc-contaminated soils is gaining interest. Use of microorganisms and excellent knowledge of their complicated association may lead to better Zn phytoremediation strategies.

Keywords

Arbuscular mycorrhizae fungi · Endophytes · Iron deficiency · Oxidative stress · Siderophores

1 Introduction

Zinc is one of the most prevalent heavy metals available, and plants from the soil primarily absorb it and are essential for their growth, development, and overall health. Zinc is indispensable for various physiological processes in plants, from essential metabolic functions to stress tolerance and defense mechanisms. While zinc is a necessary micronutrient for plant growth at low concentrations, excessive levels can lead to phytotoxicity, causing detrimental effects on plant physiology, development, and ultimately crop yield and even pose health risks to humans after consuming through bioaccumulation (Bolan et al. 2014).

The adverse impacts of heavy metal contamination on ecosystems and human health have become increasingly evident in recent years. Among the heavy metals, zinc (Zn) is a prevalent pollutant due to its widespread industrial applications and natural occurrence in soils. Anthropogenic events, such as unlimited utilization of agrochemicals, have significantly increased their concentration globally. Excess Zn accumulation in plant cytosol harms plants, so it is immediately translocated in the cell vacuoles released on plant demand (Vendi Pande et al. 2022). Traditional techniques for remediating zinc-contaminated soils, such as soil excavation and chemical treatments, are typically expensive and environmentally damaging. Several methods have been developed to remove heavy metals from sludge, water, and soil by chlorination, electrochemical processes, membrane technology, ion exchange reverse osmosis, chelating agents, acid treatment at high temperatures, etc. These technologies have been shown to be expensive, labor-intensive operations, and less efficient (Awasthi et al. 2017). To overcome all these problems, indigenous microbes and plant microbes' symbionts have gained interest. Microbes can tolerate heavy metal concentrations in different polluted sites and play a significant role in the soil's regulation of zinc and other metals (Ge et al. 2009). Soil bacteria are an important component of the ecosystem and have an intricate relationship with plants. They help plants grow directly by delivering nutrients and growth regulators. Zinc (Zn^{+2} cations) cannot diffuse across cell membranes; hence, specific Zn transporters (Zn-regulated transporter (ZRT), iron-regulated transporter (IRT)-like proteins broadly classified as ZIP protein family) are required for Zn ion homeostasis by regulating Zn uptake and transport.

Furthermore, ZIP genes are responsible for translocating, detoxifying, and storing Zn or Fe in plant cells (Jain et al. 2020). Bioremediation techniques offer a cost-effective and environmentally friendly approach to restoring polluted soil compared to traditional remediation methods. It involves using biological agents, such as microorganisms and plants, to degrade, transform, or immobilize contaminants, restoring the soil to a less toxic or nontoxic state (Romantschuk et al. 2023). High concentrations of heavy metals in the environment produce selection pressure for the establishment of metal-tolerant bacterial strains in such places. These bacteria may alter the reactivity and mobility of heavy metals and can be used to detoxify some metals, avoiding future metal pollution. Microorganisms, such as certain bacteria and fungi, typically possess the ability to reduce the mobility and bioavailability of zinc, thereby mitigating zinc phytotoxicity through various mechanisms such as zinc immobilization, phytostabilization, phytoextraction assistance, bioremediation, induced systemic resistance (ISR), and bioaugmentation.

Specific groups of microbes, including plant growth-promoting rhizobacteria (PGPR), mycorrhizal fungi, and metal-resistant bacteria, are conferred for their beneficial effects in remediation processes. This book chapter addresses the physiological effects of zinc toxicity in plants and the potential of microorganisms in decreasing zinc phytotoxicity, providing insights into their mechanisms of action and practical applications in sustainable remediation efforts.

2 Zinc Toxicity in Plants

Zinc (Zn) is an essential nutrient all species require for various physiological functions and processes. In plants, it is necessary to activate numerous plant enzymes, including alcohol dehydrogenase, carbonic anhydrase, and RNA polymerase (Castillo-González et al. 2018). Zn forms tetrahydric complexes with N and O and is notably related to S in a wide range of organic molecules. Because plants absorb it quickly, it may be quite hazardous. Zinc toxicity commonly causes growth inhibitions (Rout and Das 2009), whereas Zn shortage causes complicated plant growth and development alterations. However, certain modifications are consistent with the micronutrient's activities in specific reactions or metabolic pathways. Zn shortage causes changes in plant metabolism, affecting carbohydrates, proteins, auxins, and membrane integrity. Many zinc-dependent enzymes are involved in glucose metabolism, particularly in leaves (Hamzah Saleem et al. 2022). When Zn shortage arises, carbonic anhydrase activity drops dramatically. In addition to its role in ribosome structural integrity, zinc also performs important bio-physicochemical roles in plants, such as gene regulation and activation, protein synthesis, carbohydrate metabolism, and morphological and anatomical involvement in biomembranes (Zaheer et al. 2022). Furthermore, Zn phospholipids interact with sulfhydryl clusters of membrane proteins, which promotes membrane stability (Kareem et al. 2022). In addition, the harmful effects of metals are also studied on photosynthesis in plants (Clijsters and Van Assche 1985, Garty et al. 1992). Inhibition of photosynthesis in *Phaseolus vulgaris* was observed after treatment with toxic zinc concentration: effect on ribulose-1, 5-bisphosphate carboxylase/oxygenase (Van Assche and Clijsters et al. 1986). Recent research on hazardous metals and their interactions with critical elements has led to a better understanding of toxicity at the biochemical level.

Zinc toxicity is determined by pH, which controls the zinc concentration in the solution. High amounts of zinc can harm plants (Davis-Carter and Shuman 1993). The primary symptoms include shoot stunting, curling, and rolling of young leaves, leaf tip death, and chlorosis. In a study by Ye et al. (1997), *Typha latifolia* seedlings became chlorotic when exposed to ~80 μM zinc. Foliar chlorosis emerged on *Thlaspi ochroleucum* seedlings treated with 1 mM Zn after 4 days of growth, according to Shen et al. (1997).

3 Effect on Germination

While zinc is necessary for plant development, high quantities can limit growth and cause toxicity symptoms. Baker found that *Silene maritima* seeds germinated more quickly and efficiently in calcium nitrate solutions with varying zinc concentrations. It does not influence seed germination but promotes plumule and radicle growth. Although Zn toxicity in plants is significantly less common than Zn deficiency, most studies focus on increasing Zn levels in crop plant tissues. Zn phytotoxicity is common in acidic and sludge-amended soil. As a result, the current data provide an overview of Zn phytotoxicity studies conducted at the physiological, biochemical, and molecular levels during the previous two decades (Kaur and Garg 2021). The effect of increasing Cd, Fe, and Zn levels on wheat and bean germination and seedling development was examined. Metals had varying effects on germination and early seedling development among the investigated species. Cd and Fe had the most substantial inhibitive effect on wheat germination. However, the germination of bean seeds was slightly inhibited in Cd, indicating that these seeds were more resistant to Fe even at high metal concentrations. Studies conducted by Mathur et al. (1987), Mahmood et al. (2007), Athar and Ahmad (2002), Li et al. (2005), and Datta et al. (2011) have found that metals have a harmful impact on seed germination. The effect of metals on seed germination was studied by several researchers as a decrease in water intake and transport (Becerril and Duke 1989).

4 Effect on Root

Despite being a necessary mineral for plant development, zinc exhibited toxicity signs in more significant quantities, limiting root growth (Baker 1978; Bradshaw and McNeilly 1981). Zinc toxicity was evident in the root system, notably in root blunting and thickness, inhibiting cell division and elongation (Wainwright and Woolhouse 1975). According to Dogar and Van Hai Tang, the rate of zinc absorption in rice plants is affected by nitrogen and phosphorus concentration in the solution (Dogar and Van Hai 1980). Pearson and Rengel discovered that as plants grow older, the amount of zinc in their roots drops (Pearson and Rengel 1995).

5 Effect on Reproductive Growth

When exposed to metals, plants that are sensitive to them struggle greatly to reach the reproductive phase. Most plant species that thrive on soil enhanced with heavy metals have deficient metal concentrations in their seeds (Lorimer and Miziorko 1981; Rout and Das 2009). Increased zinc levels promoted *Silene vulgaris* genotypes resistant to zinc in vegetative growth and seed production. A threshold zinc content in soil exceeds results in decreased growth and no flower output.

6 Effect on Plant Physiology and Morphology

The physiology of metal toxicity in plants primarily focuses on metal absorption and translocation as well as metal transport from the soil to the roots. The intensity of the soil's adsorption and plant effectors, such as root exudates for metal chelation or reduction, determine a metal's availability to plants in the soil. There is a lot of research on metal homeostasis in soil and metal transfer from soil to root (Foy et al. 1978). There were very few studies on the physiological and biochemical toxicity of metals. The cytoplasmic activity of a metal was influenced by the pH and chelating carboxylic and amino acids. Metal phytotoxicity occurs when metals can transfer from the soil to root systems.

7 Effect of Zinc on Metabolism

Many plant species are generally resistant to specific concentrations of metals. This is likely accomplished by the metals being trapped inside the metal-binding proteins. Zinc is an essential component of several enzymes that promote different metabolic processes, including the metabolism of nucleic acids, protein synthesis, photosynthesis, respiration, and the metabolism of carbohydrates (Shkolnik 2012). The effects of low pH and zinc on the breakdown of chlorophyll in lichen *Ramalina duriaei* were investigated by Garty et al. (1992). They discovered reduced photosynthesis in lichens with chlorophyceae photobionts at pH 2.0 and low levels of zinc. According to Meharg (1993), the cell's compartment that may, at the very least, partially control a heavy metal ion's admission into the cell is the plasma membrane. Knowing which metals are phytotoxic and how they interact with other metals is necessary to assess toxicity's physiological and biochemical impacts in a meaningful way (Cunningham et al. 1975). Before beginning a phytotoxicity experiment, it is essential to know the metal movement, including its absorption and translocation in the plant system. Zinc pollution from various sources has resulted in zinc phytotoxicity in acidic soils. Seldom are agricultural soils significantly polluted to result in phytotoxicity from zinc (Holmgren et al. 1993). Zinc phytotoxicity is more severe. Therefore, if soils rich in zinc become acidic, there may be a risk of zinc phytotoxicity. Rainfall leaching and N fertilizers gradually acidify soils, increasing the danger of future zinc phytotoxicity if soil pH is not controlled. Since

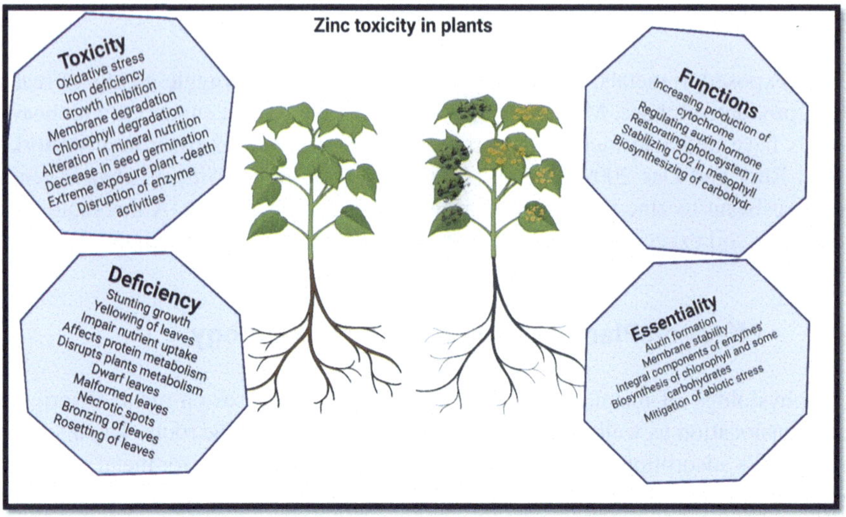

Fig. 1 Effect of zinc on plants: showing deficiency, excess, and optimum concentration of zinc micronutrient and physiological effects

sewage sludge will be one of the most significant long-term sources of soil Zn enrichment, the recent massive body of research on the land application of sewage sludge provides valuable information on possible Zn phytotoxicity. The effects of zinc on plant growth, including deficiency, excess, and optimum concentrations, along with their physiological impacts, are illustrated in Fig. 1.

8 Role of Microorganisms in Ameliorating the Zinc Phytotoxicity

Among the various strategies used to mitigate the harmful effects of zinc phytotoxicity on plants, one emerging approach is the utilization of microorganisms in the remediation process. Rhizosphere microorganisms, PGPR, endophytic bacteria, and mycorrhizae, as well as the use of biofertilizers and exogenous chelators, can improve Zn mobility and bioavailability and effectively aid phytoremediation.

9 Plant Growth-Promoting Rhizobacteria (PGPR)

The rhizosphere, or region of soil impacted by root secretions, is home to a variety of bacteria collectively called growth-promoting rhizobacteria (PGPR) that are capable of actively colonizing the plant's root system and improving their growth and yield (Sharma et al. 2021). These bacteria can potentially solubilize insoluble zinc complexes, increasing zinc availability to plants while lowering its toxicity and aiding in the absorption of vital nutrients by plants, reducing zinc toxicity indirectly.

They possess plant growth-promoting qualities, such as the formation of ACC deaminase and IAA, nitrogen (N2) fixation, phosphate solubilization, and siderophore secretion (Kour et al. 2024). The microbial colonizing rhizosphere includes bacteria, fungi, actinomycetes, protozoa, and algae. However, bacteria are the most abundant microbial present in the rhizosphere.

PGPR enhances plant growth through a wide variety of mechanisms, although the mode of action of different PGPR varies depending on the type of host plants. In addition, some PGPR have heavy metal tolerance mechanisms, and they can improve phytoremediation and increase the metal resistance in plants (Rajkumar et al. 2012). Significant Zn-tolerant PGPR strains include the genus *Cupriavidus*, *Pseudomonas*, *Streptomyces*, *Micrococcus*, *Sphingomonas*, *Klebsiella*, *Serratia*, *Proteus*, etc. Devendra Jain et al. (2020) studied that zinc-tolerant plant growth-promoting bacteria alleviate phytotoxic effects of zinc on maize through zinc immobilization. Four strains, viz., ZTB15, ZTB24, ZTB28, and ZTB29 belonging to *Serratia*, were isolated from Zn-contaminated soil, exhibiting high tolerance to Zn up to 62.5 Mm. Pot experiment using maize (*Zea mays*) showed that injection of ZTB strains alleviated the Zn toxicity and enhanced the plant growth parameters. Another study investigated plant growth-promoting rhizobacteria (PGPR)-assisted phytoremediation using a combination of rice and Zn-tolerant PGPR strain *Bacillus* sp. ZC3–2-1 for the remediation of Zn-contaminated soil (Liu et al. 2022). The zinc-tolerant plant growth-promoting bacteria absorb the zinc content from the soil and thereby reduce the phytotoxic levels of these heavy metals in the soil. The ability of these zinc-tolerant bacteria to tolerate the toxic level of zinc can serve as suitable candidates for developing microbial formulations for the growth of crop plants in Zn-contaminated areas. Rhizosphere microbial communities also affect the bioavailability of Zn in the soil (Hrynkiewicz et al. 2012). Among the bacteria, species belonging to the genera *Acinetobacter*, *Bacillus*, *Gluconacetobacter*, and *Pseudomonas* have been described as Zn solubilizers (Ramesh et al. 2014).

Rhizosphere microflora can increase Zn solubilization through a variety of methods, including soil pH decrease, Zn chelation, and improved root development. These processes vary from one microbe to another. Some species may employ one of these methods, whereas others can boost plant Zn uptake via many ways (Srithaworn et al. 2023).

10 Endophyte Bacteria

Bacterial endophytes reside within the tissues of living plant species without causing any harm or disease to their hosts. These endophytes play various roles in plant health growth and environmental adaptation, enhancing the uptake of essential nutrients such as nitrogen, phosphorus, and micronutrients by facilitating their mobilization and assimilation within the plant. They help plants tolerate various environmental stresses, including drought, salinity, heavy metals, and pathogens, by producing stress-related compounds such as antioxidants and osmoprotectants and inducing systemic resistance in host plants. Endophytes colonizing plant tissues can degrade or detoxify environmental pollutants, including heavy metals, pesticides, and organic contaminants, and hence contribute to the phytoremediation of contaminated soils and water bodies. The properties of endophytes that promote nutrition uptake and stress tolerance have been studied for their extensive application in agriculture to improve crop productivity. Studies have revealed the potential of specific endophytes in mitigating zinc toxicity in plants. Examples of endophytes that help mitigate zinc toxicity include *Bacillus* spp., *Pseudomonas* spp., *Trichoderma* spp., and *Enterobacter* (He et al. 2020).

Endophytic bacteria enhance zinc tolerance in plants through mechanisms such as the production of chelating agents, stimulation of antioxidant enzymes, and modulation of plant hormone levels (Mosa et al. 2016). Trichoderma spp., an endophytic fungus, enhances zinc detoxification in plants through the production of metallothioneins, which sequester heavy metals, and stimulation of plant growth-promoting hormones. Ma et al. (2015) studied that the metal-resistant endophytic bacteria improve its phytoextraction capacity in multi-metal contaminated soil. Results demonstrated that the endophytes can be used to enhance phytoextraction of soils contaminated with multiple heavy metals by inoculating metal hyperaccumulating plants with their own selected functional endophytic bacterial strains.

Endophyte relationships are important in natural and controlled ecosystems because they provide nutritional and non-nutritional advantages to their hosts. Endophytes have the ability to operate as biofertilizers, which can affect plant productivity. Endophytes appear to have favorable effects on their hyperaccumulators via mechanisms similar to those identified for PGPR; hence, endophytic bacteria give tolerance to plants cultivated in heavy metal-contaminated soil. Hyperaccumulating plants accumulate massive amounts of heavy metals in their shoots, far exceeding those present in the surrounding soils or non-accumulating plant species nearby, and they provide a specific environment for bacterial endophytes that could adapt to survive in high metal concentrations (Rajkumar et al. 2009). Saravanan and coauthors studied the zinc-solubilizing potential of *Gluconacetobacter diazotrophicus* under in vitro conditions with different Zn compounds using glucose or sucrose as carbon sources. *G. diazotrophicus* showed variations in their solubilization potential with the strains and the Zn compounds tested. *Gluconacetobacter diazotrophicus*, an endophytic diazotroph encountered as a rhizosphere bacterium, is reported to possess different plant growth-promoting characteristics. These studies provide insights into the interactions between endophytes and plants under zinc stress conditions and highlight the potential of endophytes in phytoremediation and plant tolerance to heavy metals.

11 Arbuscular Mycorrhizae Fungi (AMF)

Arbuscular mycorrhiza (AM) is a symbiotic association between plants and members of the Glomeromycota, an ancient fungal phylum that provides the host plant with water and nutrients such as phosphate and nitrogen. In exchange, the fungus, which lives in the roots of 70–90% of terrestrial plants, receives up to 20% of the plant's fixed carbon. AM fungi are considered "living fossils" and "ancient asexuals" since structurally identical fungi were discovered with the earliest land plant fossils, yet sexual stages or processes remain unknown. Audet and Charset (2007) studied the dynamics of arbuscular mycorrhizal symbiosis in heavy metal phytoremediation through conceptual perspectives. An extensive literature survey was carried out, and it was found that AM responses for parameters at low and high soil-HM concentration intervals could be determined that the roles of AM symbiosis are characterized by (1) increased HM phytoextraction via mycorrhizosphere "enhanced uptake" at low soil-HM concentration and (2) a reduced HM bioavailability via AM fungal "metal-binding' processes at high soil-HM levels, resulting in increased plant biomass and enhanced plant tolerance through HM stress avoidance. Endophytes can provide tolerance to the host plant against Zn-induced stress. Inoculation of AMF helps in the balanced uptake of Zn and enhances the growth and yield of crops (Saboor et al. 2021).

12 Mycorrhizae

ZnO nanoparticles (NPs) threaten the environment and human health since they are used in various applications and commonly accumulate in the environment. The roots of higher plant have symbiotic relationships with arbuscular mycorrhizal (AM) fungus (AMF), which aid in the host plants, and uptake of phosphorus (P). AMF can decrease ZnO NP toxicity, although the advantages of AMF for host plants greatly depend on the amount of P in the soil. Several experiments have suggested that hyphae. Arbuscular mycorrhizal plants, which were initially developed in environments with mild zinc contamination, have the potential to produce larger plant yields and significantly lower zinc shoot concentrations when compared to non-mycorrhizal plants on the host plant, thereby protecting against potential zinc toxicity (Saleem et al. 2022). The effects of arbuscular mycorrhiza on Zn absorption and translocation by the host in moderately polluted soils or substrates are summarized in this study based on some recent studies. The ultimate goal is to re-vegetate moderately contaminated soils by using arbuscular mycorrhizas in their ability to shield plants against Zn toxicity. This can be accomplished by growing mycotrophic plant species and making sure that the soil's physicochemical conditions are favorable for forming arbuscular mycorrhizal associations or by introducing efficient fungal strains into the soil.

13 Exogenous Chelators (Siderophores)

Exogenous chelators, such as siderophores, are important in mitigating heavy metal phytotoxicity by facilitating the detoxification and sequestration of metal ions in the rhizosphere and within plant tissues. The formation of metal-chelator complexes promotes plant tolerance and resistance toward metals or stimulates metal translocation and accumulation within the plants, strengthening the effectiveness of phytoremediation. Sedum plumbizincicola has been found to accumulate both Zn and Cd (Montreemuk et al. 2023). Soil amendments with chelating agents can improve heavy metal phytoextraction efficiency (Kafle et al. 2022). Applying ethylenediaminetetraacetic acid (EDTA) to sludge soil significantly promoted the accumulation of Zn, Cd, Pb, Ni, and Cu in *Bryophyllum laetivirens* roots by enhancing metal ion solubility in the soil (Li et al. 2020). Exogenous chelators, particularly siderophores, play crucial roles in alleviating zinc phytotoxicity by sequestering zinc ions, mobilizing zinc in the soil, regulating zinc uptake in plants, inducing plant defense responses, and facilitating beneficial interactions between plants and microorganisms. Harnessing the properties of exogenous chelators offers promising strategies for improving plant growth and productivity in zinc-contaminated environments. Though the primary role of siderophores is to update iron nutrients, studies also show that they help alleviate metal toxicity in plants.

Enterobactin is a well-studied siderophore produced by various bacterial species, including *Enterobacter* and *Escherichia coli.* While its primary role is iron chelation, enterobactin has also been found to bind other metals, including zinc. In a study by Upadhayay et al. (2022), it was demonstrated that enterobactin-producing bacteria isolated from the rhizosphere of a zinc hyperaccumulator plant (*Sedum alfredii*) could enhance zinc tolerance in the host plant. The researchers suggested that enterobactin-mediated zinc sequestration contributes to zinc detoxification in the rhizosphere. *Pseudomonas aeruginosa and Pseudomonas fluorescens* produce siderophores called pyoverdine. The study by Ghssein and Ezzedine (2022) demonstrated that pyoverdine-producing *Pseudomonas* strains isolated from the rhizosphere of metal-tolerant plants could promote plant growth and alleviate zinc toxicity in contaminated soils. The researchers proposed that pyoverdine-mediated zinc chelation contributes to the enhanced zinc tolerance observed in the plants. Desferrioxamine is a siderophore produced by various soil bacteria and fungi. While its primary function is iron chelation, desferrioxamine has been shown to bind other metals, including zinc. A study by Dimkpa et al. (2008) demonstrated that desferrioxamine-producing bacteria isolated from the rhizosphere of maize plants could enhance zinc tolerance in the host plant. The researchers suggested that desferrioxamine-mediated zinc sequestration might contribute to zinc detoxification in the rhizosphere.

14 Role of Genetically Engineered Microorganisms (GEMs) in Ameliorating the Zinc Phytotoxicity

GEMs offer promising strategies for mitigating zinc phytotoxicity and improving plant growth and health in zinc-contaminated environments. Genetically engineered microorganisms (GEMs) have been explored for their potential to alleviate zinc phytotoxicity in plants through various mechanisms. To address the zinc phytotoxicity, microorganisms can be engineered for bioremediation of zinc-contaminated soils, where GEMs can express enzymes or proteins involved in zinc detoxification pathways, such as metallothioneins or zinc-binding peptides. By colonizing the rhizosphere or soil, GEMs can promote the immobilization, sequestration, or transformation of excess zinc, reducing its availability to plants and mitigating phytotoxicity.

They can enhance metal accumulation and tolerance in plants growing in zinc-contaminated environments. For example, GEMs can be designed to produce chelators, such as siderophores or organic acids, that bind and sequester zinc ions in the rhizosphere, preventing their uptake by plants (Verma et al. 2020). Additionally, GEMs can be engineered to enhance the expression of transporters or enzymes involved in zinc uptake, sequestration, or detoxification in plant roots, thereby increasing plant tolerance to zinc toxicity.

Facilitation of Plant-Microbe Interactions GEMs can be engineered to establish beneficial interactions with plants and promote their growth and health under zinc stress conditions. GEMs can be modified to produce plant growth-promoting hormones or compounds that stimulate plant defense responses against zinc toxicity. By colonizing the rhizosphere or forming symbiotic associations with plant roots, GEMs can enhance nutrient uptake, improve water-use efficiency, and protect plants against zinc-induced oxidative stress and damage (Omotola et al. 2016).

Biofortification of Plants with Zinc GEMs can be utilized to biofortify plants with zinc, enhancing their nutritional quality and improving human health. GEMs can be engineered to enhance zinc uptake, transport, or accumulation in edible plant tissues, such as grains, fruits, or vegetables. By expressing zinc transporters or regulatory genes, GEMs can increase the zinc content of food crops, addressing zinc deficiency in populations dependent on zinc-poor diets.

Engineering Microbial Consortia for Synergistic Effects GEMs can be combined with other beneficial microorganisms to form microbial consortia with complementary functions for addressing zinc phytotoxicity. By engineering microbial consortia consisting of GEMs and plant growth-promoting rhizobacteria or mycorrhizal fungi, synergistic effects can be achieved, enhancing plant growth, nutrient uptake, and stress tolerance in zinc-contaminated soils. Consortium *Bacillus* spp. *and Serratia marcescens* can be recommended as a possible bioinoculant for the rice crop grown on zinc-deficient soil after assessment on farm trials at multiple locations (Cao et al. 2022). However, further optimization of GEM-based approaches, their safety and environmental compatibility, and their long-term effectiveness in agricultural and environmental applications are essential for their application in agriculture.

Mechanism of Zinc Bioremediation Studies over the years have reported the effectiveness of *Brevibacterium* spp. in bioremediating zinc from polluted environments, demonstrating its capability to remove zinc even at concentrations as low as 0.1 mm (Taniguchi et al. 2000). Gram-negative and Gram-positive bacteria exhibit general resistance mechanisms encompassing a range of agents, with significant overlap across species for specific metals. Various species of bacteria including

Bacillus spp., *Staphylococcus* spp., *Streptococcus* spp., *Escherichia coli*, *Pseudomonas* spp., *Klebsiella* spp., and *Enterobacter* spp. exhibit general resistance mechanisms. These mechanisms include Zn sequestration, mechanism of zinc homeostasis, efflux pump system, chelation, phytostimulation, bioaccumulation, biomineralization, and uptake, which collectively contribute to metal resistance in bacteria. Several pathways for Zn have been widely described in numerous bacterial species (Hussain et al. 2022). However, the effectiveness of bioremediation depends on factors such as concentration of pollutants, type and soil properties, climate, and microbial activity.

Zinc Sequestration Zinc sequestration refers to the process by which zinc ions are bound and immobilized, reducing their bioavailability and potential environmental toxicity. Microorganisms and plants employ various mechanisms for zinc sequestration (Subramanian et al. 2011). Resistance to hazardous zinc levels can be attributed to extracellular accumulation or sequestration by metallothioneins (MT), intracellular physical sequestration, or efflux. Synechocystis PCC6803 and Synechococcus PCC7942 have closely similar zinc sensors, ZiaR and SmtB, respectively. However, they regulate distinct structural proteins, leading to divergent consequences for zinc's cellular biology. ZiaR triggers the expulsion of excess zinc through ZiaA-mediated efflux into the periplasm, while SmtB facilitates internal sequestration by metallothionein (MT), notably SmtA (Choudhury and Srivastava 2007). Some bacteria produce extracellular ligands, such as siderophores, exopolysaccharides, and small organic acids, with high affinity for zinc ions. These ligands can form stable complexes with zinc, preventing its interaction with cellular components and reducing its availability for uptake by other organisms (Bhagat et al. 2021).

Some bacteria can induce the precipitation of zinc ions as insoluble mineral phases, such as zinc sulfides or oxides. This biomineralization process results in mineral precipitates or nanoparticles that effectively immobilize zinc in the environment, reducing its mobility and bioavailability (Heim 2011). This process can reduce the mobility and toxicity of zinc in contaminated soils or sediments. By immobilizing zinc ions through various cellular and extracellular processes, bacteria contribute to the natural attenuation of zinc pollution and play a significant role in environmental remediation. Microorganisms involved in zinc immobilization include *Pseudomonas* spp., *Bacillus*, *Enterobacter. Fungi Arbuscular mycorrhizal fungi (AMF), Aspergillus* spp., *Penicillium* spp., *Algae: Microalgae and cyanobacteria, Actinomycetes, Streptomyces* spp., *Yeast: Saccharomyces cerevisiae* (Jain et al. 2020). The efficacy of these microorganisms in zinc immobilization may vary depending on environmental conditions, soil properties, and the specific microbial strains used. Additionally, applying these microorganisms for remediation should be approached cautiously, considering ecological impact and regulatory considerations. Table 1 presents few types of microorganisms and their resistance mechanisms to zinc.

Table 1 Microbe-mediated remediation and resistance mechanisms in microorganisms for zinc

Microbial group	Microorganism	Microbial/resistance mechanism	References
Bacteria	*Desulfovibrio desulfuricans*	Extracellular sequestration	Yue et al. (2015)
	Pseudomonas aeruginosa	Extracellular sequestration, zinc solubilization	Saravanan et al. (2011), and Dimkpa et al. (2019)
	Synechococcus sp.	Intracellular sequestration	Blindauer et al. (2008)
	Escherichia coli	Active export	Lerebours et al. (2016)
	Bacillus spp.	Active export	
	Saccharomyces cerevisiae	Biosorption	
	Trichoderma spp.	Phytases, biomineralization	Contreras-Cornejo et al. (2016)
Algae	*Asparagopsis armata*	Biosorption	Romera et al. (2007)
	Codium vermilara	Biosorption	Romera et al. (2007)
Fungi	*Arbuscular mycorrhizal fungi (AMF)*	Symbiotic associations with plant roots, increasing root surface area, and nutrient absorption	Gao et al. (2020)

15 Mechanism of Zinc Homeostasis and Resistance Mechanisms in Bacteria

The bacteria have developed several mechanisms for zinc uptake and efflux. The complex network of zinc homeostasis mechanisms in bacteria permits them to maintain optimal zinc levels for cellular functions while protecting them from zinc toxicity or deficiency.

Zinc regulators in cells control the transcriptional expression of zinc transporters, which aids in the import and export of zinc into and out of the cell. The Zur/SlyA/MarR family and the MerR (ZntR) and ArsR/SmtB families regulate intracellular Zn^{2+} ion transport across the cell membrane (Mikhaylina et al. 2018). Znu is an ABC transporter that enhances Zn 2+ ion absorption, which is inhibited by zur at typical Zn^{2+} concentrations in the cell (Hantke 2001). However, when Zn^{2+} is scarce, Zur reverses its actions and causes the transporter to absorb Zn^{2+} from the surrounding environment. Under Zn toxicity circumstances, it triggers the transcription of a P-type ATPase, which facilitates the efflux of Zn^{2+} out of the cell (Capdevila et al. 2016).

16 Mechanisms of Zinc Efflux in Bacteria

Three families of exporters are available in the bacterial cell, namely, cation diffusion facilitator (CDF), P-type ATPases, and resistance nodulation division (RND) efflux pumps majorly reported (Hussain et al. 2022). Among all these types, CDF proteins are commonly found in many bacteria. Zit B and YiiP proteins in the *E. coli* are of the CDF family that regulate the efflux of Zn (Porcheron et al. 2013). In *Staphylococcus aureus* proteins present ZntA. In case of *B. subtilis*, zinc efflux is mediated by CadA, which is a CPx-type ATPase efflux system (Gaballa et al. 2002). RND efflux pump system is found mainly in Gram-negative bacteria involved in the active efflux of antibiotics and some therapeutic agents. In *Caulobacter crescentus*, the protein CzrCBA exports cadmium and Zn ions from the cytoplasm of the cell. In *Ralstonia metallidurans*, it is facilitated by CzcABC system regulated by a two-component system (Hussain et al. 2022).

17 Biochar Amendment

Carbon-enriched biochar promotes the proliferation of indigenous microbial communities by balancing pH levels and offering nutritional support. Stimulated by biochar, phytoremediation holds significant promise to immobilize cationic heavy metals and metalloids in mine sludge soils and other metal-contaminated soils, especially those with high acidity. Biochar significantly lowers the bioavailability and leaching potential of cationic metals and metalloids in soils while enhancing soil structure, physicochemical characteristics, fertility, revegetation potential, and microbial communities (Narayanan and Ma 2022). Adding biochar, a carbon-rich material derived from the pyrolysis of organic waste, to zinc-contaminated soil can help immobilize zinc and reduce its bioavailability. Biochar acts as a sorbent for zinc ions, preventing their uptake by plants and reducing leaching into groundwater. Additionally, biochar amendments can enhance microbial activity and soil fertility, facilitating zinc bioremediation (Mukherjee et al. 2022).

18 Phytoextraction-Assisted Bioleaching

Phytoextraction-assisted bioleaching combines hyperaccumulator plants and metal-mobilizing microorganisms to enhance zinc removal from contaminated soil. Hyperaccumulator plants uptake zinc from the soil into their biomass, which is harvested and subjected to bioleaching processes. Metal-mobilizing microorganisms, such as acidophilic bacteria, solubilize and extract zinc from the plant biomass, facilitating its recovery and recycling. Enhancing chemical amendment capacity is crucial for reducing the time and cost required for remediating heavy metal-contaminated soil. This involves increasing the bioaccumulation index of heavy metals in plants (Hasan et al. 2019).

19 Genetic Engineered Microbes for Zinc Bioremediation

Genetic engineering techniques can be employed to enhance the zinc tolerance and remediation capabilities of plants and microorganisms. These techniques are economically plausible and environmentally safe (Saravanan et al. 2022). Native microbes often exhibit limited capability or are unable to effectively remove toxic compounds. Interdisciplinary efforts have been undertaken to enhance their pollutant removal potential, such as integrating molecular methods, genomics, bioinformatics, and microbiology that are used to change their genetic composition, resulting in genetically engineered creatures (Hussain et al. 2018; Wu et al. 2021). Researchers can develop engineered organisms with improved zinc bioremediation potential by introducing genes involved in zinc transport, sequestration, or detoxification (Rafeeq et al. 2023). However, releasing genetically modified organisms into the environment requires careful consideration of potential ecological risks and regulatory requirements. Recent research has emphasized gene overexpression in zinc uptake, transport, and accumulation in plant tissues, including rice seeds (Lonnerdal 2003).

20 Phytostabilization

Phytostabilization is a technique used in environmental remediation to stabilize contaminants and hefty metals in soil by using plants. Metal-accumulating plants can immobilize and reduce the bioavailability of contaminants in the soil. Certain soil bacteria help in phytostabilization, such as rhizosphere bacteria facilitating the immobilization and transformation of pollutants. For example, some bacteria produce exopolysaccharides that bind heavy metals, reducing their mobility in the soil. Other bacteria can facilitate the precipitation of metal ions into less bioavailable forms, such as metal sulfides or oxides, through processes like biomineralization. AMF can also contribute to phytostabilization by increasing soil aggregate stability and influencing the contaminants' speciation in the rhizosphere. They may also produce extracellular enzymes that degrade organic pollutants and promote the formation of metal-organic complexes. Actinomycetes are a group of filamentous bacteria commonly found in soil environments. They can produce extracellular enzymes and organic acids that influence the availability and mobility of contaminants in the rhizosphere. Actinomycetes may also participate in the degradation of organic pollutants and the stabilization of soil aggregates, contributing to the overall effectiveness of phytostabilization.

Fungi, including saprophytic fungi and mycorrhizal fungi, can influence the fate of contaminants in the rhizosphere through various mechanisms. Some fungi produce enzymes that degrade organic pollutants, while others can form complexes with heavy metals, reducing their bioavailability to plants. These microorganisms interact with plant roots and soil particles in the rhizosphere, forming complex microbial communities that contribute to the phytostabilization of contaminants. Two strains, *Bacillus megaterium* BolR EW3_A03 and *Stenotrophomonas maltophilia* BolN EW3_B03, isolated from a Zn–Pb mine waste heap tested positive for all examined plant growth-promoting traits and showed co-tolerance to Zn, Cu, Cd, and Pb can be considered as potential facilitators of phytostabilization.

21 Induced Systemic Resistance (ISR)

Plants have a variety of active defense mechanisms that can be activated in response to biotic stressors (pathogens and parasites) at various sizes (from tiny viruses to phytophagous insects). Induced resistance is a condition of increased defensive capability that a plant develops when adequately stimulated. Selected zinc-tolerant microbial strains or consortia can be put into polluted soil to improve zinc bioremediation. These microorganisms can be supplemented with organic amendments or biostimulants to boost their survival and activity in soil. Selected strains of plant growth-promoting rhizobacteria (PGPR) decrease illnesses by antagonizing bacteria and soil-borne pathogens, as well as developing systemic resistance in plants to both root and foliar pathogens. Several rhizobacteria produce SA at the root surface, which activates the salicylic acid (SA)-dependent SAR pathway; however, other rhizobacteria activate signaling pathways that are not dependent on SA (Choudhary et al. 2007). *Arabidopsis thaliana* was used to study the SA-independent ISR pathway, which is reliant on jasmonic acid (JA) and ethylene signaling. Specific *Pseudomonas* strains cause systemic resistance in carnation, cucumber, radish, tobacco, and *Arabidopsis*, as indicated by increased defensive capability following challenge inoculation. Aside from *Pseudomonas strains, additional Bacillus species, such as B. amyloliquefaciens*, *B. subtilis*, *B. pasteurii*, *B. cereus*, *B. pumilus*, *B. mycoides*, *and B. sphaericus*, significantly reduce the frequency or severity of many illnesses in a variety of hosts.

22 Conclusion

The persistent threat of Zn contamination in soil, water poses significant challenges to agricultural production and environmental sustainability. Traditional remediation methods have proven insufficient in addressing the complex and widespread nature of this issue. However, promising alternatives, such as harnessing the power of beneficial microorganisms, offer sustainable solutions to mitigate zinc phytotoxicity and restore soil health. By exploring the diverse roles of microbes, including plant growth-promoting rhizobacteria, mycorrhizal fungi, and metal-resistant bacteria, innovative remediation strategies can be developed. These microbes contribute to enhancing plant tolerance to heavy metal stress by facilitating several mechanisms such as metal solubilization, immobilization, and detoxification processes. Leveraging microbial diversity and metabolic versatility holds immense potential for sustainable zinc phytoremediation practices, thereby promoting both agricultural productivity and environmental well-being. Continued research for further understanding the intricate interactions between microorganisms and their environment is the need of the hour. By integrating this knowledge into practical applications, we can effectively combat zinc contamination, safeguarding agricultural ecosystems and ensuring a healthier future.

References

Athar R, Ahmad M (2002) Heavy metal toxicity: effect on plant growth and metal uptake by wheat, and on free living Azotobacter. Water Air Soil Pollut 138:165–180. https://doi.org/10.1023/A:1015594815016

Audet P, Charest C (2007) Heavy metal phytoremediation from a meta-analytical perspective. Environ Pollut 147(1):231–237

Awasthi MK, Selvam A, Lai KM, Wong JW (2017) Critical evaluation of post-consumption food waste composting employing thermophilic bacterial consortium. Bioresour Technol 245:665–672. https://doi.org/10.1016/j.biortech.2017.09.014

Baker AJ (1978) Ecophysiological aspects of zinc tolerance in Silene maritima With. New Phytol 80(3):635–642. https://doi.org/10.1111/j.1469-8137.1978.tb01596.x

Becerril JM, Duke SO (1989) Protoporphyrin IX content correlates with activity of photobleaching herbicides. Plant Physiol 90(3):1175–1181. https://doi.org/10.1104/pp.90.3.1175

Bhagat N, Raghav M, Dubey S, Bedi N (2021) Bacterial exopolysaccharides: insight into their role in plant abiotic stress tolerance. J Microbiol Biotechnol 31(8):1045–1059. https://doi.org/10.4014%2Fjmb.2105.05009

Blindauer C, Harrison M, Parkinson J, Robinson N, Sadler P (2008) Isostructural replacement of zinc by cadmium in bacterial metallothionein. Metal Ions Biol Med 10:167–173

Bolan N, Kunhikrishnan A, Thangarajan R, Kumpiene J, Park J, Makino T, Kirkham MB, Scheckel K (2014) Remediation of heavy metal (loid) s contaminated soils–to mobilize or to immobilize? J Hazard Mater 266:141–166. https://doi.org/10.1016/j.jhazmat.2013.12.018

Bradshaw AD, McNeilly T (1981) Evolution and pollution. (No Title). https://lccn.loc.gov/81173979

Cao Z, Yan W, Ding M, Yuan Y (2022) Construction of microbial consortia for microbial degradation of complex compounds. Front Bioeng Biotechnol 10:1051233. https://doi.org/10.3389/fbioe.2022.1051233

Capdevila DA, Wang J, Giedroc DP (2016) Bacterial strategies to maintain zinc metallostasis at the host-pathogen interface. J Biol Chem 291(40):20858–20868. https://doi.org/10.1074/jbc.R116.742023

Castillo-González J, Ojeda-Barrios D, Hernández-Rodríguez A, González-Franco AC, Robles-Hernández L, López-Ochoa GR (2018) Zinc metalloenzymes in plants. Interciencia 43(4):242–248

Choudhary DK, Prakash A, Johri BN (2007) Induced systemic resistance (ISR) in plants: mechanism of action. Indian J Microbiol 47:289–297. https://doi.org/10.1007/s12088-007-0054-2

Choudhury R, Srivastava S (2007) Zinc resistance mechanisms in bacteria. Curr Sci 81(7):768–775. http://www.jstor.org/stable/24106396

Clijsters H, Van Assche F (1985) Inhibition of photosynthesis by heavy metals. Photosynth Res 7:31–40

Contreras-Cornejo HA, Macías-Rodríguez L, del-Val E, Larsen J (2016) Ecological functions of Trichoderma spp. and their secondary metabolites in the rhizosphere: interactions with plants. FEMS Microbiol Ecol 92: fiw036

Cunningham JD, Keeney DR, Ryan JA (1975) Phytotoxicity and uptake of metals added to soils as inorganic salts or in sewage sludge. American Society of Agronomy, crop science society of America, and soil science society of America

Datta JK, Bandhyopadhyay A, Banerjee A, Mondal NK (2011) Phytotoxic effect of chromium on the germination, seedling growth of some wheat (Triticum aestivum L.) cultivars under laboratory condition. J Agric Technol 7(2):395–402

Davis-Carter JG, Shuman LM (1993) Influence of texture and pH of kaolinitic soils on zinc fractions and zinc uptake by peanuts. Soil Sci 155(6):376–384. https://journals.lww.com/soilsci/toc/1993/06000

Dimkpa CO, Svatos A, Dabrowska P, Schmidt A, Boland W, Kothe E (2008) Involvement of siderophores in the reduction of metal-induced inhibition of auxin synthesis in Streptomyces spp. Chemosphere 74(1):19–25. https://doi.org/10.1016/j.chemosphere.2008.09.079

Dimkpa CO, Singh U, Bindraban PS, Elmer WH, Gardea-Torresdey JL, White JC (2019) Zinc oxide nanoparticles alleviate drought-induced alterations in sorghum performance, nutrient acquisition, and grain fortification. Sci Total Environ 688:926–934. https://doi.org/10.1016/j.scitotenv.2019.06.392

Dogar MA, Van Hai T (1980) Effect of P, N and HCO3-levels in the nutrient solution on rate of Zn absorption by Rice Roots and Zn content in plants. Z Pflanzenphysiol 98(3):203–212. https://doi.org/10.1016/S0044-328X(80)80270-4

Foy CD, Chaney RT, White MC (1978) The physiology of metal toxicity in plants. Annu Rev Plant Physiol 29(1):511–566

Gaballa A, Wang T, Ye RW, Helmann JD (2002) Functional analysis of the Bacillus subtilis Zur regulon. J Bacteriol 184(23):6508–6514. https://doi.org/10.1128/jb.184.23.6508-6514.2002

Gao X, Guo H, Zhang Q et al (2020) Arbuscular mycorrhizal fungi (AMF) enhanced the growth, yield, fiber quality and phosphorus regulation in upland cotton (Gossypium hirsutum L.). Sci Rep 10:2084. https://doi.org/10.1038/s41598-020-59180-3

Garty J, Karary Y, Harel J (1992) Effect of low pH, heavy metals and anions on chlorophyll degradation in the lichen Ramalina duriaei (De Not.). Bagl Environ Exper Botany 32(3):229–241. https://doi.org/10.1016/0098-8472(92)90006-N

Ge HW, Lian MF, Wen FZ, Yun YF, Jian FY, Ming T (2009) Isolation and characterization of the heavy metal resistant bacteria CCNWRS33-2 isolated from root nodule of Lespedeza cuneata in gold mine tailings in China. J Hazard Mater 162:50–56. https://doi.org/10.1016/j.jhazmat.2008.05.040

Ghssein G, Ezzeddine Z (2022) A review of pseudomonas aeruginosa metallophores: pyoverdine, pyochelin and pseudopaline. Biology (Basel). 11:1711

Hamzah Saleem M, Usman K, Rizwan M, Al Jabri H, Alsafran M (2022) Functions and strategies for enhancing zinc availability in plants for sustainable agriculture. Front Plant Sci 13:1033092. https://doi.org/10.3389/fpls.2022.1033092

Hantke K (2001) Bacterial zinc transporters and regulators. Zinc biochemistry, physiology, and homeostasis: recent insights and current trends. Biometals:53–63. https://doi.org/10.1007/978-94-017-3728-9_4

Hasan MM, Uddin MN, Ara-Sharmeen I et al (2019) Assisting phytoremediation of heavy metals using chemical amendments. Plants (Basel) 8(9):295. https://doi.org/10.3390/plants8090295

He W, Megharaj M, Wu CY, Subashchandrabose SR, Dai CC (2020) Endophyte-assisted phytoremediation: mechanisms and current application strategies for soil mixed pollutants. Crit Rev Biotechnol 40(1):31–45. https://doi.org/10.1080/07388551.2019.1675582

Heim C (2011) Microbial biomineralization. In: Reitner J, Thiel V (eds) Encyclopedia of geobiology. Encyclopedia of Earth Sciences Series. Springer, Dordrecht. https://doi.org/10.1007/978-1-4020-9212-1_33

Holmgren GGS, Meyer MW, Chaney RL, Daniels RB (1993) Cadmium, lead, zinc, copper, and nickel in agricultural soils of the United States of America. J Environ Qual 22:335–348. https://doi.org/10.2134/jeq1993.00472425002200020015x

Hrynkiewicz K, Dabrowska G, Baum C, Niedojadlo K, Leinweber P (2012) Interactive and single effects of ectomycorrhiza formation and Bacillus cereus on metallothionein MT1 expression and phytoextraction of Cd and Zn by willows. Water Air Soil Pollut 223:957–968. https://doi.org/10.1007/s11270-011-0915-5

Hussain I, Aleti G, Naidu R et al (2018) Microbe and plant assisted-remediation of organic xenobiotics and its enhancement by genetically modified organisms and recombinant technology: a review. Sci Total Environ 628–629:1582–1599. https://doi.org/10.1016/j.scitotenv.2018.02.037

Hussain S, Khan M, Sheikh TM, Mumtaz MZ, Chohan TA, Shamim S, Liu Y (2022) Zinc essentiality, toxicity, and its bacterial bioremediation: A comprehensive insight. Front Microbiol 13:900740. https://doi.org/10.3389/fmicb.2022.900740

Jain D, Kour R, Bhojiya AA et al (2020) Zinc tolerant plant growth promoting bacteria alleviates phytotoxic effects of zinc on maize through zinc immobilization. Sci Rep 10(1):13865. https://doi.org/10.1038/s41598-020-70846-w

Kafle A, Timilsina A, Gautam A, Adhikari K, Bhattarai A, Aryal N (2022) Phytoremediation: mechanisms, plant selection and enhancement by natural and synthetic agents. Environ Adv 8:100203. https://doi.org/10.1016/j.envadv.2022.100203

Kareem HA, Hassan MU, Zain M, Irshad A, Shakoor N, Saleem S, Niu J, Skalicky M, Chen Z, Guo Z, Wang Q (2022) Nanosized zinc oxide (n-ZnO) particles pretreatment to alfalfa seedlings alleviate heat-induced morpho-physiological and ultrastructural damages. Environ Pollut 303:119069. https://doi.org/10.1016/j.envpol.2022.119069

Kaur H, Garg N (2021) Zinc toxicity in plants: a review. Planta 253:129. https://doi.org/10.1007/s00425-021-03642-z

Kour D, Khan SS, Kour H et al (2024) ACC deaminase producing phytomicrobiomes for amelioration of abiotic stresses in plants for agricultural sustainability. J Plant Growth Regul 43:963–985. https://doi.org/10.1007/s00344-023-11163-0

Lerebours A, To VV, Bourdineaud JP (2016) Danio rerio ABC transporter genes abcb3 and abcb7 play a protecting role against metal contamination. J Appl Toxicol 36(12):1551–1557. https://doi.org/10.1002/jat.3313

Li W, Khan MA, Yamaguchi S, Kamiya Y (2005) Effects of heavy metals on seed germination and early seedling growth of Arabidopsis thaliana. Plant Growth Regul 46:45–50. https://doi.org/10.1007/s10725-005-6324-2

Liu A, Wang W, Zheng X, Chen X, Fu W, Wang G, Guan C (2022) Improvement of the Cd and Zn phytoremediation efficiency of rice (Oryza sativa) through the inoculation of a metal-resistant PGPR strain. Chemosphere 302:134900. https://doi.org/10.1016/j.chemosphere.2022.134900

Lönnerdal B (2003) Genetically modified plants for improved trace element nutrition. J Nutr 133(5):1490S–1493S. https://doi.org/10.1093/jn/133.5.1490S

Lorimer GH, Miziorko HM (1981) RuBP carboxylase: the mechanism of activation and its relation to catalysis. Regulation of carbon metabolism. Phtosynthesis. IV. Balaban Int Sci Serv Philadelphia:3–16

Ma Y, Oliveira RS, Nai F, Rajkumar M, Luo Y, Rocha I, Freitas H (2015) The hyperaccumulator Sedum plumbizincicola harbors metal-resistant endophytic bacteria that improve its phytoextraction capacity in multi-metal contaminated soil. J Environ Manag 1(156):62–69. https://doi.org/10.1016/j.jenvman.2015.03.024

Mahmood T, Islam KR, Muhammad S (2007) Toxic effects of heavy metals on early growth and tolerance of cereal crops. Pak J Bot 39(2):451

Mathur KC, Srivastava RK, Chaudhary K (1987) Effect of Cd and Cr metals on germination and early growth performance of Allium cepa seeds. Proc Nat Acad Sci India Sect B (Biol Sci) 57:191–196

Meharg AA (1993) The role of the plasmalemma in metal tolerance in angiosperms. Physiol Plant 88(1):191–198. https://doi.org/10.1111/j.1399-3054.1993.tb01777.x

Mikhaylina A, Ksibe AZ, Scanlan DJ, Blindauer CA (2018) Bacterial zinc uptake regulator proteins and their regulons. Biochem Soc Trans 46(4):983–1001. https://doi.org/10.1042/BST20170228

Montreemuk J, Stewart TN, Prapagdee B (2023) Bacterial-assisted phytoremediation of heavy metals: Concepts, current knowledge, and future directions. Environ Technol Innov 33:103488. https://doi.org/10.1016/j.eti.2023.103488

Mosa KA, Saadoun I, Kumar K, Helmy M, Dhankher OP (2016) Potential biotechnological strategies for the cleanup of heavy metals and metalloids. Front Plant Sci 7:183531. https://doi.org/10.3389/fpls.2016.00303

Mukherjee S, Sarkar B, Aralappanavar VK, Mukhopadhyay R, Basak BB, Srivastava P, Marchut-Mikołajczyk O, Bhatnagar A, Semple KT, Bolan N (2022) Biochar-microorganism interactions for organic pollutant remediation: Challenges and perspectives. Environ Pollut 308:119609. https://doi.org/10.1016/j.envpol.2022.119609

Narayanan M, Ma Y (2022) Influences of biochar on bioremediation/phytoremediation potential of metal-contaminated soils. Front Microbiol 13:929730

Omotola FM, Ngole-Jeme VM, Babalola OO (2016) Heavy metal pollution from gold mines: environmental effects and bacterial strategies for resistance. Int J Environ Res Public Health 13(11):1047. https://doi.org/10.3390/ijerph13111047

Pande V, Pandey SC, Sati D, Bhatt P, Samant M (2022) Microbial interventions in bioremediation of heavy metal contaminants in agroecosystem. Front Microbiol 13:824084. https://doi.org/10.3389/fmicb.2022.824084

Pearson JN, Rengel Z (1995) Uptake and distribution of 65Zn and 54Mn in wheat grownat sufficient and deficient levels of Zn and Mn: I. During vegetative growth. J Exp Bot 46(7):833–839. https://doi.org/10.1093/jxb/46.7.833

Porcheron G, Garénaux A, Proulx J, Sabri M, Dozois CM (2013) Iron, copper, zinc, and manganese transport and regulation in pathogenic Enterobacteria: correlations between strains, site of infection and the relative importance of the different metal transport systems for virulence. Front Cell Infect Microbiol 5(3):90. https://doi.org/10.3389/fcimb.2013.00090

Rafeeq H, Afsheen N, Rafique S, Arshad A, Intisar M, Hussain A, Bilal M, Iqbal HM (2023) Genetically engineered microorganisms for environmental remediation. Chemosphere 310:136751. https://doi.org/10.1016/j.chemosphere.2022.136751

Rai KK, Pandey N, Meena RP, Rai SP (2021) Biotechnological strategies for enhancing heavy metal tolerance in neglected and underutilized legume crops: A comprehensive review. Ecotoxicol Environ Saf 208:111750. https://doi.org/10.1016/j.ecoenv.2020.111750

Rajkumar M, Ae N, Freitas H (2009) Endophytic bacteria and their potential to enhance heavy metal phytoextraction. Chemosphere 77(2):153–160. https://doi.org/10.1016/j.chemosphere.2009.06.047

Rajkumar M, Sandhya S, Prasad MNV, Freitas H (2012) Perspectives of plant-associated microbes in heavy metal phytoremediation. Biotechnol Adv 30(6):1562–1574

Ramesh A, Sharma SK, Sharma MP, Yadav N, Joshi OP (2014) Inoculation of zinc solubilizing Bacillus aryabhattai strains for improved growth, mobilization and biofortification of zinc in soybean and wheat cultivated in Vertisols of central India. Appl Soil Ecol 73:87–96

Romantschuk M, Lahti-Leikas K, Kontro M, Allen JA, Sinkkonen A (2023) Bioremediation of contaminated soil and groundwater by in situ biostimulation. Front Microbiol 14:1258148. https://doi.org/10.3389/fmicb.2023.1258148

Romera E, González F, Ballester A, Blázquez ML, Munoz JA (2007) Comparative study of biosorption of heavy metals using different types of algae. Bioresour Technol 98(17):3344–3353. https://doi.org/10.1016/j.biortech.2006.09.026

Rout GR, Das P (2009) Effect of Metal Toxicity on Plant Growth and Metabolism: I. Zinc. In: Lichtfouse E, Navarrete M, Debaeke P, Véronique S, Alberola C (eds) Sustainable Agriculture. Springer, Dordrecht. https://doi.org/10.1007/978-90-481-2666-8_53

Saboor A, Ali MA, Hussain S, El Enshasy HA, Hussain S, Ahmed N, Gafur A, Sayyed RZ, Fahad S, Danish S, Datta R (2021) Zinc nutrition and arbuscular mycorrhizal symbiosis effects on maize (Zea mays L.) growth and productivity. Saudi J Biol Sci 28(11):6339–6351. https://doi.org/10.1016/j.sjbs.2021.06.096

Saleem HM, Usman K, Rizwan M, Al Jabri H, Alsafran M (2022) Functions and strategies for enhancing zinc availability in plants for sustainable agriculture. Front Plant Sci 13:1033092

Saravanan VS, Kumar MR, Sa TM (2011) Microbial zinc solubilization and their role onplants. In Bacteria in agrobiology: Plant nutrient management, Springer, Berlin, Heidelberg, pp 47–63

Saravanan A, Kumar PS, Ramesh B, Srinivasan S (2022) Removal of toxic heavy metals using genetically engineered microbes: molecular tools, risk assessment and management strategies. Chemosphere. 298:134341. https://doi.org/10.1016/j.chemosphere.2022.134341

Sharma P et al (2021) Rhizosphere, rhizosphere biology, and rhizospheric engineering. In: Mohamed HI, El-Beltagi HEDS, Abd-Elsalam KA (eds) Plant growth-promoting microbes for sustainable biotic and abiotic stress management. Springer, Cham. https://doi.org/10.1007/978-3-030-66587-6_21

Shen ZG, Zhao FJ, McGrath SP (1997) Uptake and transport of zinc in the hyperaccumulator Thlaspi caerulescens and the non-hyperaccumulator Thlaspi ochroleucum. Plant Cell Environ 20(7):898–906. https://doi.org/10.1046/j.1365-3040.1997.d01-134.x

Shkolnik MY (2012) Trace elements in plants. Elsevier

Srithaworn M, Jaroenthanyakorn J, Tangjitjaroenkun J, Suriyachadkun C, Chunhachart O (2023) Zinc solubilizing bacteria and their potential as bioinoculant for growth promotion of green soybean (*Glycine max* L. Merr.). Peer J 11:e15128

Subramanian KS, Virgine Tenshia JS, Jayalakshmi K, Ramachandran V (2011) Antioxidant enzyme activities in arbuscular mycorrhizal (Glomus intraradices) fungus inoculated and non-inoculated maize plants under zinc deficiency. Indian J Microbiol 51(1):37–43. https://doi.org/10.1007/s12088-011-0078-5

Taniguchi J, Hemmi H, Tanahashi K, Amano N, Nakayama T, Nishino T (2000) Zinc biosorption by a zinc-resistant bacterium, Brevibacterium sp. strain HZM-1. Appl Microbiol Biotechnol 54(4):581–588. https://doi.org/10.1007/s002530000415

Upadhayay VK, Singh AV, Khan A (2022) Cross talk between zinc-solubilizing bacteria and plants: a short tale of bacterial-assisted zinc biofortification. Front Soil Sci 1:788170

Van Assche F, Clijsters H (1986) Inhibition of photosynthesis in Phaseolus vulgaris by treatment with toxic concentration of zinc: effect on ribulose-1, 5-bisphosphate carboxylase/oxygenase. J Plant Physiol 125(3–4):355–360. https://doi.org/10.1016/S0176-1617(86)80157-2

Verma PK, Verma S, Meher AK, Tripathi RD, Chakrabarty D, Pandey N (2020) Bioremediation of heavy metals using the interaction between plants and genetically engineered microbes. Int J Plant Envir 6(04):241–252. https://doi.org/10.18811/ijpen.v6i04.02

Wainwright SJ, Woolhouse HW (1975) Physiological mechanism of heavy metal tolerance in plants. Br Ecol Soc Symp;(United Kingdom) 15. https://www.osti.gov/etdeweb/biblio/6350612

Wu JY, Lay CH, Chen CC, Wu SY, Zhou D, Mohamed Abdula P (2021) Textile wastewater bioremediation using immobilized Chlorella sp. Wu-G23 with continuous culture. Clean Technol Environ Policy 23:153–161

Ye ZH, Baker AJ, Wong MH, Willis AJ (1997) Zinc, lead and cadmium tolerance, uptake and accumulation by Typha latifolia. New Phytol 136(3):469–480. https://doi.org/10.1046/j.1469-8137.1997.00759.x

Yue ZB, Li Q, Li CC, Chen TH, Wang J (2015) Component analysis and heavy metal adsorption ability of extracellular polymeric substances (EPS) from sulfate reducing bacteria. Bioresour Technol 194:399–402. https://doi.org/10.1016/j.biortech.2015.07.042

Zaheer IE, Ali S, Saleem MH, Yousaf HS, Malik A et al (2022) Combined application of zinc and iron-lysine and its effects on morpho-physiological traits, antioxidant capacity and chromium uptake in rapeseed (Brassica napus L.). PLoS ONE 17(1):e0262140. https://doi.org/10.1371/journal.pone.0262140

Brassinosteroids Contribute to Zinc Stress Tolerance: Dissecting the Responses in Crops

Ynglety Cascaes Pereira, Camille Ferreira Maia, Maria Antonia Machado Barbosa, and Allan Klynger da Silva Lobato

Abstract

Zinc (Zn) is a mineral element recognized as a micronutrient in plant nutrition. Brassinosteroids are a class of over 70 polyhydroxy steroid derivatives, widely dispersed across the plant kingdom, exhibiting a diverse range of physiological effects when administered exogenously at nanomolar to micromolar concentrations. This chapter seeks to determine if exogenous EBR treatment might mitigate stress induced by Zn in plants, assessing the potential effects on physiological, biochemical, nutritional, and morphological responses. This chapter demonstrated that the external administration of BRs alleviated the adverse effects induced by Zn stress on plant metabolism. The advantageous impacts on gas exchange, chlorophyll fluorescence, and oxidant substances validate the merits of this steroid. Decreases in stress markers imply that the function of BRs is associated with antioxidant metabolism and the preservation of chloroplast integrity.

Keywords

24-Epibrassinolide · Micronutrient · Plant growth regulator · Required element

Y. C. Pereira · C. F. Maia · A. K. da. S. Lobato (✉)
Núcleo de Pesquisa Vegetal Básica e Aplicada, Universidade Federal Rural da Amazônia, Paragominas, Pará, Brazil

M. A. M. Barbosa
Grupo de Estudos em Fitotecnia, Universidade do Estado de Minas Gerais, Ituiutaba, Minas Gerais, Brazil

S. Srivastava, H. Kaur (eds.), *Zinc in Soil-Plant Continuum*,
https://doi.org/10.1007/978-981-96-4253-3_12

1 Introduction

Zinc (Zn) is a mineral element classified as a micronutrient, required for plant growth. Zn is a metal that, in contrast to copper, iron, and manganese, is divalent (Sturikova et al. 2018). In nutrient solution for plants, Zn is a cation devoid of redox activity, due to the absence of valence alterations (Kumar et al. 2022). Zn is a transition metal (Zlobin et al. 2021) with an atomic number of 30 and ranks as the 23rd most prevalent element on Earth, occurring in the +2 oxidation state in the natural environment (Broadley et al. 2007; Gondal 2021).

2 Zinc in the Soil

Zinc is naturally released into the soil from minerals containing zinc oxides, sulfates, sulfides, carbonates, silicates, and phosphates, which in turn are released from the parent rock (Sturikova et al. 2018). Very minute quantity of Zn is present in soil, and on average it varies between 0.2–0.002 μg Zn per gram of soil (Tayyiba et al. 2021). In soils, zinc can be present in different forms, mainly as a free ion (Zn^{2+} and $ZnOH^{+}$) or complexed with organic matters (Balafrej et al. 2020). Zn has low mobility in the soil, which is an important factor when deciding on the fertilization of crops, since it has a low leaching rate, which can contribute to its residual effect in the soil (Han et al. 2011).

Concomitantly, Zn enters soils through various sources, such as rock weathering, forest fires, mining activities, manure, sewage sludge, and phosphate fertilizers (Kaur and Garg 2021). Studies indicate that the critical levels of zinc in the soil are between 0.8 and 1.0 mg dm^{-3} (Khoshgoftarmansh et al. 2012). About Zn supplementation, the fertilizers most used as a source of zn are ZnO and $ZnSO_4$ (Montalvo et al. 2016).

Zinc is naturally released into the soil from minerals that include zinc oxides and sulfates, sulfides, carbonates, silicates, and phosphates, which are subsequently liberated from the parent rock (Sturikova et al. 2018). A negligible quantity of Zn is found in soil, averaging between 0.2–0.002 μg Zn per gram of soil (Tayyiba et al. 2021). In soils, zinc may exist as many forms, primarily as a free ion (Zn^{2+} and $ZnOH^{+}$) or complexed with organic substances. Zinc exhibits minimal mobility in soil, a critical consideration in decision-making (Balafrej et al. 2020). Regarding the fertilization of crops, due to its minimal leaching rate, which may enhance its residual effect within the soil (Han et al. 2011).

Simultaneously, zinc infiltrates soils from multiple sources, including rock weathering, forest fires, mining operations, manure, sewage sludge, and phosphate fertilizers (Kaur and Garg 2021). In general, adequate concentrations of zinc in the soil range from 0.8 to 1.0 mg dm^{-3} (Khoshgoftar Mansh et al. 2012). Regarding Zn supplementation, the most commonly utilized fertilizers as a source of Zn are ZnO and ZnSO (montalvo et al. 2016).

3 Zinc in Plants

The optimal concentration of zinc required for the proper growth of most plants ranges between 30 to 200 μg Zn per gram of dry weight. Plant roots absorb Zn in its divalent form (Zn^{2+}). Additionally, it is absorbed as organic Zn-binding complexes (Hassan et al. 2020). Concerning Zn, in the context of redistribution in plants, metallothioneins (MTs), which are involved in metal homeostasis, were deemed partially accountable for the distribution of Zn in plants (Barbosa et al. 2016).

This element is abundantly found in proteins in plants, either in the active site or elsewhere, as a fundamental element of proteins (Singh et al. 2019; Jomova et al. 2022), including carbonic anhydrase, alcohol dehydrogenase, and copper (Cu)/zinc superoxide dismutase (Zeng et al. 2019). The maintenance of biological membrane integrity is also crucial in pollen production (Suganya et al. 2020), including the carbohydrate synthesis and co-factor of auxins (Hassan et al. 2020). Furthermore, zinc finger protein (ZFP) contains one or more "zinc fingers" that bind to one or increased zinc ions due to its residues histidine (His) and cysteine (Cys) (Li et al. 2013a, b). "Zinc finger" proteins facilitate the binding of transcription factors to DNA and protein-protein interactions (Takatsuji et al. 1998; Cabot et al. 2019). Research indicates that Zn also contributes to the metabolism of reactive species.

Reactive oxygen species and zinc-finger proteins facilitating a sequence of physiological and biochemical responses (Stanton et al. 2022; Wu et al. 2015). Noman et al. (2019) assert that zinc finger proteins are essential have functions linked to phytohormone responsiveness, plant growth and development, and stress resilience. According to Rizwan et al. (2019), zinc can enhance plant resilience to excessive cadmium. Simultaneously, Shemi et al. (2021) indicated that the use of Zn can enhance gas exchange, photosynthesis, and pigments in *Zea mays* subjected to hydric stress. Wu et al. (2015) examined the effects of zinc. Research on cotton plants subjected to drought stress indicated that zinc supplementation enhances drought resistance by promoting the formation of antioxidant defenses and facilitating osmotic adjustment.

4 Zinc Deficiency

The accessibility of Zn to plants is affected by soil characteristics, biological processes in the rhizosphere, and interactions with other nutrients (Moreno-Lora et al. 2020). Elements such as elevated pH greater than 7, reduced redox potential (Riaz et al. 2020), extended inundation, microbial communities in the rhizosphere, elevated organic matter content, and high carbonate and calcium bicarbonate levels (Sturikova et al. 2018) indicate that elevated iron levels can diminish the availability of zinc in the soil. Simultaneously, practices aimed at diminishing soil acidity may adversely affect the availability of zinc in soil (Ghasemi-FASAEI et al. 2008).

Zinc deficiency in plants hinders photosynthesis and nitrogen metabolism, in addition to resulting in diminished blooming and fruit development (Das and Green 2013). Zinc deficiency can impact photochemical activities in thylakoids and the inhibition of associated photosynthetic processes (Hajiboland et al. 2010). Plants with zinc shortage also demonstrate reduced levels of indole-3-acetic acid (IAA), a crucial component for plant development (Rehman et al. 2017).

Symptoms resulting from the shortage of this vitamin, associated with plant growth, are frequently attributable to a malfunction in the synthesis of "zinc finger" proteins, which are crucial for DNA transcription (Mattiello et al. 2015; Rudani et al. 2018). Producing an impact on diverse flora activities was reliant on particular proteins, including photosynthesis and respiration (Marschner 2012). Zn insufficiency reduces the concentration of Zn in cereal grains and compromises their nutritional value quality (Sadeghzadeh 2013).

5 Zinc Toxicity

Zinc, while a crucial element for plants, becomes toxic in excessive amounts, potentially hindering plant development (Fukao et al. 2011; Pereira-Matos et al. 2023). In general, most plants contain 30–100 mg Zn/kg of dry matter; on the other hand, contents greater than 300 mg/kg are generally toxic (Noulas et al. 2018).

Excess zinc in plants induces morphological, biochemical, and physiological abnormalities (Balafrej et al. 2020). Zinc toxicity causes symptoms similar to those of iron deficiency such as chlorosis and reduced growth (Mccauley et al. 2009). Excess Zn can provoke oxidative damage by enhancing the levels of reactive radicals (Natasha et al. 2022). This led the researchers to relate the symptoms to the inhibition of Fe activity due to excess zinc in the soil, which may be due to the composition of that soil or the mining activity that can release zinc into the soil Ghasemi-FASAEI et al. 2008).

Insufficient Zn supplementation adversely impacts leaf development and photosynthetic efficiency, hindering plant growth (Pereira-Matos et al. 2023). Investigating the effects of high Zn concentration on sugar beet plants (*Beta vulgaris* L.), Sagardoy et al. (2009) observed that plants treated with high concentrations of Zn sulfate in the nutrient solution (50, 100 e 300 μm) showed a reduction in the fresh and dry mass of the root and shoot, in addition to a decrease in the levels of N, Mg, K, and Mn in all parts of the plant. To Balafrej et al. (2020), the impacts of excess Zn depend on its concentration available in the external environment, the duration of exposure, the plant genotype, and its development phase.

6 Growth and Development Induced by Brassinosteroids

Brassinosteroids (BRs) are a class of roughly 70 polyhydroxy steroid derivatives, which are seemingly ubiquitous in the plant world (Clouse 2011). The discovery of endogenous steroid chemicals in plants emerged from over 30 years of study aimed at identifying novel growth-promoting components in pollen extracts from various plant species (Farriduddin et al. 2014). The initial identification of BRs occurred in a study utilizing *Brassica napus* pollen extract, which emerged as the most potent growth promoter, demonstrating enhancements in stem elongation and cellular division inside bean internodes (Oklestkova et al. 2015). The findings of the study by Mitchell et al. (1970) prematurely asserted that BRs are specific transposable organic chemicals derived from a plant, which facilitated quantifiable growth regulation when administered in small quantities to another plant (Clouse 2011).

Brassinolide is one of the most physiologically active naturally occurring forms of brassinosteroids (Joo et al. 2015; Azhar et al. 2017). Clouse and Sasse (1998) asserted that campesterol is anticipated to be the plant steroidal precursor of brassinolide due to its lateral chain structure and that brassinolide is synthesized from campesterol, as indicated by the relative biological activities, co-occurrence, and molecular structures of the intermediates in teasterone, tifasterol, and castasterone. Numerous researches have been conducted based on these facts.

As a result, BRs are recognized substances that play a significant role in processes primarily associated with the promotion of vegetative organ growth through their combined effects on cell expansion and division (Gudesblat and Russinova 2011). Oh et al. (2012) indicated that BRs exhibit a wide range of physiological effects when applied externally at nanomolar to micromolar concentrations. Numerous research efforts in agriculture are underway to assess the potential of BRs in promoting and regulating various biological responses (Divi and Krishna 2009).

Zhiponova et al. (2013) conducted a study utilizing the *Arabidopsis* BR-deficient mutant, which exhibits constitutive photomorphogenesis and dwarfism, to investigate the role of BR in leaf growth. Their findings indicate that this compound is crucial for proliferation, expansion, and cell division and that the temporal balance between proliferation and differentiation is contingent upon BR levels.

Pereira-Netto et al. (2009) assert that BR is associated with shooting growth. In their study involving grafted apple plants, it was noted that brassinolide had a differential impact on elongation and the formation of central and primary lateral shoots, which subsequently led to a reduction in the apical domain. Wei and Li (2016) presented another example of BR research, indicating that recent studies have enhanced the understanding of this steroid's role in root growth and development. It was noted that BRs regulate root meristem size and lateral root development, functioning in a cell type-specific manner throughout these processes.

7 Interferences on Chlorophyll Fluorescence

For q_P, BR-deficient plants exhibited a 30% increase when supplemented with BR, compared to BR-deficient plants without BR (Table 1). In NPQ, BR-deficient plants exhibited a 13% reduction when treated with BR. Concerning ETR/PN, both genotypes subjected to BR exhibited substantial reductions compared to plants of the same genotypes devoid of BR. The interventions for Φ_{PSII}, ETR, and EXC yielded little alterations.

The BR application enhanced Φ_{PSII}, q_P, and ETR. The elevation of q_P following BR spray facilitated the dissociation of electron charges in the reaction centers,

Table 1 Chlorophyll fluorescence in tomato genotypes sprayed with BR

Genotype	Treatment	Φ_{PSII}	q_P	NPQ	ETR (μmol m^{-2} s^{-1})	EXC (μmol m^{-2} s^{-1})	ETR/P_N
BR-efficient	- BR	0.53 ± 0.03a	0.59 ± 0.03b	1.47 ± 0.11b	78.4 ± 3.8a	0.25 ± 0.02a	10.3 ± 0.5c
BR-efficient	+ BR	0.56 ± 0.05a	0.63 ± 0.02b	1.38 ± 0.09b	81.5 ± 6.8a	0.23 ± 0.02a	9.0 ± 0.4d
BR-deficient	- BR	0.51 ± 0.05a	0.51 ± 0.03c	1.72 ± 0.09a	75.5 ± 6.7a	0.24 ± 0.02a	14.3 ± 0.4a
BR-deficient	+ BR	0.55 ± 0.05a	0.67 ± 0.01a	1.49 ± 0.12b	81.2 ± 7.5a	0.21 ± 0.02a	12.5 ± 1.0b

Φ_{PSII} represents the effective quantum yield of PSII photochemistry; q_P denotes the photochemical quenching coefficient; NPQ indicates nonphotochemical quenching; ETR represents the electron transport rate; EXC indicates the relative energy excess at the PSII level; ETR/PN denotes the ratio between the electron transport rate and the net photosynthetic rate. Columns marked with distinct letters denote significant differences as determined by the Scott-Knott test ($P < 0.05$). The values represent the means and standard deviations derived from five replicates enhancing Φ_{PSII} and ETR. The results validate the enhancements associated with chlorophyll fluorescence, including an augmentation in photochemical dissipation and efficiency in energy acquisition, attributable to the sustained openness of reaction centers (Guo et al. 2006; Zhang et al. 2013; Thussagunpanit et al. 2015a). A study by Dobrikova et al. (2014) on *Pisum sativum* indicated that the application of BR optimized Φ_{PSII}, q_P, and ETR, enhancing the photosynthetic apparatus's ability to sustain oxidized Q_A and facilitate electron transfer through PSII.

The BR spray resulted in reductions in NPQ, EXC, and ETR/PN. Concerning NPQ, the decrease noted in plants treated with BR resulted in reduced heat dissipation and sustained passage of excitation energy from the antenna system to the PSII reaction centers (Calatayud and Barreno 2004; Thussagunpanit et al. 2015b). The decline in ETR/PN in plants subjected to BR application is inherently linked to the reduction of EXC, indicating that surplus electrons were utilized less for secondary processes, such as photorespiration, and were likely available for primary functions, such as NADP+ reduction during CO_2 fixation (Silva et al. 2012). Ogweno et al. (2008) demonstrated that the application of BR in *Lycopersicon esculentum* under high-temperature stress dramatically diminished NPQ, hence enhancing the protection of PSII from potential damage caused by excessive excitation and improving the integrity of the thylakoid membranes.

8 Benefits Promoted by the BR on Gas Exchange

BR-efficient plants with BR expressed increases of 19%, 16%, 40%, and 37% in the variables P_N, E, g_s, and P_N/C_i, respectively, when compared to plants of the same genotype that did not have BR (Table 2). The combination of BR-deficient and BR resulted in increases in P_N, E, g_s, and P_N/C_i of 25%, 11%, 15%, and 48%, respectively, when compared to BR-deficient alone. Plants of both genotypes treated with BR exhibited reductions in C_i.

Beneficial effects on P_N, E, and g_s were observed in plants administered with BR. The gains are associated with the advantages conferred by BR on the photosynthetic machinery, as evidenced by the enhancements observed in qP in our study. The rise in P_N/C_i can be attributed to BR enhancing RuBisCO activity (resulting in a decrease in C_i) and the fixation of CO_2 (leading to an increase in P_N) during photosynthesis (Farooq et al. 2009; Xia et al. 2009; Shu et al. 2016). Hu et al. (2010) reported that during cold stress, the application of BR in *Cucumis sativus* enhanced P_N and g_s, mitigating the adverse effects of stress on these variables.

9 BR Application Caused Maximization of Plant Growth

BR-efficient plants exhibited superior growth metrics. Moreover, BR-efficient plants + BR exhibited increases of 146%, 123%, 102%, and 122% in LDM, RDM, SDM, and TDM, respectively, in comparison to plants of the same genotype devoid

Table 2 Gas exchange in tomato genotypes sprayed with BR

Genotype	Treatment	P_N (μmol m^{-2} s^{-1})	E (mmol m^{-2} s^{-1})	g_s (mol m^{-2} s^{-1})	C_i (μmol mol^{-1})	WUE (μmol $mmol^{-1}$)	P_N/C_i (μmol m^{-2} s^{-1} Pa^{-1})
BR-efficient	- BR	7.63 ± 0.42b	2.00 ± 0.08b	0.17 ± 0.008b	275 ± 10a	3.82 ± 0.10a	0.028 ± 0.002b
BR-efficient	+ BR	9.05 ± 0.56a	2.31 ± 0.14a	0.24 ± 0.016a	236 ± 14b	3.93 ± 0.21a	0.038 ± 0.001a
BR-deficient	- BR	5.27 ± 0.51d	1.61 ± 0.09c	0.15 ± 0.005c	294 ± 15a	3.28 ± 0.22b	0.018 ± 0.001c
BR-deficient	+ BR	6.59 ± 0.34c	1.74 ± 0.10c	0.17 ± 0.009b	256 ± 7b	3.78 ± 0.15a	0.026 ± 0.002b

P_N = Net photosynthetic rate; E = Transpiration rate; g_s = Stomatal conductance; C_i = Intercellular CO_2 concentration; WUE = Water-use efficiency; P_N/C_i = Carboxylation instantaneous efficiency. Columns with different letters indicate significant differences from the Scott-Knott test ($P < 0.05$). Values correspond to means and standard deviations from five replicates (Maia et al. 2018)

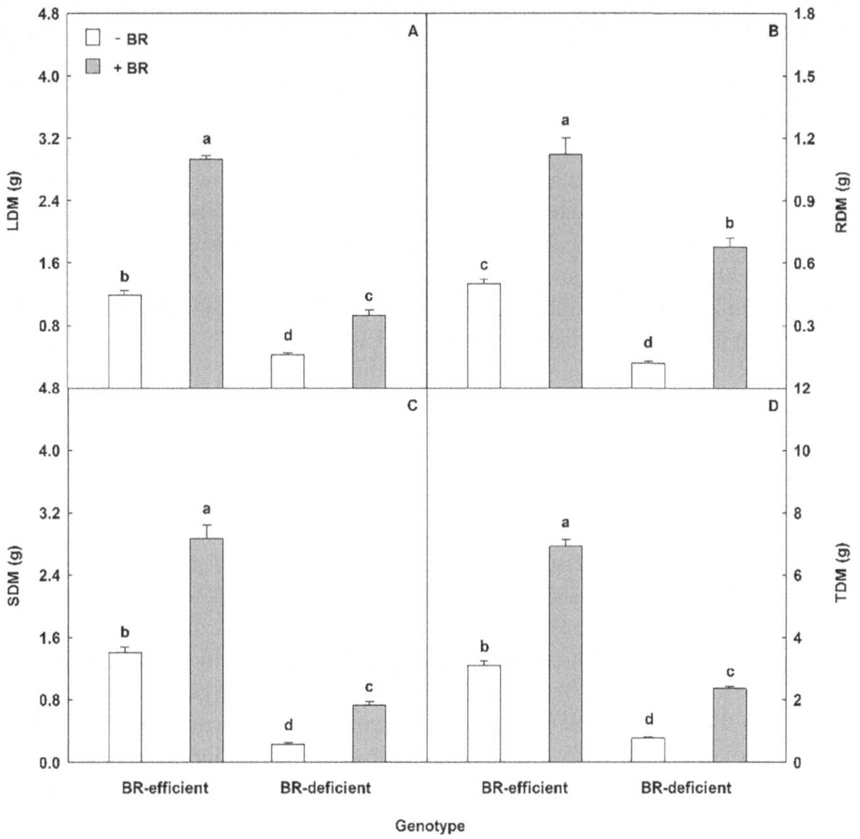

Fig. 1 Leaf dry matter (LDM), root dry matter (RDM), stem dry matter (SDM), and total dry matter (TDM) in tomato genotypes sprayed with BR. Different letters indicate significant differences from the Scott-Knott test ($P < 0.05$). Values correspond to means and standard deviations from five replicates (Maia et al. 2018)

of BR (Fig. 1). BR-deficient plants with BR expression exhibited increases of 120%, 469%, 219%, and 203% for the variables LDM, RDM, SDM, and TDM, respectively.

The BR application resulted in enhanced plant growth (LDM, RDM, SDM, and TDM). These reactions are inherently linked to the increases observed in q_P and P_N in our study, signifying the enhancements facilitated by the BR on chlorophyll fluorescence and gas exchange. The biomass expansion in plants subjected to this chemical correlates with enhanced photosynthetic rates, driven by elevated energy absorption and CO_2 fixation (Shahbaz et al. 2008). Zheng et al. (2016) assessed *Lycopersicon esculentum* under saline stress and found that the application of BR enhanced leaf, root, stem, and total biomass.

10 Beneficial Interaction Between Steroid and Zn

Both deficient and excessive Zn inputs induced an imbalance in Zn concentrations within the root, stem, and leaf tissues of tomato plants (Table 3). Plants subjected to low zinc levels and treated with EBR exhibited increases in zinc concentrations of 48% (leaf), 10% (stem), and 6% (root) relative to those with low zinc without EBR application. The enhancements for the control treatment and EBR spray were 16%, 29%, and 3% for leaf, stem, and root, respectively, in comparison to the equivalent treatment without EBR. Conversely, plants subjected to high Zn + EBR exhibited reductions in Zn concentrations of 19% in leaves, 19% in stems, and 36% in roots compared to the identical treatment devoid of EBR.

The EBR application enhanced Zn levels in the roots, stems, and leaves of tomato plants under low and regulated Zn availability. The findings stem from the steroids augmenting the absorption and distribution of this metallic nutrient. During Zn absorption, plants employ many mechanisms, including the interaction of the Zn^{2+} ion with organic molecules, inorganic acids, and specialized organelles (Sbartai et al. 2012). There is probably an advantageous interaction between steroids and metallic ions. Conversely, EBR spray mitigated the harmful effects of excess Zn, as this natural compound enhances the plant's protection against ionic toxicity. This steroid promotes the synthesis and accumulation of glutathione and phytochelatins, which are crucial to the detoxification process of metals, including Zn (Adediran et al. 2016; Arora et al. 2010). The synthesis of phytochelatins enhances zinc accumulation in the leaves of plants cultivated in soil with sufficient zinc availability, hence improving plant performance (Kuhnlenz et al. 2016). Talarek-Karwel et al. (2019) examined the effects of EBR in *Acutodesmus obliquus* under lead (Pb) stress and found that the application of 1 μM EBR enhanced the synthesis and activation of phytochelatins. Santos et al. (2020) investigated the effects of EBR on the root anatomy and antioxidant system of soybean plants exposed to Zn stress, noting that 100 nM EBR resulted in elevated Zn concentrations in the leaves, stems, and roots of plants receiving minimal Zn supplementation (0.2 μM Zn).

Table 3 Zn contents in tomato plants sprayed with EBR and exposed to different Zn supplies

EBR	Zn supply	Zn in root (μg g DM^{-1})	Zn in stem (μg g DM^{-1})	Zn in leaf (μg g DM^{-1})
–	Low	26.79 ± 1.80^{Ba}	8.25 ± 0.65^{Ba}	14.16 ± 1.06^{Cb}
–	Control	28.06 ± 0.74^{Ba}	8.39 ± 0.35^{Bb}	18.97 ± 0.77^{Bb}
–	High	3901.35 ± 86.28^{Aa}	1099.29 ± 40.11^{Aa}	1117.70 ± 72.53^{Aa}
+	Low	28.50 ± 0.84^{Ba}	9.10 ± 0.46^{Ca}	21.00 ± 0.29^{Ba}
+	Control	28.91 ± 0.65^{Ba}	10.85 ± 0.48^{Ba}	22.03 ± 0.92^{Ba}
+	High	2511.17 ± 131.79^{Ab}	891.88 ± 17.85^{Ab}	907.05 ± 13.25^{Ab}

Zn = zinc. Columns with different uppercase letters between Zn supplies (low, control, and high Zn supply under equal EBR level) and lowercase letters between EBR level (with and without EBR under equal Zn supply) indicate significant differences from the Scott-Knott test ($P < 0.05$). Means ± SD, n = 5 (Pereira-Matos et al. 2023)

11 EBR Mitigating Oxidative Damages in Plants Exposed to Zn Inadequate

Concerning stress markers (Fig. 2), both low and high levels of Zn supplementation stimulated the excessive formation of oxidant chemicals. Nonetheless, plants treated with EBR alleviated these effects. We observed reductions of 49%, 7%, and 30% in O_2^- levels for the low, control, and high Zn treatments, respectively, compared to the identical treatments without EBR. Regarding H_2O_2, plants treated with EBR exhibited reductions of 33%, 24%, and 25% under low, control, and high Zn conditions, respectively. Concerning MDA, EBR spray facilitated decreases of 10%, 9%, and 9% in low, control, and high Zn treatments, respectively. Regarding EL, there were reductions of 12%, 7%, and 14% in plants subjected to low, control, and high Zn, respectively, in comparison to identical treatment without EBR.

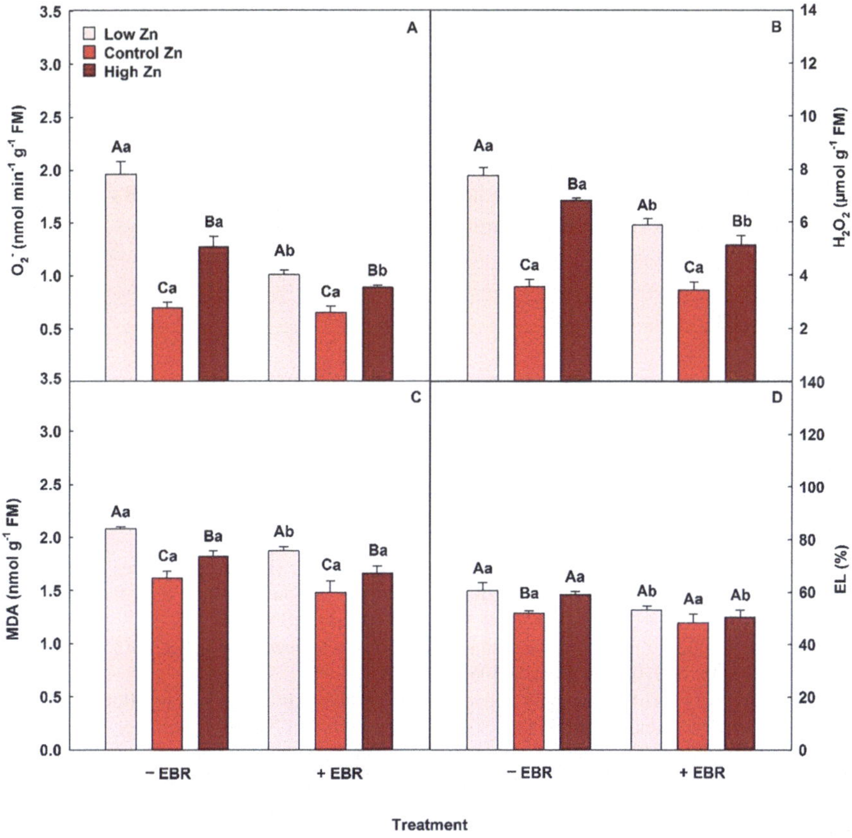

Fig. 2 Superoxide (O_2^-; **a**), hydrogen peroxide (H_2O_2; **b**), malondialdehyde (MDA; **c**), and electrolyte leakage (EL; **d**) in tomato plants treated with EBR and subjected to varying Zn levels. Columns marked with distinct uppercase letters among Zn supplies (low, control, and high Zn supply at a uniform EBR level) and lowercase letters between EBR levels (with and without EBR at a constant Zn supply) denote significant differences as determined by the Scott-Knott test ($P < 0.05$). Mean ± Standard Deviation, $n = 5$ (Pereira-Matos et al. 2023)

The EBR application alleviated the detrimental impacts of both low and high zinc levels on O_2^-, H_2O_2, MDA, and EL. Both zinc toxicity and deficiency induce oxidative stress, marked by the excessive generation of reactive oxygen species (ROS), and adversely affect the antioxidant defense system (Vázquez et al. 2013). Our findings indicate that the EBR enhanced the activities of antioxidant enzymes associated with ROS removal, hence diminishing oxidative damage, as seen by the reductions in O_2^- and H_2O_2, attributable to an increase in SOD activity. This enzyme catalyzes the dismutation from O_2^- to H_2O_2, subsequently converted to H_2O by the enzymes CAT APX and POX, mitigating the production of oxidant compounds (Bhaduri and Fulekar 2012; Li et al. 2013a, b). Consistent with our research, Arora et al. (2010) examined the effects of EBR interference on the antioxidant system in 30-day-old *Brassica juncea* plants subjected to Zn stress, reporting decreases in MDA, EL, O_2^-, and H_2O_2 levels.

12 Conclusions

This chapter revealed that the exogenous application of BRs mitigated the negative impacts promoted by Zn stresses in plant metabolism. The positive effects on gas exchange, chlorophyll fluorescence, and oxidant compounds confirm the benefits of this steroid. Reductions in stress indicators suggest that the role of BRs are linked to antioxidant metabolism and the maintenance of chloroplast integrity.

References

Adediran GA, Ngwenya BT, Mosselmans JFW, Heal KV (2016) Bacteria–zinc co-localization implicates enhanced synthesis of cysteine-rich peptides in zinc detoxification when Brassica juncea is inoculated with rhizobium leguminosarum. New Phytol 209(280–293):527. https://doi.org/10.1111/nph.13588

Arora P, Bhardwaj R, Kumar Kanwar M (2010) 24-epibrassinolide induced antioxidative defense system of Brassica juncea L. under Zn metal stress. Physiol Mol Biol Plants 16:285–293. https://doi.org/10.1007/s12298-010-0031-9

Azhar N et al (2017) Exogenously Applied 24-Epibrassinolide (EBL) Ameliorates Detrimental Effects of Salinity by Reducing K+ Efflux via Depolarization-Activated K+ Channels. Plant Cell Physiol 58(4):802–810. https://doi.org/10.1093/pcp/pcx026

Balafrej H et al (2020) Zinc Hyperaccumulation in Plants: a review. Plants, [S.L.] 9(5):562. MDPI AG. https://doi.org/10.3390/plants9050562

Barbosa BCF et al (2016) Zinc supply impacts on the relative expression of a metallothionein-like gene in Coffea arabica plants. Plant Soil, [S.L.] 411(1–2):179–191, . Springer Science and Business Media LLC. https://doi.org/10.1007/s11104-016-2983-

Bhaduri AM, Fulekar MH (2012) Assessment of arbuscular mycorrhizal fungi on the phytoremediation potential of Ipomoea aquatica on cadmium uptake. 3 Biotech 2:193–198. https://doi.org/10.1007/s13205-012-0046-8

Broadley MR et al (2007) Zinc in plants. New Phytol, [S.L.] 173(4):677–702, Wiley. https://doi.org/10.1111/j.1469-8137.2007.01996.x

Cabot C et al (2019) A role for zinc in plant defense against pathogens and herbivores. Frontiers Plant Sci [S.L.] 10(4) Frontiers Media SA:1171. https://doi.org/10.3389/fpls.2019.01171

Calatayud A, Barreno E (2004) Response to ozone in two lettuce varieties on chlorophyll a fluorescence, photosynthetic pigments and lipid peroxidation. Plant Physiol Biochem 42:549–555.https://doi.org/10.1016/j.plaphy.2004.05.002
Clouse SD (2011) Brassinosteroids. The Arabidopsis Book. pp e0151. https://doi.org/10.1199/tab.0151
Clouse SD, Sasse JM (1998) Brassinosteroids: essential regulators of plant growth and development. Annu Rev Plant Biol 49(1):427–451. https://doi.org/10.1146/annurev.arplant.49.1.427
Das S, Green A (2013) Importance of zinc in crops and human health. J SAT Agric Res 11:1–7
Divi UK. Krishna P (2009) Brassinosteroid: a biotechnological target for enhancing crop yield and stress tolerance. New Biotechnol 26:131–136. https://doi.org/10.1016/j.nbt.2009.07.006
Dobrikova AG, Vladkova RS, Rashkov GD et al (2014) Effects of exogenous 24-epibrassinolide on the photosynthetic membranes under non-stress conditions. Plant Physiol Biochem 80:75–82. https://doi.org/10.1016/j.plaphy.2014.03.022
Farooq M, Wahid A, Basra SMA, Islam-ud-Din (2009) Improving water relations and gas exchange with brassinosteroids in rice under drought stress. J Agron Crop Sci 195:262–269. https://doi.org/10.1111/j.1439-037X.2009.00368.x
Farriduddin Q et al (2014) Brassinosteroids and their role in response of plants to abiotic stresses. Biol Plant 58:9–17. https://doi.org/10.1007/s10535-013-0374-5
Fukao Y et al (2011) ITRAQ analysis reveals mechanisms of growth defects due to excess zinc in Arabidopsis. Plant Physiol, [SL] 155(4):1893–1907, . Oxford University Press (OUP). https://doi.org/10.1104/pp.110.169730
Ghasemi-FASAEI R et al (2008) Interaction of iron with copper, zinc, and manganese in wheat as affected by iron and manganese in a calcareous soil. J Plant Nutrition, [S.L.] 31(5):839–848, . Informa UK Limited. https://doi.org/10.1080/01904160802043148
Gondal AH (2021) A detailed review study of zinc involvement in animal, plant and human nutrition. Indian J Pure Appl Biosci, [S.L.] 9(2):262–271. Vital Biotech. https://doi.org/10.18782/2582-2845.8652
Gudesblat GE, Russinova E (2011) Plants grow on brassinosteroids. Curr Opin Plant Biol 14:530–537. https://doi.org/10.1016/j.pbi.2011.05.004
Guo HX, Liu WQ, Shi YC (2006) Effects of different nitrogen forms on photosynthetic rate and the chlorophyll fluorescence induction kinetics of flue-cured tobacco. Photosynthetica 44:140–142. https://doi.org/10.1007/s11099-005-0170-3
Hajiboland R et al (2010) Growth, photosynthesis and antioxidant systém in Zn-deficient red cabbage plants. Plant Soil Environ, [S.L.] 56(5):209–217, . Czech Academy of Agricultural Sciences. https://doi.org/10.17221/207/2009-pse
Han X, Li X, Uren N, Tang C (2011) Zinc fractions and availability to soybeans in representative soils of Northeast China. J Soils Sediments 11:596–606
Hassan MU et al (2020) The critical role of zinc in plants facing the drought stress. Agriculture, [S.L.] 10(9):396. MDPI AG. https://doi.org/10.3390/agriculture10090396
Hill J, Polasky S, Nelson E, Tilman D, Huo H, Ludwig L, Neumann J, Zheng H, Bonta D (2009) Climate change and health costs of air emissions from biofuels and gasoline. Proceedings of the National academy of sciences of the United states of America 106(6):2077–2082. https://doi.org/10.1073/pnas.0812835106
Hu WH, Wu Y, Zeng JZ et al (2010) Chill-induced inhibition of photosynthesis was alleviated by 24-epibrassinolide pretreatment in cucumber during chilling and subsequent recovery. Photosynthetica 48:537–544. https://doi.org/10.1007/s11099-010-0071-y
Jomova K et al (2022) Essential metals in health and disease. Chem Biol Interact:110173
Joo S et al (2015) Biosynthetic relationship between C 28-brassinosteroids and C 29brassinosteroids in rice (Oryza sativa) seedlings. Phytochemistry 111: 84–90. https://doi.org/10.1016/j.phytochem.2014.11.006
Kaur H, Garg N (2021) Zinc toxicity in plants: a review. Planta 253(6):129
Khoshgoftarmansh AH et al (2012) Critical deficiency level of zinc for corn on calcareous salt-affected soils in Central Iran. J Plant Nutr 35(12):1806–1818

Kuhnlenz T, Hofmann C, Uraguchi S, Schmidt H, Schempp S, Weber M, Lahner B, Salt DE, Clemens S (2016). Phytochelatin synthesis promotes leaf Zn accumulation of arabidopsis thaliana plants grown in soil with adequate Zn supply and is essential for survival on Zn-contaminated soil. Plant Cell Physiol 57:2342–2352. https://doi.org/10.1093/pcp/pcw148

Kumar S et al (2022) Fortification of micronutrients for sustainable development in field crops: a review. Pharma Innov J 11:363–373

Li W-T et al (2013a) Zinc finger protein (ZFP) in plants-a review. Plant Omics 6(6)

Li D, Sun CJ, Ye JF (2013b) Removal efficiency of organic substance in municipal sludge by supercritical water oxidation. Adv Mater Res 726–731:1732–1738. https://doi.org/10.4028/www.scientific.net/AMR.726-731.1732

Maia CF, Silva BRS, Lobato AKS (2018) Brassinosteroids positively modulate growth: physiological, biochemical and anatomical evidence using two tomato genotypes contrasting to dwarfism. J Plant Growth Regul 37:1099–1112. https://doi.org/10.1007/s00344-018-9802-2

Marschner P (2012) Marschner's mineral nutrition of hight plants, 3rd edn. Elsevier, Oxford-Inglaterra, p 643

Mattiello EM et al (2015) Zinc deficiency affects physiological and anatomical characteristics in maize leaves. J Plant Physiol 183:138–143

Mccauley A, Jones C, Jacobsen J (2009) Plant nutrient functions and deficiency and toxicity symptoms. Nutrient Manag Module 9:1–16

Mitchell JW et al (1970) Brassins—a new family of plant hormones from rape pollen. Nature 225:1065. https://doi.org/10.1038/2251065a0

Montalvo D et al (2016) Agronomic effectiveness of zinc sources as micronutrient fertilizer. In: Advances in agronomy, [S.L.]. Elsevier, pp 215–267. https://doi.org/10.1016/bs.agron.2016.05.004

Moreno-Lora A et al (2020) Factors determining Zn availability and uptake by plants in soils developed under Mediterranean climate. Geoderma, [S.L.] 376:114509 . Elsevier BV. https://doi.org/10.1016/j.geoderma.2020.114509

Natasha N et al (2022) Zinc in soil-plant-human system: a data-analysis review. Sci Total Environ, [S.L.] 808:152024. Elsevier BV. https://doi.org/10.1016/j.scitotenv.2021.152024

Noman A et al (2019) Zinc finger protein transcription factors: integrated line of action for plant antimicrobial activity. Microbial Pathogenesis, [S.L.] 132:141–149. Elsevier BV. https://doi.org/10.1016/j.micpath.2019.04.042

Noulas C et al (2018) Zinc in soils, water and food crops. J Trace Elem Med Biol, [S.L.] v. 49:252–260. Elsevier BV. https://doi.org/10.1016/j.jtemb.2018.02.009

Ogweno JO, Song XS, Shi K et al (2008) Brassinosteroids alleviate heatinduced inhibition of photosynthesis by increasing carboxylation efficiency and enhancing antioxidant systems in Lycopersicon esculentum. J Plant Growth Regul 27:49–57. https://doi.org/10.1007/s00344-007-9030-7

Oh K et al (2012) Synthesis of novel brassinosteroid biosynthesis inhibitors based on the ketoconazole scaffold. Bioorg Med Chem Lett 22:1625–1628. https://doi.org/10.1016/j.bmcl.2011.12.120

Oklestkova J et al (2015) Brassinosteroids: synthesis and biological activities. Phytochem Rev 14:1053–1072. https://doi.org/10.1007/s11101-015-9446-9

Pereira-Matos YC et al (2023) Exogenous 24-epibrassinolide reverses disturbances in zinc-stressed tomato by synergistically stimulating leaf structures, photosynthesis and growth. South Afr J Botany, [S.L.] 159:447–460. Elsevier BV. https://doi.org/10.1016/j.sajb.2023.06.034

Pereira-Netto AB et al (2009) Shooting control by brassinosteroids: metabolomic analysis and effect of brassinazole on Malus prunifolia, the Marubakaido apple rootstock. Tree Physiol 29(4):607–620. https://doi.org/10.1093/treephys/tpn052

Rehman A et al (2017) Zinc nutrition in wheat-based cropping systems. Plant Soil, [S.L.] 422(1–2):283–315. Springer Science and Business Media LLC. https://doi.org/10.1007/s11104-017-3507-3

Riaz MU et al (2020) Fate of micronutrients in alkaline soils. In: Resources use efficiency in agriculture, [S.L.]. Springer Singapore, pp 577–613. https://doi.org/10.1007/978-981-15-6953-1_16

Rizwan M et al (2019) A critical review on the effects of zinc at toxic levels of cadmium in plants. Environ Sci Pollution Res, [S.L.] 26(7):6279–6289. Springer Science and Business Media LLC. https://doi.org/10.1007/s11356-019-04174-6

Rudani K, Vishal P, Kalavati P (2018) The importance of zinc in plant growth-a review. Int Res J Nat Appl Sci 5(2):38–48

Sadeghzadeh B (2013) A review of zinc nutrition and plant breeding. J Soil Sci Plant Nutrition, [S.L.]:0. Springer Science and Business Media LLC. https://doi.org/10.4067/s0718-95162013005000072

Sagardoy R et al (2009) Effects of zinc toxicity on sugar beet (Beta vulgaris L.) plants grown in hydroponics. Plant Biol [SL] 11(3):339–350

Santos LR d, Paula L d S, Pereira YC, Silva BRS d, Batista BL, Alsahli AA, Lobato AK d S (2020) Brassinosteroids-mediated amelioration of iron deficiency in soybean plants: beneficial effects on the nutritional status, photosynthetic pigments and chlorophyll 821 fluorescence. J Plant Growth Regul. https://doi.org/10.1007/s00344-020-10232-y

Sbartai H, Reda M, Sbartai I, Berrabbah H (2012) Comptes Rendus Biologies Bioaccumulation du Cd et du Zn chez les plants de tomates (Lycopersicon esculentum L.) Bioaccumulation of cadmium and zinc in tomato (Lycopersicon esculentum L.). Comptes Rendus – Biol 335:585–593. 828. https://doi.org/10.1016/j.crvi.2012.08.001

Shahbaz M, Ashraf M, Athar H-R (2008) Does exogenous application of 24-epibrassinolide ameliorate salt induced growth inhibition in wheat (Triticum aestivum L.)? Plant Growth Regul 55:51–64. https://doi.org/10.1007/s10725-008-9262-y

Shemi R, Wang R, Gheith ESMS et al (2021) Effects of salicylic acid, zinc and glycine betaine on morpho-physiological growth and yield of maize under drought stress. Sci Rep 11:3195. https://doi.org/10.1038/s41598-021-82264-7

Shu S, Tang Y, Yuan Y et al (2016) The role of 24-epibrassinolide in the regulation of photosynthetic characteristics and nitrogen metabolism of tomato seedlings under a combined low temperature and weak light stress. Plant Physiol Biochem 107:344–353. https://doi.org/10.1016/j.plaphy.2016.06.021

Silva EN, Ribeiro RV, Ferreira-Silva SL et al (2012) Coordinate changes in photosynthesis, sugar accumulation and antioxidative enzymes improve the performance of Jatropha curcas plants under drought stress. Biomass Bioenerg 45:270–279. https://doi.org/10.1016/j.biombioe.2012.06.009

Singh P et al (2019) Zinc application enhances superoxide dismutase and carbonic anhydrase activities in zinc-efficient and zinc-inefficient wheat genotypes. J Soil Sci Plant Nutr 19:477–487

Stanton C et al (2022) Zinc in plants: integrating homeostasis and biofortification. Molecular Plant, [SL] 15(1):65–85

Sturikova H et al (2018) Zinc, zinc nanoparticles and plants. J Hazard Mater [S.L.] 349:101–110. Elsevier BV. https://doi.org/10.1016/j.jhazmat.2018.01.040

Suganya A et al (2020) Role of zinc nutrition for increasing zinc availability, uptake, yield, and quality of maize (Zea mays L.) grains: an overview. Commun Soil Sci Plant Anal 51(15):2001–2021

Takatsuji H et al (1998) Zinc-finger transcription factors in plants. Cell Mol Life Sci (Cmls), [S.L.] 54(6):582–596. Springer Science and Business Media LLC. https://doi.org/10.1007/s000180050186

Talarek-Karwel M, Bajguz A, Piotrowska-Niczyporuk A (2019) 24-Epibrassinolide modulates primary metabolites, antioxidants, and phytochelatins in Acutodesmus obliquus exposed to lead stress. J Appl Phycol 14. https://doi.org/10.1007/s10811-019-01966-8

Tayyiba L et al (2021) Efficiency of zinc in plants, its deficiency and sensitivity for different crops. Curr Res Agric Sci, [S.L.] 8(2):128–134, . Conscientia Beam. https://doi.org/10.18488/journal.68.2021.82.128.134

Thussagunpanit J, Jutamanee K, Kaveeta L et al (2015a) Comparative effects of brassinosteroid and brassinosteroid mimic on improving photosynthesis, lipid peroxidation, and rice seed set under heat stress. J Plant Growth Regul 34:320–331. https://doi.org/10.1007/s00344-014-9467-4

Thussagunpanit J, Jutamanee K, Sonjaroon W et al (2015b) Effects of brassinosteroid and brassinosteroid mimic on photosynthetic efficiency and rice yield under heat stress. Photosynthetica 53:312–320. https://doi.org/10.1007/s11099-015-0106-5

Vázquez MN, Guerrero YR, González LM, de la Noval WT (2013) Brassinosteroids and plant responses to heavy metal stress. Overview Open J Met 03:34–41. https://doi.org/10.4236/ojmetal.2013.32A1005

Wei Z, Li J (2016) Brassinosteroids regulate root growth, development, and symbiosis. Mol Plant 9:86–100. https://doi.org/10.1016/j.molp.2015.12.003

Wu S, Hu C, Tan Q et al (2015) Drought stress tolerance mediated by zinc-induced antioxidative defense and osmotic adjustment in cotton (*Gossypium Hirsutum*). Acta Physiol Plant 37:167. https://doi.org/10.1007/s11738-015-1919-3

Xia X-J, Huang L-F, Zhou Y-H et al (2009) Brassinosteroids promote photosynthesis and growth by enhancing activation of Rubisco and expression of photosynthetic genes in Cucumis sativus. Planta 230:1185–1196. https://doi.org/10.1007/s00425-009-1016-1

Zeng H, Zhang X, Ding M, Zhu Y (2019) Integrated analyses of miRNAome and transcriptome reveal zinc deficiency responses in rice seedlings. Bmc Plant Biol, [S.L.] 19(1):585. Springer Science and Business Media LLC. https://doi.org/10.1186/s12870-019-2203-2

Zhang YP, Zhu XH, Ding HD et al (2013) Foliar application of 24-epibrassinolide alleviates high-temperature induced inhibition of photosynthesis in seedlings of two melon cultivars. Photosynthetica 51:341–349. https://doi.org/10.1007/s11099-013-0031-4

Zheng Q, Liu J, Liu R et al (2016) Temporal and spatial distributions of sodium and polyamines regulated by brassinosteroids in enhancing tomato salt resistance. Plant Soil 400:147–164. https://doi.org/10.1007/s11104-015-2712-1

Zhiponova MK et al (2013) Brassinosteroid production and signaling differentially control cell division and expansion in the leaf. New New Phytol 197(2):490502. https://doi.org/10.1111/nph.12036

Zlobin IE et al (2021) Current understanding of plant zinc homeostasis regulation mechanisms. Plant Physiol Biochem, [S.L.] 162:327–335. Elsevier BV. https://doi.org/10.1016/j.plaphy.2021.03.003

The manufacturer's authorised representative in the EU is Springer
Nature Customer Service Centre GmbH, Europaplatz 3, 69115 Heidelberg,
Germany. If you have any concerns regarding our products, please
contact ProductSafety@springernature.com

Printed and bound by CPI Group (UK) Ltd, Croydon, CR0 4YY
01/10/2026
02225929-0005